Oxford Lecture Series in
Mathematics and its Applications 12

Series editors
John Ball   Dominic Welsh

OXFORD LECTURE SERIES IN MATHEMATICS AND ITS APPLICATIONS

1. J. C. Baez (ed.): *Knots and quantum gravity*
2. I. Fonseca and W. Gangbo: *Degree theory in analysis and applications*
3. P.-L. Lions: *Mathematical topics in fluid mechanics, Vol. 1: Incompressible models*
4. J. E. Beasley (ed.): *Advances in linear and integer programming*
5. L. W. Beineke and R. J. Wilson (eds): *Graph connections: Relationships between graph theory and other areas of mathematics*
6. I. Anderson: *Combinatorial designs and tournaments*
7. G. David and S. W. Semmes: *Fractured fractals and broken dreams*
8. Oliver Pretzel: *Codes and algebraic curves*
9. M. Karpinski and W. Rytter: *Fast parallel algorithms for graph matching problems*
10. P.-L. Lions: *Mathematical topics in fluid mechanics, Vol. 2: Compressible models*
11. W. T. Tutte: *Graph theory as I have known it*
12. Andrea Braides and Anneliese Defranceschi: *Homogenization of multiple integrals*
13. Thierry Cazenave and Alain Haraux: *An introduction to semi-linear evolution equations*
14. J. Y. Chemin: *Perfect incompressible fluids*
15. Giuseppe Buttazzo, Mariano Giaquinta and Stefan Hildebrandt: *One dimensional variational problems: an introduction*

Homogenization of Multiple Integrals

Andrea Braides

SISSA, Trieste, Italy

and

Anneliese Defranceschi

Parma University, Parma, Italy

CLARENDON PRESS • OXFORD

1998

Oxford University Press, Great Clarendon Street, Oxford OX2 6DP

Oxford New York

*Athens Auckland Bangkok Bogota Bombay Buenos Aires Calcutta
Cape Town Chennai Dar es Salaam Delhi Florence Hong Kong Istanbul
Karachi Kuala Lumpur Madrid Melbourne Mexico City Mumbai
Nairobi Paris São Paolo Singapore Taipei Tokyo Toronto Warsaw*

*and associated companies in
Berlin Ibadan*

Oxford is a registered trade mark of Oxford University Press

*Published in the United States
by Oxford University Press, Inc., New York*

A catalogue record for this book is available from the British Library

*Library of Congress Cataloging in Publication Data
(Data available)
ISBN 0 19 850246 X*

Typeset by the authors

*Printed in India
Thomson Press (India) Ltd*

In memoria di Ennio De Giorgi

PREFACE

In these lecture notes we present some basic results in the mathematical theory of homogenization of multiple integrals. We present a general approach using weak convergence methods which are particularly suited for the investigation of problems of a non-linear nature.

The purpose of the book is to provide a common introduction to various fundamental concepts which have been studied and developed in recent years to deal with problems involving weak convergence, such as quasiconvexity, poly-convexity and Γ-convergence, and to specialize these notions to homogenization problems. The need for a book of this kind arose naturally from the lack of a text unifying the basic results of non-linear theories which provide the background for developing research. It was conceived in fact after giving various courses at Ph D level on the subject, and trying to answer the requests of young researchers. The book is almost completely self-contained, requiring as prerequisites only a basic knowledge of Sobolev spaces, standard functional analysis and measure theory. The only exceptions are some regularity results for the solutions of partial differential equations, collected together in Appendix C. Moreover, we sometimes use some elementary notions of the calculus of variations for smooth functionals, such as the computation of Euler equations.

The book is organized in four parts. First we present recent developments on weak convergence and on the study of the structure of weakly lower semicontinuous integral functionals, which provide the essential framework for the study of non-linear variational problems. Various notions of convexity ($W^{1,p}$-quasiconvexity, polyconvexity, rank-1-convexity) are introduced and analysed, following the work of Morrey (1952), Ball (1977), Ball and Murat (1984), Dacorogna (1989), Šverák (1991) and (1992).

The second part of the book contains an outline of the theory of De Giorgi's Γ-convergence, which is the main tool for describing the asymptotic behaviour of sequences of variational functionals. We specialize the so-called direct methods of Γ-convergence to the framework of Sobolev spaces, with particular regard to the functionals of homogenization, which have driven our choices for the notation. Moreover, many results are proved in a context suited to integrals satisfying degenerate growth conditions, completing the more extensive introduction of Dal Maso (1993).

The third part describes the application of the concepts introduced in the previous parts to highly oscillating integrals, providing the main convergence theorems. First we deal with the periodic case, showing the essential differences between the scalar- and the vector-valued case (Chapter 14). Subsequently, we study various types of generalizations, for which a natural framework is that of almost-periodic functions (Chapters 15 and 17). We show how these results apply to the study of oscillating Hamilton Jacobi equations and to the limit of periodic Riemannian metrics (Chapter 16). A closure theorem shows how the homogenization process is stable under perturbations (Chapter 17); on the other hand, regularity and polyconvexity properties are not maintained after homogenization (Chapters 15 and 18).

The fourth part contains the finer homogenization results. In Chapters 19–21 we treat functionals with various types of degeneracy, which correspond to complex, or degenerate, growth conditions. First, we deal with functionals defined on periodic connected Lipschitz sets, in which case an extension theorem provides coerciveness for the limit. In Chapter 20 we study functionals describing media with stiff and soft inclusions, and, in Chapter 21, functionals satisfying some non-standard growth conditions. Chapter 22 is devoted to the study of media with multiple scales of homogenization and Chapter 23 provides a correctors result in the strictly convex case. Finally, in Chapter 24 we outline some results for the homogenization of multi-dimensional structures.

Exercises at the ends of most of the chapters in the book complete the presentation.

We gratefully acknowledge contributions to the contents of this work, through direct collaboration or fruitful discussion with, in chronological order, Ennio De Giorgi, Giuseppe Buttazzo, Gianni Dal Maso, Emilio Acerbi, Luigi Ambrosio, Serguei Kozlov, Stefan Müller, Valeria Chiadò Piat, John Ball, Vladimir Šverák, Enrico Vitali, Adriana Garroni, Micol Amar, Paolo Dall'Aglio, Vincenzo Nesi, Irene Fonseca, Gilles Francfort and Nadia Ansini.

The core of this book is based on lectures given by both authors at the School on Homogenization held at ICTP Trieste in September 1993, and on two courses given at SISSA Trieste in 1993/94 and 1995/96 by the first author, which were also elaborated in a series of lectures at the Tata Institute of Fundamental Research in Bangalore during summer 1994. Part of the book was improved during a visit of the first author to the Max Planck Institute for Mathematics in the Sciences on a Marie Curie fellowship of the EU program 'Training and Mobility of Researchers'. The authors are grateful to these institutions for providing a stimulating environment, and to the Department of Mathematics of the University of Trento for hospitality granted during the final writing of the manuscript.

This work was supported by *Consiglio Nazionale delle Ricerche* through the project *Equazioni alle derivate parziali e calcolo delle variazioni*.

Trieste
April 1998

A.B., A.D.

CONTENTS

NOTATION

$a \otimes b$ matrix whose entries are $(a \otimes b)_{ij} = a_i b_j$

$\mathcal{A}_n$ family of all bounded open subsets of $\mathbf{R}^n$

Ax ($A \in \mathbf{M}^{m \times n}$ and $x \in \mathbf{R}^n$) the vector $y \in \mathbf{R}^m$ with $y_i = \sum_{j=1}^n A_{ij} x_j$

Ax (with a slight abuse of notation) the linear function $x \mapsto Ax$ ($A \in \mathbf{M}^{m \times n}$)

$\mathcal{A}(\Omega)$ family of all open subsets of Ω

$B(x, r)$ open ball of centre x and radius r

c (unless otherwise stated) a strictly positive constant whose value may change from line to line and which depends only on the fixed parameters of the problem

$e_1, \ldots, e_N$ canonical base of $\mathbf{R}^N$ (N-dimensional Euclidean space)

$|E|$ Lebesgue measure of the set E

$\mathbf{M}^{m \times n}$ space of $m \times n$ real matrices (identified with $\mathbf{R}^{mn}$)

$\lim_j$ ($\liminf_k$, etc.) limit as the discrete parameter j tends to $+\infty$

$[t]$ integer part of $t \in \mathbf{R}$

$\langle x, y \rangle$, $|x|$ scalar product of x and y, Euclidean norm of x

$U \subset\subset V$ U is compactly contained in V, i.e. $\overline{U}$ is compact and $\overline{U} \subset U$

χ_E characteristic function of E ($\chi_E(x) = 1$ if $x \in E$, $\chi_E(x) = 0$ if $x \notin E$)

$\mu \llcorner E$ restriction of the measure μ to E, i.e. $\mu \llcorner E(B) = \mu(B \cap E)$

Ω (unless otherwise stated) bounded open subset of $\mathbf{R}^n$

In the following notation Ω is not necessarily bounded, and the target space $\mathbf{R}^m$ is omitted if $m = 1$.

$C^k(\Omega; \mathbf{R}^m)$ k-times continuously differentiable functions $u : \Omega \to \mathbf{R}^m$

$C_0^k(\Omega; \mathbf{R}^m)$ functions in $C^k(\Omega; \mathbf{R}^m)$ with compact support in Ω

Du weak gradient of u, distributional derivative of u (Chapter 24)

$L_\mu^p(\Omega; \mathbf{R}^m)$ Lebesgue space with respect to the measure μ (μ is omitted if it is the Lebesgue measure)

p', p^* dual and Sobolev exponents of p, respectively

$u_j \to u$ u_j converges strongly to u

$u_j \rightharpoonup u$, $u_j \rightharpoonup^* u$ u_j converges weakly to u, u_j converges weakly* to u

$W^{1,p}(\Omega; \mathbf{R}^m)$ Sobolev space of $L^p(\Omega; \mathbf{R}^m)$ functions with $Du \in L^p(\Omega; \mathbf{M}^{m \times n})$

$W_0^{1,p}(\Omega; \mathbf{R}^m)$ closure of $C_0^\infty(\Omega; \mathbf{R}^m)$ in $W^{1,p}(\Omega; \mathbf{R}^m)$

$W^{-1,p'}(\Omega; \mathbf{R}^m)$ dual space of $W_0^{1,p}(\Omega; \mathbf{R}^m)$

$X_{\mathrm{loc}}(\Omega; \mathbf{R}^m)$ $\{u : \Omega \to \mathbf{R}^m : u \in X(U; \mathbf{R}^m) \text{ for all open } U \subset\subset \Omega\}$ (X a generic notation for a function space)

$X_\#((0, k)^n; \mathbf{R}^m)$ k-periodic functions in $X_{\mathrm{loc}}(\mathbf{R}^n; \mathbf{R}^m)$

$\triangle u$ Laplacian of u

INTRODUCTION

The object of homogenization theory is the description of the macroscopic properties of media with fine microstructure, covering a wide range of applications
that run from the study of the properties of composites to optimal design. The
structures under consideration may model composites, fiber materials, stratified
or porous media, finely damaged materials, compliance problems, materials with
many holes or cracks. In mathematical terms, this study can be translated into
the asymptotic analysis of fast-oscillating ordinary or partial differential equations and systems, or integral functionals, depending on a small-scale parameter
ε, as this parameter tends to 0.

In this book, we study the basic topics in the homogenization of non-linear
integral functionals, with particular regard to those general results that do not
rely on smoothness or convexity assumptions. As an illustrating example, we can
consider a cellular (possibly) non-linearly hyperelastic material, with a bounded
open subset Ω of $\mathbf{R}^n$ as the reference configuration (the physical case being
$n = 3$), whose periodicity cell has side-length ε. The elastic energy of such a
material subject to a displacement u can be written as an integral of the form

$$F_\varepsilon(u) = \int_\Omega f\left(\frac{x}{\varepsilon}, Du\right) dx. \tag{0.1}$$

If the scale ε is very small, the fine behaviour of this material may not be interesting, and only averaged quantities at a larger scale are relevant. The overall
properties of such a medium can be described by replacing F_ε by a 'homogenized'
simpler integral

$$F_{\text{hom}}(u) = \int_\Omega f_{\text{hom}}(Du)\, dx, \tag{0.2}$$

if for every choice of external force g, $\omega \subseteq \partial\Omega$, and boundary datum φ, the
solutions and minimum values of the problems

$$\min\left\{ \int_\Omega \left(f(\tfrac{x}{\varepsilon}, Du) - \langle g, u \rangle \right) dx : u = \varphi \text{ on } \omega \right\}, \tag{0.3}$$

which at present we suppose exist for the sake of simplicity, converge in some
sense to the corresponding solution and minimum value of

$$\min\left\{ \int_\Omega \left(f_{\text{hom}}(Du) - \langle g, u \rangle \right) dx : u = \varphi \text{ on } \omega \right\}, \tag{0.4}$$

as $\varepsilon \to 0$. Clearly, the study of the latter problem is much simpler than the first
one. In this framework, major tasks of homogenization theory are the proof of this

convergence, the description of f_{hom} through the solution of auxiliary problems (homogenization formulae), and the study of its qualitative and quantitative properties (such as convexity conditions, coerciveness, optimal bounds for f_{hom}, etc.).

Since the solutions u_ε to (0.3) are in general wildly oscillating as ε vanishes, their convergence must be understood in the weak sense of Sobolev spaces. The natural framework for the study of these kinds of problems is then the setting of weakly lower semicontinuous functionals on $W^{1,p}(\Omega; \mathbf{R}^m)$, to whose study the first part of the book is devoted. As lower semicontinuity conditions will be essential only for the limit problem, we will focus our analysis on integrands which depend only on the gradient Du and not on the space variable x. In this case, the notion of $W^{1,p}$-quasiconvexity (see Ball and Murat (1984)) turns out to be a necessary condition for weak lower semicontinuity: a function $\psi : \mathbf{M}^{m\times n} \to \overline{\mathbf{R}}$ ($\mathbf{M}^{m\times n}$ denotes the space of $m \times n$ matrices) is $W^{1,p}$-quasiconvex if

$$\psi(A)|E| \le \int_E \psi(A + Du)\,dx \tag{0.5}$$

for all open bounded smooth subsets E of $\mathbf{R}^n$, $A \in \mathbf{M}^{m\times n}$ and $\psi \in W_0^{1,p}(E; \mathbf{R}^m)$ (Chapter 4). Moreover, under proper growth conditions this property is equivalent to Morrey's (1952) quasiconvexity, and turns out also to be sufficient. Clearly, Jensen's inequality implies that convex functions are $W^{1,p}$-quasiconvex. An important sufficient condition is polyconvexity, introduced by Morrey and studied by Ball (1977), while a crucial necessary property is rank-1-convexity, from which we can deduce for instance that quasiconvex functions are locally Lipschitz (Chapter 5). Important examples of non-polyconvex quasiconvex functions and of non-quasiconvex rank-1-convex functions have been presented by Šverák (1991) and (1992) (Chapter 6).

The right notion of variational convergence of functionals F_ε to F_{hom} has proved to be De Giorgi's Γ-convergence (see De Giorgi and Franzoni (1975)), whose natural framework is that of lower semicontinuous functionals, and which implies, and in most cases is equivalent to, the convergence of problems (0.3) to (0.4) in the sense specified above (Chapters 7 and 8). For this reason, in most cases only Γ-convergence results are stated throughout this book, the convergence of minimum problems being an immediate consequence. The *direct methods* of Γ-convergence, a quite general version of which is presented in Chapters 9–11 and which have been most fruitful in the study of limits of integral functionals, were first outlined by De Giorgi (1975) and successively elaborated by many authors (see Dal Maso (1993)). They consist first of a localization procedure that highlights the dependence of integrals as set functions. In the case of functionals of the form (0.1), the localized functionals are of the form

$$F_\varepsilon(u, U) = \int_U f\left(\frac{x}{\varepsilon}, Du(x)\right) dx, \tag{0.6}$$

with U varying on all open subsets of Ω. Next, general abstract compactness results are proven that assure the existence of Γ-converging subsequences. The Γ-limit will be an abstract functional $F_0(u, U)$ defined for all $u \in W^{1,p}(\Omega; \mathbf{R}^m)$ and U open subset of Ω. It is crucial then to recover enough information on F_0, regarded both as a functional on $W^{1,p}(\Omega; \mathbf{R}^m)$ and as a set function, to obtain a representation in an integral form

$$F_0(u, U) = \int_U \psi(x, Du(x)) \, dx. \tag{0.7}$$

The main technical ingredient is the so-called 'fundamental estimate', whose aim is to 'join functions without introducing a large error'. Under some growth conditions, in the case of the functionals F_ε as above it asserts that given U, U' and V open subsets of Ω with $U' \subset\subset U$, and $u, v \in W^{1,p}(\Omega; \mathbf{R}^m)$, for ε sufficiently small it is possible to find a suitable interpolation w between u and v, coinciding with u on U' and with v on $V \setminus U$, such that

$$F_\varepsilon(w, U' \cup V) \le F_\varepsilon(u, U) + F_\varepsilon(v, V) + c \int_{U \cap V} |u - v|^p \, dx \tag{0.8}$$

(c a positive constant), up to an error which vanishes as $\varepsilon \to 0$. Using (0.8), it is possible to show that $F_0(u, \cdot)$ is a measure, and eventually obtain (0.7). The fundamental estimate holds for instance if f satisfies a standard growth condition of order p, i.e.

$$\alpha |A|^p \le f(x, A) \le \beta(1 + |A|^p) \tag{0.9}$$

for all $A \in \mathbf{M}^{m \times n}$ and $x \in \mathbf{R}^n$ (Chapter 12).

In the case of integrands satisfying (0.9) the problem of the Γ-convergence of F_ε to F_{hom} is reduced to the identification of the integrand ψ in (0.7), which a priori depends on a subsequence (ε_j). If f is 1-periodic in the first variable, the function ψ can be chosen to be independent of x, and is $W^{1,p}$-quasiconvex. Hence, it can be expressed as a minimum problem:

$$\psi(A) = \min\left\{ \int_{(0,1)^n} \psi(A + Du) \, dx : u \in W_0^{1,p}((0,1)^n; \mathbf{R}^m) \right\}. \tag{0.10}$$

The crucial point is then to prove the existence of the limit

$$\lim_{t \to +\infty} \inf\left\{ \frac{1}{t^n} \int_{(0,t)^n} f(x, A + Du) \, dx : u \in W_0^{1,p}((0,t)^n; \mathbf{R}^m) \right\}. \tag{0.11}$$

This limit defines the function f_{hom} at A and is called the *asymptotic homogenization formula*. From the Γ-convergence's property of convergence of minima (and infima), after rewriting the minimum problems in (0.11) on $(0,1)^n$ by a change of variables, we can deduce that $\psi(A)$ equals the limit $f_{\text{hom}}(A)$ in (0.11), and hence we have the convergence of the whole family (F_ε) to F_{hom} as defined in (0.2). In the convex case, an alternative description can be given, which takes into account only one minimum problem: the so-called *cell-problem formula*

$$f_{\mathrm{hom}}(A) = \min\left\{ \int_{(0,1)^n} f(x, A + Du)\, dx : u \in \mathrm{W}^{1,p}_{\#}((0,1)^n; \mathbf{R}^m) \right\}, \qquad (0.12)$$

where $\mathrm{W}^{1,p}_{\#}((0,1)^n; \mathbf{R}^m)$ stands for the set of $\mathrm{W}^{1,p}_{\mathrm{loc}}(\mathbf{R}^n; \mathbf{R}^m)$ 1-periodic functions. In this case f_{hom} is convex, and, moreover, the growth conditions can be weakened, requiring only that

$$0 \leq f(x, A) \leq \beta(1 + |A|^p) \qquad (0.13)$$

for all $A \in \mathbf{M}^{m \times n}$ and $x \in \mathbf{R}^n$. Such weak conditions are useful to deal with degenerate media, such as fiber or porous materials. It is worth noting that in the non-convex case formula (0.12) does not describe the homogenized integrand, as an example by Müller (1987) shows (Chapter 14). Moreover, the structure of the homogenized integrand may differ from that of f: for example, even if $f(x, \cdot)$ is polyconvex for all x the function f_{hom} may not be polyconvex (Chapter 18).

The homogenization result in the periodic case cannot be generalized using periodic techniques, even to the simple case of the sum of two periodic integrands $\int_\Omega (f(\frac{x}{\varepsilon}, Du) + g(\frac{x}{\varepsilon}, Du))\, dx$, if f and g have no common period, or to integrands of the form $f(\frac{x}{\varepsilon}, \frac{u}{\varepsilon}, Du)$. Note that in the latter case, even if f is 1-periodic in the first two variables the function $x \mapsto f(x, Ax, A)$ is in general non-periodic. An easy way to treat problems of this type is to use almost-periodic methods (Chapter 15). In this case the Γ-limit of functionals of the form

$$F_\varepsilon(u) = \int_\Omega f\left(\frac{x}{\varepsilon}, \frac{u}{\varepsilon}, Du\right) dx \qquad (0.14)$$

can be computed using a suitable generalization of the asymptotic homogenization formula, under hypotheses of uniform almost periodicity in the first two variables. This result can be applied to describe the asymptotic behaviour of viscosity solutions of oscillating Hamilton Jacobi equations, since integrals of the form (0.14) appear in the Lax formula. Another application is the study of periodic Riemannian metrics, related to functionals of the form

$$F_\varepsilon(u) = \int_0^1 \sum_{i,j=1}^m a\left(\frac{u}{\varepsilon}\right) u_i' u_j'\, dt, \qquad (0.15)$$

which may converge to Finsler metrics related to

$$F_{\mathrm{hom}}(u) = \int_0^1 \psi(u')\, dt, \qquad (0.16)$$

where ψ may not be a quadratic form. Notice that in both applications the limit integrand may lose smoothness (Chapters 15 and 16).

The process of homogenization is stable under small perturbations: for example, if we consider a compact support perturbation of f (even not periodic) in the homogenization processes described above, the Γ-limit remains unchanged. This observation can be generalized to a closure theorem: if (f_j) is a sequence of 'homogenizable functions' (i.e. there exists the Γ-limit of the related functionals constructed as in (0.1), with a limit integrand $(f_j)_{\text{hom}}$), and

$$\lim_j \fint \sup_{|A| \leq M} |f_j(x, A) - f(x, A)| \, dx = 0 \qquad (0.17)$$

for all $M \geq 0$ (where $\fint$ denotes the upper mean value on $\mathbf{R}^n$), then f also is 'homogenizable', and $(f_j)_{\text{hom}}$ converges to f_{hom}. As a consequence, this result allows a homogenization theorem under very weak (almost) periodicity for f, through a density argument (Chapter 17).

The methods for the homogenization above can also be applied to describe some finer properties in the case of degenerate structures. A first example is that of functionals describing porous media. In this case we are led to consider energies of the form

$$F_\varepsilon(u) = \int_{\Omega \cap \varepsilon E} f\left(\frac{x}{\varepsilon}, Du\right) dx. \qquad (0.18)$$

Under the hypothesis that E is a connected periodic open subset of $\mathbf{R}^n$, and f satisfies a growth condition of order $p > 1$, the homogenized functional exists and is coercive on $W^{1,p}(\Omega; \mathbf{R}^m)$. Furthermore, it is possible to find compact minimizing sequences of Neumann or Dirichlet boundary value problems, thanks to an extension theorem by Acerbi *et al.* (1992) (Chapter 19). Note that the functional F_ε can be written as in (0.1) if we extend f to 0 on $\mathbf{R}^n \setminus E$ (which can itself be a connected domain, if $n \geq 3$). At the same time we can allow f to have 'stiff' degeneracies, i.e. zones where the growth condition from above is violated. For example, we can even allow f to be of the form

$$f(x, A) = \begin{cases} +\infty & \text{if } A \neq 0 \\ 0 & \text{if } A = 0, \end{cases} \qquad (0.19)$$

for x belonging to a set composed of well-separated domains (Chapter 20). A further type of degeneracy can be that of energy densities satisfying a non-standard growth condition of the form

$$\alpha|A|^p \leq f(x, A) \leq \beta(1 + |A|^q), \qquad (0.20)$$

which can be dealt with under some restrictions on p and q. As an example, these energies may model mixtures of materials with different non-linearities. In this case f_{hom} may not satisfy a growth condition of order r for any r; an example by Kozlov (1989) shows that the domain of F_{hom} may be an Orlicz Sobolev space (Chapter 21).

Functionals F_{hom} defined on Sobolev spaces can also be obtained as a result of a homogenization process starting from energies depending on lower-dimensional or multi-dimensional structures. Two complementary approaches can be followed, and are outlined in Chapter 24. The smooth one considers energies of the form

$$F_\varepsilon(u) = \int_\Omega f\left(\frac{x}{\varepsilon}, Du\right) d\mu_\varepsilon, \tag{0.21}$$

defined on $C^1(\Omega; \mathbf{R}^m)$, where μ_ε is the ε-periodic measure given by $\mu_\varepsilon(B) = \varepsilon^n \mu(\frac{1}{\varepsilon}B)$, and μ is a fixed 1-periodic measure. A second approach is based on the notion of Sobolev space with respect to a measure introduced by Ambrosio *et al.* (1996). In this case,

$$F_\varepsilon(u) = \int_\Omega f\left(\frac{x}{\varepsilon}, \frac{dDu}{d\mu_\varepsilon}\right) d\mu_\varepsilon \tag{0.22}$$

is defined on functions whose distributional derivative is a measure absolutely continuous with respect to μ_ε. An example of the application of the first approach is the study of 1-dimensional networks or fibers in $\mathbf{R}^n$, while a simple case in which the second approach can be applied is to finite-difference approximations of Sobolev functions. Chapter 24 also contains a homogenization result for thin structures.

A further interesting issue is the study of structures with multiple scales of homogenization, whose energies can be written as

$$F_\varepsilon(u) = \int_\Omega f\left(\frac{x}{\varepsilon}, \frac{x}{\varepsilon^2}, \ldots, \frac{x}{\varepsilon^k}, Du\right) dx. \tag{0.23}$$

The homogenized functional F_{hom} can be computed by iterating the homogenization formula (Chapter 22). The process of iterated homogenization can be useful in the computation of the effective behaviour of multi-layered materials, as pointed out, for example, by Avellaneda (1987).

Another problem is the description of the oscillating solutions u_ε to problems (0.3) in terms of the simpler solution u to (0.4). Clearly, we cannot expect Du_ε to converge strongly to Du. However, in the case of f strictly convex and smooth, it is possible to construct a family of 'correctors' depending only on the function $D_\xi f(x, \xi)$, by which Du_ε can be expressed in terms of Du up to an error which converges to 0 strongly in $L^p(\Omega; \mathbf{R}^n)$. These correctors can be described easily from the cell-problem formula in the periodic case (Chapter 23).

Some important issues among the applications of homogenization that have not been dealt with in this book, but to which these notes may be of partial introduction, are: the problem of optimal bounds for mixtures of materials with prescribed proportions (see Hashin and Strickman (1963), Lurie and Cherkaev (1984), Murat and Tartar (1985)), relaxed Dirichlet problems (i.e. the asymptotic behaviour of Dirichlet problems in perforated domains; see Cioranescu and Murat (1982), Dal Maso and Mosco (1987), Buttazzo *et al.* (1987)), and problems in optimal design whose solution can be obtained using a homogenization method (see Murat and Tartar (1985a), Kohn and Strang (1986)).

Finally, we would like to point out that even though the Γ-convergence approach is most suited to the non-linear and non-smooth setting, it is necessarily limited in depth by its generality. Various other techniques have been developed for the homogenization in smoother situations, especially in the linear case, and have themselves often drawn inspiration from issues in weak convergence and the study of lower semicontinuous functionals. We would like to mention G-convergence (Spagnolo (1968), De Giorgi and Spagnolo (1973)), H-convergence and H-measures (Tartar (1990)), compensated compactness (Murat (1978), Tartar (1979)), two-scale convergence (Allaire (1992)), and the perturbed test-function method (Evans (1989)). We think that these techniques are complementary to Γ-convergence, and knowledge of them is advisable for a complete understanding of the homogenization processes, and at the same time that this book may be used as an introduction to most of the problems dealt with by those other methods.

Part I

Lower semicontinuity of integral functionals

1

LOWER SEMICONTINUITY AND COERCIVENESS

The direct methods of the calculus of variations can be summarized in the equation

$$\text{lower semicontinuity} \ + \ \text{coerciveness} \ = \ \text{existence},$$

meaning that, in order to obtain the existence of minimizers for a certain functional, it suffices to find a suitable topology in which a compact set can be found where a minimizing sequence lies (coerciveness), and at the same time the functional under consideration is lower semicontinuous. In this chapter we present a general introduction to lower semicontinuity and coerciveness. In Chapters 2–6 these two concepts will be specialized to the case of integral functionals defined on Sobolev spaces of vector-valued functions.

In what follows (X, d) is a metric space. We will consider functions defined on X with values in the extended real line $\overline{\mathbf{R}} = \mathbf{R} \cup \{-\infty, +\infty\}$. Such a function f will be called *proper* if there exists $x \in X$ such that $f(x) \in \mathbf{R}$, and f never takes the value $-\infty$. We use the notation $\{f \leq t\} = \{x \in X : f(x) \leq t\}$, and similar.

1.1 Lower semicontinuity

Definition 1.1 *A function $f : X \to \overline{\mathbf{R}}$ will be said to be (sequentially) lower semicontinuous (l.s.c. for short) at $x \in X$, if for every sequence (x_j) converging to x we have*

$$f(x) \leq \liminf_j f(x_j), \tag{1.1}$$

or in other words

$$f(x) = \min\{\liminf_j f(x_j) : \ x_j \to x\}. \tag{1.2}$$

We will say that f is lower semicontinuous (on X) if it is l.s.c. at all $x \in X$.

Remark 1.2 The following conditions are equivalent:

(i) f is lower semicontinuous;

(ii) if we define the *lower limit* of f at x as

$$\liminf_{y \to x} f(y) = \sup_{U \in \mathcal{N}(x)} \inf_{y \in U} f(y) \ \left(= \sup_{\delta > 0} \inf_{d(x,y) < \delta} f(y) \right),$$

where $\mathcal{N}(x)$ denotes the family of all neighbourhoods of x, then we have

$$f(x) = \liminf_{y \to x} f(y) \tag{1.3}$$

for all $x \in X$;

    (iii) the *epigraph* of f, $epi(f) = \{(x,t): \ x \in X, \ t \in \mathbf{R}, \ t \geq f(x)\}$, is closed in $X \times \mathbf{R}$;

    (iv) for all $t \in \mathbf{R}$ the sublevel set $\{f \leq t\}$ is closed;

    (v) for all $x \in X$ and for all $t < f(x)$ there exists $U \in \mathcal{N}(x)$ such that $f(y) > t$ for all $y \in U$.

Indeed, (i) $\Leftrightarrow$ (ii) as (1.2) $\Leftrightarrow$ (1.3) since X is a metric space; (i) $\Rightarrow$ (iii) since if $epi(f) \ni (x_j, t_j) \to (x,t)$ then $t = \lim_j t_j \geq \liminf_j f(x_j) \geq f(x)$; hence $(x,t) \in epi(f)$; (iii) $\Rightarrow$ (iv) since if $(x_j) \subset \{f \leq t\}$ and $x_j \to x$ then $epi(f) \ni (x_j, t) \to (x,t) \in epi(f)$, and $x \in \{f \leq t\}$; (iv) $\Rightarrow$ (v) since $\{f > t\}$ is open; hence if $x \in \{f > t\}$ then there exists $U \in \mathcal{N}(x)$ such that $U \subset \{f > t\}$; finally (v) $\Rightarrow$ (ii): from (v) $\sup_{U \in \mathcal{N}(x)} \inf_{y \in U} f(y) \geq t$ for all $t < f(x)$, i.e. $\sup_{U \in \mathcal{N}(x)} \inf_{y \in U} f(y) \geq f(x)$.

Definition 1.1 makes sense also if we only have a notion of convergence of sequences on X. Any of the equivalent conditions (ii)–(v) can be taken as the definition of lower semicontinuity for functions defined on arbitrary topological spaces. In general topological spaces these notions differ from *sequential* lower semicontinuity as introduced in Definition 1.1.

Remark 1.3 (i) If f and g are l.s.c. at x, then so is $f + g$.

    (ii) Let $\{f_i : \ i \in I\}$ be a family of l.s.c. functions (I an arbitrary set of indices). Then the function defined by $f(x) = \sup_i f_i(x)$ is l.s.c. In fact, for fixed $x \in X$ and $x_j \to x$, we have

$$f_i(x) \leq \liminf_j f_i(x_j) \leq \liminf_j f(x_j).$$

By taking the supremum for $i \in I$ we obtain $f(x) \leq \liminf_j f(x_j)$. In particular, the supremum of a family of continuous functions is l.s.c.

    (iii) If $f = \chi_E$ is the characteristic function of the set E, then f is l.s.c. if and only if E is open, by Remark 1.2(iv).

    (iv) A function $f : X \to \overline{\mathbf{R}}$ is called *upper semicontinuous* if $-f$ is l.s.c. All the results of this section have an obvious counterpart for upper semicontinuous functions. In particular $f = \chi_E$ is upper semicontinuous if and only if E is closed.

Definition 1.4 *Let $f : X \to \overline{\mathbf{R}}$ be a function. Its* lower semicontinuous envelope *$\overline{f}$ is the greatest lower semicontinuous function not greater than f, i.e. for every* $x \in X$

$$\overline{f}(x) = \sup\{g(x): \ g \ l.s.c., \ g \leq f\}. \tag{1.4}$$

Notice that (1.4) gives an l.s.c. function by (ii) of the previous remark. The function $\overline{f}$ can be described as follows: for every $x \in X$

$$\overline{f}(x) = \liminf_{y \to x} f(y) = \min\{\liminf_j f(x_j) : \ x_j \to x\}. \tag{1.5}$$

To verify (1.5) we check first that the function $h(x) = \liminf_{y \to x} f(y)$ is l.s.c., i.e. by (1.3), that

$$h(x) \le \liminf_{y \to x} h(y).$$

Now, we have

$$
\begin{aligned}
\liminf_{y \to x} h(y) &= \sup_{U \in \mathcal{N}(x)} \inf_{y \in U} h(y)\\
&= \sup_{U \in \mathcal{N}(x)} \inf_{y \in U} \sup_{V \in \mathcal{N}(y)} \inf_{z \in V} f(z)\\
&\ge \sup_{U \in \mathcal{N}(x)} \inf_{y \in U} \sup_{U \supseteq V \in \mathcal{N}(y)} \inf_{z \in V} f(z)\\
&\ge \sup_{U \in \mathcal{N}(x)} \inf_{y \in U} \sup_{U \supseteq V \in \mathcal{N}(y)} \inf_{z \in U} f(z)\\
&\ge \sup_{U \in \mathcal{N}(x)} \inf_{z \in U} f(z) = h(x).
\end{aligned}
$$

If $g \le f$ is l.s.c. then $g(x) = \liminf_{y \to x} g(y) \le \liminf_{y \to x} f(y) = h(x)$ for all $x \in X$, and the first equality in (1.5) is proven. As X is metric

$$\liminf_{y \to x} f(y) = \inf\{\liminf_j f(x_j) : x_j \to x\}.$$

It is easy to check by a diagonal argument that indeed there exists a sequence (x_j) such that $x_j \to x$ and $\overline{f}(x) = \lim_j f(x_j)$; hence

$$\overline{f}(x) = \min\{\lim_j f(x_j) : \ x_j \to x, \ \exists \lim_j f(x_j)\}, \tag{1.6}$$

and in particular the second equality in (1.5) holds.

1.2 Yosida transforms

We address now the problem of the approximation of l.s.c. functions by continuous functions.

Definition 1.5 *Let $\psi : [0, +\infty) \to [0, +\infty)$ be a strictly increasing continuous function with $\psi(0) = 0$. If $f : X \to \overline{\mathbf{R}}$, then for all $\lambda \ge 0$ we define the* Yosida transform *of f, $T_\lambda^\psi f : X \to \overline{\mathbf{R}}$, by*

$$T_\lambda^\psi f(x) = \inf\{f(y) + \lambda \psi(d(x,y)) : \ y \in X\}$$

for every $x \in X$. If $\psi(t) = t^p$ with $p > 0$, then we use the notation $T_\lambda^p f$ for $T_\lambda^\psi f$; moreover, if $p = 1$ we write $T_\lambda f$ in place of $T_\lambda^p f$:

$$T_\lambda^p f(x) = \inf\{f(y) + \lambda(d(x,y))^p : \ y \in X\},$$

$$T_\lambda f(x) = \inf\{f(y) + \lambda d(x,y) : \ y \in X\}.$$

Remark 1.6 (a) Whatever ψ and f may be, if $0 \leq \lambda \leq \eta$ then

$$T_0^\psi f = \inf_X f \leq T_\lambda^\psi f \leq T_\eta^\psi f \leq f, \quad \text{and} \quad T_\lambda^\psi f \leq T_\lambda^\psi g \quad \text{if } f \leq g.$$

(b) Let f be bounded from below and proper; then

(i) for all $x_1, x_2 \in X$

$$|T_\lambda f(x_1) - T_\lambda f(x_2)| \leq \lambda\, d(x_1, x_2).$$

Moreover, if $d(x, y) < +\infty$ for all $x, y \in X$ and $p \geq 1$, then we have the following local Lipschitz condition: for all $x_1, x_2 \in X$

$$|T_\lambda^p f(x_1) - T_\lambda^p f(x_2)| \leq (1 + \lambda)c\big(1 + (d(x_1, x_0))^{p-1} + (d(x_2, x_0))^{p-1}\big)\, d(x_1, x_2),$$

where c is a suitable constant and x_0 is any point in X fixed; in particular, if $X = \mathbf{R}^n$, and we take $x_0 = 0$, then we have

$$|T_\lambda^p f(x_1) - T_\lambda^p f(x_2)| \leq (1 + \lambda)c(1 + |x_1|^{p-1} + |x_2|^{p-1})|x_1 - x_2|;$$

(ii) if $0 < p \leq 1$ then $T_\lambda^p f$ is the greatest function not greater than f satisfying the Hölder condition

$$|T_\lambda^p f(x_1) - T_\lambda^p f(x_2)| \leq \lambda(d(x_1, x_2))^p;$$

in particular, $T_\lambda f$ is the greatest Lipschitz function with Lipschitz constant λ not greater than f.

As for the proof of (i), we begin by noticing that it is not restrictive to suppose $f \geq 0$. Fix $\overline{x} \in X$ such that $f(\overline{x}) < +\infty$. It is not restrictive to prove (i) with $x_0 = \overline{x}$. For fixed $\varepsilon \in (0, 1)$, we can find $y_\varepsilon \in X$ such that $f(y_\varepsilon) + \lambda(d(x_1, y_\varepsilon))^p \leq T_\lambda^p f(x_1) + \varepsilon$. Then

$$\begin{aligned} T_\lambda^p f(x_2) &\leq f(y_\varepsilon) + \lambda(d(x_2, y_\varepsilon))^p \\ &\leq f(y_\varepsilon) + \lambda(d(x_1, x_2) + d(x_1, y_\varepsilon))^p. \end{aligned}$$

If $p = 1$ we get $T_\lambda f(x_2) \leq T_\lambda f(x_1) + \varepsilon + \lambda d(x_1, x_2)$, and, by a symmetry argument and the arbitrariness of ε, $|T_\lambda f(x_2) - T_\lambda f(x_1)| \leq \lambda d(x_1, x_2)$. In general, note that

$$\begin{aligned} \lambda(d(x_1, y_\varepsilon))^p &\leq T_\lambda^p f(x_1) + 1 \\ &\leq f(\overline{x}) + \lambda(d(x_1, \overline{x}))^p + 1 \leq c + \lambda(d(x_1, x_0))^p, \end{aligned}$$

hence $(d(x_1, y_\varepsilon))^{p-1} \leq c(\lambda^{(1/p)-1} + (d(x_1, x_0))^{p-1})$. Thus, using the simple inequality $||a + b|^p - |a|^p| \leq c(1 + |a|^{p-1} + |b|^{p-1})|b|$,

$$T_\lambda^p f(x_2) \leq f(y_\varepsilon) + \lambda(d(x_1, y_\varepsilon))^p + \lambda\Big((d(x_1, x_2) + d(x_1, y_\varepsilon))^p - (d(x_1, y_\varepsilon))^p\Big)$$

$$\leq T_\lambda^p f(x_1) + \varepsilon + \lambda c\Big(1 + (d(x_1, x_2))^{p-1} + (d(x_1, y_\varepsilon))^{p-1}\Big) d(x_1, x_2)$$
$$\leq T_\lambda^p f(x_1) + \varepsilon$$
$$+ (1 + \lambda)c\Big(1 + (d(x_1, x_0))^{p-1} + (d(x_2, x_0))^{p-1}\Big) d(x_1, x_2).$$

Letting $\varepsilon \to 0$ we obtain

$$T_\lambda^p f(x_2) - T_\lambda^p f(x_1) \leq (1 + \lambda)c\Big(1 + (d(x_1, x_0))^{p-1} + (d(x_2, x_0))^{p-1}\Big) d(x_1, x_2),$$

and the same inequality for $T_\lambda^p f(x_1) - T_\lambda^p f(x_2)$ follows by a symmetry argument.

We prove (ii) only in the case $p = 1$. The case $p < 1$ is left as an exercise (see also Dal Maso (1993) p. 107). By (i) $T_\lambda f$ is Lipschitz with constant λ. Note that if g is Lipschitz with constant λ, then $g(x) \leq g(y) + \lambda d(x, y)$ for all x, y in X. Hence $T_\lambda g = g$. If in addition $g \leq f$, then $g = T_\lambda g \leq T_\lambda f$. Hence, we have (ii).

(c) If $X = \mathbf{R}^n$, and ψ and f are convex, then $T_\lambda^\psi f$ is also convex.

Proposition 1.7 *Let $f : X \to \overline{\mathbf{R}}$ be bounded from below; then for all $x \in X$*

$$\overline{f}(x) = \sup_{\lambda \geq 0} T_\lambda^\psi f(x).$$

Proof We suppose without loss of generality that $f \geq 0$. Fix $x_0 \in X$; if there exists $\eta > 0$ such that $f(x) = +\infty$ for all $x \in X$ with $d(x, x_0) < \eta$, then trivially

$$\overline{f}(x_0) = \sup_{\lambda \geq 0} T_\lambda^\psi f(x_0) = +\infty.$$

Hence we fix $\varepsilon > 0$ and suppose that there exists $\overline{x} \in X$ with $\psi(d(x_0, \overline{x})) \leq \frac{\varepsilon}{2}$ such that $f(\overline{x}) < +\infty$. Let $\varepsilon\lambda \geq 2f(\overline{x})$. If $\psi(d(x, x_0)) \geq \varepsilon$, then

$$f(x) + \lambda\psi(d(x, x_0)) \geq \lambda\varepsilon = \lambda\frac{\varepsilon}{2} + \lambda\frac{\varepsilon}{2}$$
$$\geq f(\overline{x}) + \lambda\frac{\varepsilon}{2} \geq f(\overline{x}) + \lambda\psi(d(\overline{x}, x_0)),$$

so that

$$T_\lambda^\psi f(x_0) = \inf_{x \in X} \{f(x) + \lambda\psi(d(x, x_0))\}$$
$$= \inf\{f(x) + \lambda\psi(d(x, x_0)) : \psi(d(x, x_0)) \leq \varepsilon\},$$

and eventually

$$\sup_\lambda T_\lambda^\psi f(x_0) = \sup_{\varepsilon > 0} \sup_\lambda \inf\{f(x) + \lambda\psi(d(x, x_0)) : \psi(d(x, x_0)) \leq \varepsilon\}$$
$$= \sup_\lambda \sup_{\delta > 0} \inf_{d(x, x_0) \leq \delta} \{f(x) + \lambda\psi(d(x, x_0))\}$$

$$= \sup_{\lambda} \liminf_{x \to x_0}(f(x) + \lambda\psi(d(x, x_0)))$$

$$= \liminf_{x \to x_0} f(x) = \overline{f}(x_0),$$

as required. $\qquad\qquad\qquad\qquad\qquad\qquad\qquad\qquad\qquad\qquad\qquad\qquad\qquad$ $\square$

Proposition 1.7 shows that any l.s.c. function f bounded from below can be seen as the pointwise supremum of a family of continuous functions (take $\psi(t) = t$ and recall Remark 1.6). The condition of boundedness from below can be replaced by $f(x) \geq c - \lambda_0\psi(d(x, x_0))$ for some $\lambda_0 \geq 0$ and $x_0 \in X$, for all $x \in X$. Moreover, if f is convex then it is the pointwise supremum of a family of Lipschitz and smooth convex functions (see Exercise 1.5).

1.3   Coerciveness conditions. The direct method

By a *compact* subset of X we mean a sequentially compact set $K \subset X$, i.e.

$$\forall(x_j) \subset K \; \exists x \in K, \; \exists(x_{j_k}) : \; x_{j_k} \to x.$$

A set $K \subset X$ is *precompact* if its closure is compact, i.e.

$$\forall(x_j) \subset K \; \exists(x_{j_k}) : \; x_{j_k} \text{ converges in } X.$$

Definition 1.8 *A function $f : X \to \overline{\mathbf{R}}$ is* coercive *if for all $t \in \mathbf{R}$ the set $\{f \leq t\}$ is precompact. A function $f : X \to \overline{\mathbf{R}}$ is* mildly coercive *if there exists a non-empty compact set $K \subset X$ such that* $\inf_X f = \inf_K f$.

Remark If f is l.s.c. and coercive, then for all $t \in \mathbf{R}$ the set $\{f \leq t\}$ is compact. If f is coercive then it is mildly coercive. In fact, if f is not identically $+\infty$ (in which case we take K as any compact subset of X), then there exists $t \in \mathbf{R}$ such that $\{f \leq t\}$ is not empty, and we take K as the closure of $\{f \leq t\}$ in X. An example of a non-coercive, mildly coercive function is given by any periodic function $f : \mathbf{R}^n \to \mathbf{R}$.

An intermediate condition between coerciveness and mild coerciveness is the requirement that there exists $t \in \mathbf{R}$ such that $\{f \leq t\}$ is not empty and precompact.

Combined lower semicontinuity and coerciveness assure the existence of minimum points, as specified by the following version of a well-known theorem.

Theorem 1.9 (Weierstrass' Theorem) *If $f : X \to \overline{\mathbf{R}}$ is mildly coercive, then there exists the minimum value $\min\{\overline{f}(x) : x \in X\}$, and it equals the infimum $\inf\{f(x) : x \in X\}$. Moreover, the minimum points for $\overline{f}$ are exactly all the limits of converging sequences (x_j) such that $\lim_j f(x_j) = \inf_X f$.*

Proof Clearly $\inf_X \overline{f} \leq \inf_X f$ since $\overline{f} \leq f$. On the other hand the constant function $g = \inf_X f$ is trivially l.s.c. and $g \leq f$; hence $g \leq \overline{f}$ so that $\inf_X f \leq \inf_X \overline{f}$.

Let K be a compact set such that $\inf_K f = \inf_X f$ $(= \inf_X \overline{f})$. By possibly extracting a converging subsequence, we find a sequence $(x_j) \subset K$ such that $f(x_j) \to \inf_X f$, $x_j \to \overline{x} \in K$. By the lower semicontinuity of $\overline{f}$

$$\inf_X \overline{f} \leq \overline{f}(\overline{x}) \leq \liminf_j \overline{f}(x_j) \leq \liminf_j f(x_j) = \inf_X f = \inf_X \overline{f}.$$

In the same way we see that each limit of a convergent minimizing sequence is a minimum point for the function $\overline{f}$. Conversely, if $\overline{x}$ is a minimum point for $\overline{f}$, by (1.6) we can find a sequence (x_j) converging to $\overline{x}$ such that $\lim_j f(x_j) = \overline{f}(\overline{x}) = \inf_X f$. $\qquad\square$

The application of this theorem to prove the existence of solutions to minimum problems is usually referred to as the 'direct method' of the calculus of variations.

1.4 Exercises

Exercise 1.1 Prove statement (i) of Remark 1.3 (it suffices to recall that we have $\liminf_j (f(x_j)+g(x_j)) \geq \liminf_j f(x_j)+\liminf_j g(x_j)$). Note that in general $\overline{f+g} \neq \overline{f}+\overline{g}$.

Exercise 1.2 Prove statement (ii) of Remark 1.6(b) in the case $0 < p < 1$.

Exercise 1.3 Prove statement (c) of Remark 1.6.

Exercise 1.4 Prove that $f : \mathbf{R}^n \to \mathbf{R}$ is coercive if and only if $\lim_{|x| \to +\infty} f(x) = +\infty$.

Exercise 1.5 Prove that if $f : [0, +\infty) \to [0, +\infty]$ is convex and lower semicontinuous, then it is the pointwise supremum of a family (f_λ) of smooth convex functions with f_λ Lipschitz with constant λ (use Proposition 1.7, Remark 1.6(c) and a convolution argument).

2

WEAK CONVERGENCE

The set Ω will be a bounded open subset of $\mathbf{R}^n$. If $1 \leq p \leq \infty$ we denote by $L^p(\Omega)$ the Lebesgue space of the p-summable functions over Ω, and by $W^{1,p}(\Omega)$ the Sobolev space of the functions in $L^p(\Omega)$ whose distributional first derivatives belong to $L^p(\Omega)$, i.e. for all $i = 1, \ldots, n$ a function $f_i \in L^p(\Omega)$ exists such that

$$\int_\Omega u \frac{\partial \phi}{\partial x_i} \, dx = - \int_\Omega f_i \phi \, dx \qquad \text{for all } \phi \in C_0^\infty(\Omega).$$

We denote the *weak partial derivatives* of u and the *weak gradient* of u by

$$\frac{\partial u}{\partial x_i} = f_i \qquad \text{and} \qquad Du = (f_1, \ldots, f_n),$$

respectively. We adopt the usual notation

$$\|u\|_{L^p(\Omega)}, \qquad \|u\|_{W^{1,p}(\Omega)}$$

for the $L^p(\Omega)$ and $W^{1,p}(\Omega)$ norm of the function u. We sometimes write $\|u\|_\infty$ in place of $\|u\|_{L^\infty(\Omega)}$ if no confusion is possible. The subspace $W_0^{1,p}(\Omega)$ of $W^{1,p}(\Omega)$ is the closure of $C_0^\infty(\Omega)$ with respect to the $W^{1,p}$ norm.

In the same way, $L^p(\Omega; \mathbf{R}^m)$ and $W^{1,p}(\Omega; \mathbf{R}^m)$ denote the spaces of $\mathbf{R}^m$-valued functions defined on Ω whose components belong to $L^p(\Omega)$ and $W^{1,p}(\Omega)$, respectively. The *conjugate exponent* of p is $p' = p/(p-1)$ ($p' = \infty$ if $p = 1$ and $p' = 1$ if $p = \infty$). If $1 \leq p < n$ the *Sobolev exponent* of p is $p^* = pn/(n-p)$.

2.1 Weak convergence in Lebesgue spaces

Definition 2.1 *Let* $1 \leq p < \infty$. *We define the* weak convergence *in* $L^p(\Omega)$ *as follows: we say that a sequence* (u_j) *in* $L^p(\Omega)$ *converges weakly to* $u \in L^p(\Omega)$ *(written* $u_j \rightharpoonup u$ *in* $L^p(\Omega)$*) if for all* $v \in L^{p'}(\Omega)$ *we have*

$$\lim_j \int_\Omega v(u_j - u) \, dx = 0. \tag{2.1}$$

We also define the convergence *in the sense of distributions,* $u_j \to u$ *in* $\mathcal{D}'(\Omega)$, *if* (2.1) *holds for all* $v \in C_0^\infty(\Omega)$.

The following properties hold for weak converging sequences.

Proposition 2.2 *Let $1 \le p < \infty$, and $u_j \rightharpoonup u$ in $\mathrm{L}^p(\Omega)$. Then*

$$\text{(i)} \quad \sup_j \|u_j\|_{\mathrm{L}^p(\Omega)} < +\infty; \tag{2.2}$$

$$\text{(ii)} \quad \|u\|_{\mathrm{L}^p(\Omega)} \le \liminf_j \|u_j\|_{\mathrm{L}^p(\Omega)}. \tag{2.3}$$

Remark 2.3 (a) Equation (2.1) can be read as the requirement that for all $v \in \mathrm{L}^{p'}(\Omega)$ the functions $u \mapsto \int_\Omega vu \, dx$ are *weakly continuous*, i.e. continuous for the weak convergence.

(b) Proposition 2.2(i) is an immediate consequence of the Banach-Steinhaus Theorem since

$$\|u\|_{\mathrm{L}^p(\Omega)} = \sup\left\{ \int_\Omega vu \, dx : \ \|v\|_{\mathrm{L}^{p'}(\Omega)} \le 1 \right\}.$$

The inequality in Proposition 2.2(ii) asserts the *weak lower semicontinuity* of the norm. This fact is explained by regarding the norm as the supremum of a family of weakly continuous functions, and hence weakly lower semicontinuous (as in Remark 1.3(ii)).

(c) If $p = \infty$ we use a different terminology. If (2.1) holds for all $v \in \mathrm{L}^1(\Omega)$, then we say that (u_j) converges weakly* to u ($u_j \rightharpoonup^* u$). The analogue of Proposition 2.2 holds.

(d) If $1 < p < \infty$ then part (ii) of Proposition 2.2 can be sharpened: it can be proved that if $u_j \rightharpoonup u$ in $\mathrm{L}^p(\Omega)$, and $\|u\|_{\mathrm{L}^p(\Omega)} = \lim_j \|u_j\|_{\mathrm{L}^p(\Omega)}$, then (u_j) converges to u *strongly* in $\mathrm{L}^p(\Omega)$. This fact is a consequence of the uniform convexity of the L^p-norm for $1 < p < \infty$, and can be derived from the so-called Clarkson inequalities: if $p \ge 2$

$$\int_\Omega \left(\left| \frac{u-v}{2} \right|^p + \left| \frac{u+v}{2} \right|^p \right) dx \le \frac{1}{2} \int_\Omega (|u|^p + |v|^p) \, dx$$

and, if $1 < p \le 2$,

$$\left(\int_\Omega \left| \frac{u-v}{2} \right|^p dx \right)^{1/(p-1)} + \left(\int_\Omega \left| \frac{u+v}{2} \right|^p dx \right)^{1/(p-1)} \le \left(\frac{1}{2} \int_\Omega (|u|^p + |v|^p) \, dx \right)^{1/(p-1)}$$

Indeed (suppose, for example, $1 < p \le 2$) if we set $v = u_j$ in the second inequality, recalling that by Proposition 2.2(ii)

$$\int_\Omega |u|^p \, dx \le \liminf_j \int_\Omega \left| \frac{u+u_j}{2} \right|^p dx$$

since $\frac{1}{2}(u + u_j) \rightharpoonup u$, and passing to the limit in j, we get

$$\limsup_j \left(\int_\Omega \left| \frac{u-u_j}{2} \right|^p dx \right)^{1/(p-1)} + \left(\int_\Omega |u|^p \, dx \right)^{1/(p-1)}$$

$$\leq \lim_j \left(\frac{1}{2}\int_\Omega (|u|^p + |u_j|^p)\, dx\right)^{1/(p-1)} = \left(\int_\Omega |u|^p\, dx\right)^{1/(p-1)},$$

so that $u - u_j \to 0$ in $\mathrm{L}^p(\Omega)$.

The same property does not hold in $\mathrm{L}^1(\Omega)$ and in $\mathrm{L}^\infty(\Omega)$ (see Exercise 2.1).

(e) If $u_j \rightharpoonup u$ in $\mathrm{L}^p(\Omega)$, and $v_j \to v$ in $\mathrm{L}^{p'}(\Omega)$, then $\int_\Omega u_j v_j\, dx \to \int_\Omega uv\, dx$. This is trivially false if we only have $v_j \rightharpoonup v$ in $\mathrm{L}^{p'}(\Omega)$.

By the Banach Alaoglu Bourbaki Theorem we have the following compactness result.

Theorem 2.4 (Weak compactness) *Let $1 < p \leq \infty$. If (u_j) is a norm-bounded sequence in $\mathrm{L}^p(\Omega)$, then there exists a subsequence of (u_j) weakly converging in $\mathrm{L}^p(\Omega)$ (weakly* if $p = \infty$).*

Remark 2.5 Theorem 2.4 and Proposition 2.2(i) show that a subset of $\mathrm{L}^p(\Omega)$ is compact for the weak convergence (in the sense of Chapter 1) if and only if it is norm bounded. Theorem 2.4 does not hold if $p = 1$. To overcome this fact it is customary to identify $\mathrm{L}^1(\Omega)$ as a subset of the space of Borel measures on Ω, on which similar compactness properties hold, associating to each $u \in \mathrm{L}^1(\Omega)$ the measure $\mu_u(E) = \int_E u\, dx$.

Remark 2.6 (Convergence of mean values) If $1 \leq p < \infty$, and $u_j \rightharpoonup u$ in $\mathrm{L}^p(\Omega)$, then, taking $v = \chi_E$ in (2.1) (E a measurable set), we have

$$\int_E u_j\, dx \to \int_E u\, dx, \tag{2.4}$$

which implies the convergence of the mean values on E of the functions u_j to the mean value on E of the function u. Conversely, if (u_j) is a bounded sequence in $\mathrm{L}^p(\Omega)$, and we have the convergence (2.4) for all measurable E, then $u_j \rightharpoonup u$ in $\mathrm{L}^p(\Omega)$. In fact, from (2.4) we get the convergence (2.1) for all piecewise constant functions v, and then for arbitrary $v \in \mathrm{L}^{p'}(\Omega)$ by approximation (recall Remark 2.3(e)). The same remark applies to the weak* convergence in $\mathrm{L}^\infty(\Omega)$.

Example 2.7 Let Y be an n-cube in $\mathbf{R}^n$, let $1 \leq p \leq \infty$ and let $u \in \mathrm{L}^p_{\mathrm{loc}}(\mathbf{R}^n)$ be Y-periodic. Define for $\varepsilon > 0$, $u_\varepsilon(x) = u(\frac{x}{\varepsilon})$. Then as $\varepsilon \to 0$

$$u_\varepsilon \rightharpoonup \frac{1}{|Y|}\int_Y u(y)\, dy \qquad (\rightharpoonup^* \text{ if } p = \infty) \tag{2.5}$$

in $\mathrm{L}^p(\Omega)$ for every bounded open subset Ω of $\mathbf{R}^n$.

Proof It is no restriction to suppose $Y = (0,1)^n$. We first treat the case $p = \infty$. Since the family (u_ε) is bounded in $\mathrm{L}^\infty(\Omega)$ there exist a sequence (ε_j) and $\overline{u} \in \mathrm{L}^\infty(\Omega)$ such that $u_{\varepsilon_j} \rightharpoonup^* \overline{u}$ in $\mathrm{L}^\infty(\Omega)$. We show that

$$\overline{u}(x) = \int_{(0,1)^n} u(y)\, dy \qquad \text{for } x \in \Omega \text{ a.e.} \tag{2.6}$$

Let $Q \subset \Omega$ be an n-cube with edges parallel to the coordinate axes and of side-length l. We can suppose $l > \varepsilon_j$, so that $(1/\varepsilon_j)Q$ is the union of $[(l/\varepsilon_j) - 1]^n$ unit cubes of the fundamental lattice ($[t]$ is the integer part of $t \in \mathbf{R}$), and a set E_j with $|E_j| = (l/\varepsilon_j)^n - [(l/\varepsilon_j) - 1]^n \le 2n(l/\varepsilon_j)^{n-1}$. We have

$$\left| \int_Q \left(u_{\varepsilon_j}(x) - \int_Y u(y)\, dy \right) dx \right| = \left| \varepsilon_j^n \int_{\frac{1}{\varepsilon_j}Q} \left(u(z) - \int_Y u(y)\, dy \right) dz \right|$$

$$= \left| \varepsilon_j^n \int_{E_j} \left(u(x) - \int_Y u(y)\, dy \right) dx \right|$$

$$\le \varepsilon_j^n \|u\|_{\mathrm{L}^\infty(Y)} |E_j|,$$

which tends to 0 as $j \to \infty$. Thus

$$\int_E u_{\varepsilon_j}\, dx \to |E| \left(\int_{(0,1)^n} u(y)\, dy \right) \tag{2.7}$$

whenever E is a countable union of cubes with edges parallel to the coordinate axes. By approximation (2.7) holds for all measurable E, and then (2.6) follows by Remark 2.6. Since $\overline{u}$ is uniquely determined by u the convergence in (2.5) follows.

Next, let $p < \infty$. Let u^j be a sequence of Y-periodic functions in $\mathrm{L}^\infty(\mathbf{R}^n)$ converging strongly to u in $\mathrm{L}^p(Y)$ (e.g. obtained by truncation or convolution of u). Then $u_\varepsilon^j \to u_\varepsilon$ in $\mathrm{L}^p(\Omega)$ as $j \to +\infty$ uniformly for $\varepsilon < 1$. In fact, if $I_\varepsilon = \{k \in \mathbf{Z}^n : (k + Y) \cap \frac{1}{\varepsilon}\Omega \ne \emptyset\}$ then

$$\int_\Omega |u_\varepsilon^j - u_\varepsilon|^p\, dx \le \varepsilon^n \int_{\frac{1}{\varepsilon}\Omega} |u^j - u|^p\, dz \le \varepsilon^n \sum_{k \in I_\varepsilon} \int_{k+Y} |u^j - u|^p\, dz$$

$$= \varepsilon^n \sum_{k \in I_\varepsilon} \int_Y |u^j - u|^p\, dz \le c \int_Y |u^j - u|^p\, dz.$$

By the first part of the proof we have $u_\varepsilon^j \rightharpoonup^* \int_Y u^j(y)\, dy$ in $\mathrm{L}^\infty(\Omega)$ as $\varepsilon \to 0$. If $v \in \mathrm{L}^{p'}(\Omega)$ then we write

$$\int_\Omega u_\varepsilon v\, dx = \int_\Omega u_\varepsilon^j v\, dx + \int_\Omega (u_\varepsilon - u_\varepsilon^j)v\, dx;$$

letting $\varepsilon \to 0$ and $j \to +\infty$, we obtain $u_\varepsilon \rightharpoonup \int_Y u(y)\, dy$ in $\mathrm{L}^p(\Omega)$, and conclude the proof. $\qquad \square$

2.2   Weak convergence in Sobolev spaces

Definition 2.8 *We extend the definition of* weak convergence *to sequences in* $L^p(\Omega; \mathbf{R}^k)$ *by saying that* $u_j \rightharpoonup u$ *in* $L^p(\Omega; \mathbf{R}^k)$ *if each component converges weakly in* $L^p(\Omega)$, *and to Sobolev spaces by saying that* $u_j \rightharpoonup u$ *in* $W^{1,p}(\Omega)$ *if* $u_j \rightharpoonup u$ *in* $L^p(\Omega)$ *and* $Du_j \rightharpoonup Du$ *in* $L^p(\Omega; \mathbf{R}^n)$ ($\rightharpoonup^*$ *if* $p = \infty$). *If* $u_j, u \in W^{1,p}(\Omega; \mathbf{R}^m)$ *we say that* $u_j \rightharpoonup u$ *in* $W^{1,p}(\Omega; \mathbf{R}^m)$ *if each component converges weakly in* $W^{1,p}(\Omega)$.

Remark 2.9 From every norm-bounded sequence (u_j) in $W^{1,p}(\Omega)$ ($1 < p < \infty$) we can extract a weakly convergent subsequence. In fact, by applying Theorem 2.4 twice we obtain the existence of a subsequence (u_{j_k}) and of two functions $u, f \in L^p(\Omega)$ such that $u_{j_k} \rightharpoonup u$ in $L^p(\Omega)$, and $Du_{j_k} \rightharpoonup f$ in $L^p(\Omega; \mathbf{R}^n)$. Then for all $\varphi \in C_0^\infty(\Omega)$ we have

$$
\int_\Omega f\varphi \, dx = \lim_k \int_\Omega Du_{j_k} \varphi \, dx
$$

$$
= -\lim_k \int_\Omega u_{j_k} D\varphi \, dx = - \int_\Omega u D\varphi \, dx;
$$

hence $f = Du$, $u \in W^{1,p}(\Omega)$ and $u_{j_k} \rightharpoonup u$ in $W^{1,p}(\Omega)$. The same remark applies to norm-bounded sequences in $W^{1,\infty}(\Omega)$ and weak* convergence.

Example 2.10 Let $Y \subset \mathbf{R}^n$ be an n-cube, and let $v \in W_{\mathrm{loc}}^{1,p}(\mathbf{R}^n; \mathbf{R}^m)$ be such that Dv is Y-periodic. Define $v_\varepsilon(x) = \varepsilon v(\frac{x}{\varepsilon})$. Then

$$
v_\varepsilon \rightharpoonup \left(\frac{1}{|Y|} \int_Y Dv(y) \, dy \right) x
$$

($\rightharpoonup^*$ if $p = \infty$) in $W^{1,p}(\Omega; \mathbf{R}^m)$ for every bounded open subset Ω of $\mathbf{R}^n$.

Proof We suppose $Y = (0, 1)^n$. For all $i = 1, \ldots, n$ the function $z_i(x) = v(x + e_i) - v(x)$ is constant. We define the matrix A by $Ae_i = z_i = z_i(x)$. The function $w(x) = v(x) - Ax$ is Y-periodic. Note that we have

$$
v_\varepsilon(x) = Ax + \varepsilon w\left(\frac{x}{\varepsilon} \right), \qquad Dv_\varepsilon(x) = A + Dw\left(\frac{x}{\varepsilon} \right).
$$

Thus $v_\varepsilon \to Ax$ in $L^p(\Omega; \mathbf{R}^m)$ and $Dv_\varepsilon \rightharpoonup \int_Y (A + Dw(y)) \, dy = \int_Y Dv(y) \, dy$ in $L^p(\Omega; \mathbf{R}^{nm})$ by Example 2.7. $\square$

Remark 2.11 Using the Hahn Banach theorem it is easy to see that a convex subset of $L^p(\Omega; \mathbf{R}^m)$ is closed if and only if it is closed for the weak convergence. As a consequence it is immediate to see for example that $W_0^{1,p}(\Omega; \mathbf{R}^m)$ is closed under weak convergence.

2.3   Weak* convergence of measures

We recall a few results on the weak* convergence of measures.

Definition 2.12 *Let (μ_j) be a sequence of (signed) Borel measures, and let $X(\Omega)$ be the closure of $C_0^\infty(\Omega)$ in the L^∞-topology. We say that the sequence (μ_j) converges weakly* to μ if*

$$\int_\Omega f\,d\mu = \lim_j \int_\Omega f\,d\mu_j$$

for all $f \in X(\Omega)$. To each signed measure μ on Ω we associate the positive measure $|\mu|$, called the total variation *of μ, given by*

$$|\mu|(B) = \sup\Big\{\sum_i |\mu(B_i)| : \bigcup_i B_i = B,\ B_i\ Borel\ sets\Big\}$$

for all Borel subsets B of Ω.

From the Banach Steinhaus and the Banach Alaoglu Bourbaki theorems it is easy to deduce the following result.

Theorem 2.13 *The sequence of measures (μ_j) converges weakly* to μ if and only if $\sup_j |\mu_j|(\Omega) < +\infty$ and there exists a dense subset $D \subset X(\Omega)$ such that*

$$\lim_j \int_\Omega u\,d\mu_j = \int_\Omega u\,d\mu \qquad \forall u \in D.$$

Moreover, every sequence (μ_j) such that $(|\mu_j|(\Omega))$ is bounded admits a weakly converging subsequence, and the map $\mu \mapsto |\mu|(\Omega)$ is l.s.c. with respect to weak* convergence.*

Remark 2.14 All finite positive Borel measures μ are *regular*, i.e.

$$\mu(B) = \inf\{\mu(A) : A\ \text{open},\ B \subset A\}.$$

Proposition 2.15 *If (μ_j) is a sequence of positive Borel measures converging weakly* to μ, then*

$$\int_\Omega f\,d\mu \le \liminf_j \int_\Omega f\,d\mu_j$$

for every l.s.c. function $f : \Omega \to [0, +\infty]$, and

$$\int_\Omega f\,d\mu \ge \limsup_j \int_\Omega f\,d\mu_j$$

for every upper semicontinuous function $f : \Omega \to [0, +\infty)$ with compact support. In particular

$$\mu(U) \le \liminf_j \mu_j(U) \tag{2.8}$$

for all open subsets $U \subset \Omega$, and

$$\mu(K) \ge \limsup_j \mu_j(K) \tag{2.9}$$

for all compact subsets $K \subset \Omega$.

Proof Let $f : \Omega \to [0, +\infty]$ be l.s.c., and let $f_\lambda = T_\lambda f$ be as in Definition 1.5. Then we have $f = \sup_\lambda f_\lambda$ by Proposition 1.7. Let $(\phi_k) \subset C_0^\infty(\Omega)$ be an increasing sequence converging to 1 in Ω. Then

$$\int_\Omega \phi_k f_\lambda \, d\mu = \lim_j \int_\Omega \phi_k f_\lambda \, d\mu_j \leq \liminf_j \int_\Omega f \, d\mu_j.$$

Letting $k \to +\infty$ and $\lambda \to +\infty$ Beppo Levi's Theorem yields the desired inequality.

If $f : \Omega \to [0, +\infty)$ is upper semicontinuous with compact support in Ω, then $-f$ is l.s.c. and bounded from below, so that $-f = \sup_\lambda T_\lambda(-f)$, and we can write $f = \inf_\lambda f_\lambda$, where $f_\lambda = -T_\lambda(-f)$. Moreover, the support of f_λ is compact in Ω for λ large. Then

$$\int_\Omega f_\lambda \, d\mu \geq \limsup_j \int_\Omega f_\lambda \, d\mu_j \geq \limsup_j \int_\Omega f \, d\mu_j.$$

Letting $\lambda \to +\infty$ we complete the proof. $\qquad\square$

Remark 2.16 From (2.8) and (2.9) we obtain that

$$\mu(U) = \lim_j \mu_j(U),$$

for every open set U with compact closure in Ω such that $\mu(\partial U) = 0$.

Remark 2.17 If μ is a signed Borel measure, we have

$$|\mu|(\Omega) = \sup\left\{ \int_\Omega f \, d\mu : \ f \in X(\Omega), \ \|f\|_\infty \leq 1 \right\}, \qquad (2.10)$$

which shows that the functional $\mu \mapsto |\mu|(\Omega)$ is l.s.c. with respect to the weak* convergence, as the supremum of a family of continuous functionals.

Remark 2.18 If (μ_j) is a sequence of signed measures such that $\mu_j \rightharpoonup^* \mu$ and $|\mu_j| \rightharpoonup^* \lambda$, then $|\mu| \leq \lambda$.

It suffices to check this fact on open sets, by the regularity of the measures. Let A be an open subset of Ω and let $t > 0$. Let $A_t = \{x \in A : \ \text{dist}\,(x, \partial A) > t\}$, and let $u \in X(\Omega)$ with $\chi_{A_t} \leq u \leq \chi_A$. Then

$$|\mu|(A_t) \leq \liminf_j |\mu_j|(A_t)$$
$$\leq \liminf_j \int_\Omega u \, d|\mu_j| = \int_\Omega u \, d\lambda \leq \lambda(A).$$

Letting $t \to 0$ we get $|\mu|(A) \leq \lambda(A)$ as desired.

2.4 Weak compactness criteria in L^1

We have the following compactness criterion.

Theorem 2.19 *Let (u_j) be a sequence in $\mathrm{L}^1(\Omega)$. If a function $f : [0, +\infty) \to [0, +\infty]$ exists such that*

$$\lim_{t \to +\infty} \frac{f(t)}{t} = +\infty$$

and

$$\sup_j \int_\Omega f(|u_j|)\, dx < +\infty,$$

then there exists a subsequence of (u_j) weakly converging in $\mathrm{L}^1(\Omega)$.

Proof First, we check that $\sup_j \|u_j\|_{\mathrm{L}^1(\Omega)} < +\infty$. In fact, let $K > 0$ be such that $f(t) \geq t$ for $t \geq K$. Then,

$$
\begin{aligned}
\int_\Omega |u_j|\, dx &= \int_{\{|u_j| \leq K\}} |u_j|\, dx + \int_{\{|u_j| > K\}} |u_j|\, dx \\
&\leq \int_{\{|u_j| \leq K\}} |u_j|\, dx + \int_{\{|u_j| > K\}} f(|u_j|)\, dx \\
&\leq K|\Omega| + \sup_j \int_\Omega f(|u_j|)\, dx.
\end{aligned}
$$

Note that we have $\lim_{K \to +\infty} \int_{\{|u_j| \geq K\}} |u_j|\, dx = 0$ uniformly in j, as

$$\int_{\{|u_j| \geq K\}} |u_j|\, dx \leq \left(\inf_{t \geq K} \frac{f(t)}{t} \right)^{-1} \sup_j \int_\Omega f(|u_j|)\, dx.$$

Note, moreover, that for all $\varepsilon > 0$ there exists $\delta > 0$ such that if $|E| \leq \delta$ then $\int_E |u_j|\, dx \leq \varepsilon$ for all j. To check this, for all $\eta > 0$ let $K(\eta)$ be such that

$$\frac{f(t)}{t} \geq \frac{1}{\eta} \sup_j \int_\Omega f(|u_j|)\, dx$$

for all $t \geq K(\eta)$. If $\delta = \left(K(\frac{\varepsilon}{2}) \right)^{-1} \frac{\varepsilon}{2}$ then

$$
\begin{aligned}
\int_E |u_j|\, dx &= \int_{E \cap \{|u_j| \leq K(\varepsilon/2)\}} |u_j|\, dx + \int_{E \cap \{|u_j| > K(\varepsilon/2)\}} |u_j|\, dx \\
&\leq K(\varepsilon/2)|E| + (\varepsilon/2) \leq \varepsilon.
\end{aligned}
$$

Let μ_j be the signed measure given by

$$\mu_j(B) = \int_B u_j\, dx$$

for all Borel subsets B of Ω. Up to subsequences (μ_j) converges weakly* to some measure μ, and $|\mu_j|$ converges weakly* to some measure λ. Moreover, if ε, δ are as above, if A is an open set and $|A| \leq \delta$, then

$$|\mu|(A) \le \lambda(A) \le \liminf_j |\mu_j|(A) \le \varepsilon.$$

This shows that $|\mu|$, and hence also μ, is absolutely continuous with respect to the Lebesgue measure. Hence there exists $u \in L^1(\Omega)$ such that

$$\mu(B) = \int_B u \, dx$$

for all Borel subsets B of Ω.

To check eventually that $u_j \rightharpoonup u$ weakly in $L^1(\Omega)$ it suffices to verify that $\int_A u_j \, dx \to \int_A u \, dx$ for every open set A. In fact, from this we obtain $\int_E u_j \, dx \to \int_E u \, dx$ for all measurable sets E by approximation, and we can conclude the proof by Remark 2.6. Let then $A \subset \Omega$ be an open set as above, let $A_t = \{x \in A : \operatorname{dist}(x, \partial A) > t\}$, and let $\phi_t \in X(\Omega)$ satisfy $\chi_{A_t} \le \phi_t \le \chi_A$. Then

$$\mu(A) = \int_A u \, dx = \lim_{t \to 0+} \int_\Omega u\phi_t \, dx = \lim_{t \to 0+} \lim_j \int_\Omega u_j \phi_t \, dx$$

$$= \lim_{t \to 0+} \lim_j \left(\int_A u_j \, dx - \int_{A \setminus A_t} u_j (1 - \phi_t) \, dx \right).$$

Since

$$\left| \int_{A \setminus A_t} u_j (1 - \phi_t) \, dx \right| \le \int_{A \setminus A_t} |u_j| \, dx,$$

and the latter integral converges to 0 uniformly in j as $t \to 0+$, the proof is completed. $\qquad\square$

Theorem 2.19 above can be made more precise to give a characterization of weakly compact sets of $L^1(\Omega)$ as in the result below, whose proof we do not provide (see e.g. Dellacherie and Meyer (1975)).

Theorem 2.20 (Dunford Pettis and de la Vallée Poussin) *Let (u_j) be a sequence in $L^1(\Omega)$. The following statements are equivalent:*

(i) (compactness) every subsequence of (u_j) admits a subsequence weakly converging in $L^1(\Omega)$;

(ii) (equi-integrability) for all $\varepsilon > 0$ there exists $t > 0$ such that for all $j \in \mathbf{N}$

$$\int_{\{|u_j| \ge t\}} |u_j(x)| \, dx \le \varepsilon; \tag{2.11}$$

(iii) for all $\varepsilon > 0$ there exists $\delta > 0$ such that for all $j \in \mathbf{N}$ and all measurable sets E with $|E| \le \delta$ we have $\int_E |u_j(x)| \, dx \le \varepsilon$;

(iv) there exists a positive Borel function $f : [0, +\infty) \to [0, +\infty]$ such that

$$\lim_{t \to +\infty} \frac{f(t)}{t} = +\infty \qquad and \qquad \sup_j \int_\Omega f(|u_j(x)|) \, dx < +\infty; \tag{2.12}$$

(v) *there exists a convex l.s.c. increasing function $f : [0, +\infty) \to [0, +\infty]$ such that (2.12) holds.*

Remark   The proof of the implications (iv) $\Longrightarrow$ (ii) $\Longrightarrow$ (iii) $\Longrightarrow$ (i) is actually contained in the proof of Theorem 2.19.

2.5   Exercises

Exercise 2.1 Show that the property in Remark 2.3(d) does not hold for $p = 1$ or $p = \infty$.

Solution: if $p = 1$, take for example any sequence of functions $u_j \geq 0$ converging weakly but not strongly in $L^1(\Omega)$ to u. Then, choosing $v = 1$ in (2.1),

$$\|u\|_{L^1(\Omega)} = \int_\Omega u \, dx = \lim_j \int_\Omega u_j \, dx = \lim_j \|u_j\|_{L^1(\Omega)}.$$

If $p = \infty$, for instance, take (u_j) converging weakly* but not strongly in $L^\infty(0,1)$, and consider the sequence $(v_j) \subset L^\infty(-1,1)$ defined by $v_j = u_j$ on $(0,1)$, and $v_j = \sup_i \|u_i\|_{L^\infty(0,1)}$ on $(-1,0)$.

Exercise 2.2 Show that if $u_j \rightharpoonup u$ in $L^p(\Omega)$, and $v_j \rightharpoonup v$ in $L^{p'}(\Omega)$, then we may have $\lim_j \int_\Omega u_j v_j \, dx \neq \int_\Omega uv \, dx$.

Exercise 2.3 Define the space of 'piecewise $W^{1,2}$-functions'

$$\mathbf{W} = \{u \in L^\infty(0,1) : \exists N \in \mathbf{N} \, \exists \, 0 = t_0 < t_1 < \cdots < t_N = 1 : u \in W^{1,2}(t_{i-1}, t_i)\}.$$

If $u \in \mathbf{W}$ we abuse notation and write $u' \in L^2(0,1)$ to denote the function defined by the weak derivative of u on each interval (t_{i-1}, t_i), and we set $S(u) = \{t \in (0,1) : u(t+) \neq u(t-)\}$ (we use the piecewise continuous representatives of u). Prove that if (u_j) is a sequence in $\mathbf{W}$ such that $(u_j(t))$ is bounded for $t \in (0,1)$ a.e. and $\sup_j \int_{(0,1)} |u_j'|^2 \, dt + \#(S(u_j)) < +\infty$, then there exists a subsequence (not relabelled) such that $u_j \to u \in \mathbf{W}$ a.e. Moreover, $u_j' \rightharpoonup u'$ weakly in $L^2(0,1)$ and $\#(S(u)) \leq \liminf_j \#(S(u_j))$.

Hint: it is not restrictive to suppose that $S(u_j) = \{t_1^j < \cdots < t_{N-1}^j\}$ with N independent of j and that t_i^j converge to some $t_i \in [0,1]$ (define also $t_0 = 0$ and $t_N = 1$). Note that if $t_{i-1} < t_i$ then (u_j) is precompact in $W^{1,2}(t_{i-1} + \varepsilon, t_i - \varepsilon)$ for all $\varepsilon > 0$ small enough. After passing to subsequences, define u as the weak limit of (u_j) in each $(t_{i-1} + \varepsilon, t_i - \varepsilon)$, and use the arbitrariness of ε.

Exercise 2.4 Let Y be an n-cube in $\mathbf{R}^n$. Let $v \in L^p_{\mathrm{loc}}(\mathbf{R}^n; \mathbf{R}^m)$ be such that $v - Ax$ is Y-periodic, and let $v_\varepsilon(x) = \varepsilon v(\frac{x}{\varepsilon})$. Prove that $v_\varepsilon \to Ax$ strongly in $L^p_{\mathrm{loc}}(\mathbf{R}^n; \mathbf{R}^m)$.

Hint: write $|v_\varepsilon(x) - Ax| = \varepsilon|v(\frac{x}{\varepsilon}) - A\frac{x}{\varepsilon}|$ and use Example 2.7.

3

MINIMUM PROBLEMS IN SOBOLEV SPACES

3.1 The direct method. An example of application

A fundamental tool for the application of the direct method of the calculus of variations to minimum problems for functionals defined on Sobolev spaces is the following well-known theorem (see e.g. Adams (1975)).

Theorem 3.1 (Poincaré's inequality) *Let $1 \leq p < \infty$. There exists a constant $C > 0$ depending on Ω, bounded open subset of $\mathbf{R}^n$, and p such that*

$$\int_\Omega |u|^p \, dx \leq C \int_\Omega |Du|^p \, dx \tag{3.1}$$

for all $u \in W_0^{1,p}(\Omega)$, and, if Ω is connected and with Lipschitz boundary,

$$\int_\Omega |u - u_\Omega|^p \, dx \leq C \int_\Omega |Du|^p \, dx \tag{3.2}$$

for all $u \in W^{1,p}(\Omega)$, where $u_\Omega = |\Omega|^{-1} \int_\Omega u \, dx$.

Example 3.2 (Weak solutions of the p-Laplace equation) Let Ω be a bounded smooth open set in $\mathbf{R}^n$, with outward normal ν, let $1 < p < \infty$, and let $g \in L^{p'}(\Omega)$. The variational problems related to the p-Laplace equations

$$\begin{cases} -\mathrm{div}\,(|Du|^{p-2}Du) = g \\ u = 0 \qquad \text{on } \partial\Omega \end{cases} \qquad \begin{cases} -\mathrm{div}\,(|Du|^{p-2}Du) = g \\ \dfrac{\partial u}{\partial \nu} = 0 \qquad \text{on } \partial\Omega \end{cases}$$

(if $p = 2$ we write $\mathrm{div}\,(|Du|^{p-2}Du) = \Delta u$) are

$$\min\left\{ \int_\Omega |Du|^p \, dx - p \int_\Omega gu \, dx : \; u \in W_0^{1,p}(\Omega) \right\},$$

$$\min\left\{ \int_\Omega |Du|^p \, dx - p \int_\Omega gu \, dx : \; u \in W^{1,p}(\Omega) \right\},$$

respectively. We can apply the direct methods to these problems. The functional

$$F(u) = \int_\Omega |Du|^p \, dx - p \int_\Omega gu \, dx$$

is weakly coercive on $W^{1,p}(\Omega)$ (i.e. coercive with respect to weak convergence). Note that if $u \in W_0^{1,p}(\Omega)$

$$F(u) \geq \|Du\|_{L^p(\Omega)}^p - c_1\|u\|_{L^p(\Omega)} \geq c_2\|u\|_{W^{1,p}(\Omega)}^p - c_1\|u\|_{W^{1,p}(\Omega)}$$

by Theorem 3.1 and Hölder's inequality. Hence $-\infty < \inf_{W_0^{1,p}(\Omega)} F \leq F(0) = 0$, and every minimizing sequence (u_j) is bounded in $W^{1,p}(\Omega)$. We can find a minimizing sequence converging to $\overline{u} \in W_0^{1,p}(\Omega)$. By the weak lower semicontinuity of the norm, and the weak continuity of $u \mapsto \int_\Omega gu\,dx$, we see that $\overline{u}$ is a minimum point for F on $W_0^{1,p}(\Omega)$.

As for the second problem, it is no restriction to suppose that Ω is connected. If $\int_\Omega g\,dx \neq 0$ then it can be immediately checked that $\inf_{W^{1,p}(\Omega)} F = -\infty$ (consider $u = t$ constant and let $t \to \infty$). Suppose $\int_\Omega g\,dx = 0$. Proceeding as above we see that F is weakly coercive on $W^{1,p}(\Omega)$. In fact, F is invariant under translations of u by a constant. Hence

$$\inf\{F(u): \ u \in W^{1,p}(\Omega)\} = \inf\left\{F(u): \ u \in W^{1,p}(\Omega), \ \int_\Omega u\,dx = 0\right\},$$

and this set is again weakly compact in $W^{1,p}(\Omega)$.

By Example 3.2 we see that a natural coerciveness condition for functionals defined on $W^{1,p}(\Omega)$ is of the type

$$c_1 \int_\Omega |Du|^p\,dx - c_2 \leq F(u).$$

If $F(u) = \int_\Omega f(x, u, Du)\,dx$ is an integral functional, this condition can be obtained directly from some growth condition on the integrand, but this is not always necessarily so (recall Korn's inequality in linear elasticity, for example).

3.2 Borel and Carathéodory functions

In order that a functional of the form

$$F(u) = \int_\Omega f(x, u, Du)\,dx$$

be well defined, it is necessary that f be equivalent to a Borel function, in the sense that there exists a Borel function $\widetilde{f}$ such that

$$f(x, \cdot, \cdot) = \widetilde{f}(x, \cdot, \cdot) \qquad \text{for a.e. } x.$$

In this case, we simply say that f *is* a Borel function.

An important class of Borel functions (in the sense of equivalence classes as above) is the family of *Carathéodory functions*: we say that $f : \Omega \times \mathbf{R}^N \to \overline{\mathbf{R}}$ is a Carathéodory function if

$$\begin{cases} f(x, \cdot) \text{ is continuous for a.e. } x \in \Omega \\[2mm] f(\cdot, z) \text{ is measurable for all } z \in \mathbf{R}^N. \end{cases}$$

Proposition 3.3 *Every Carathéodory function is (equivalent to) a Borel function.*

Proof Without loss of generality we suppose $0 \leq f \leq 1$. Let $\mathcal{U} = \{U_j : j \in \mathbf{N}\}$ be a countable base of open sets of $\mathbf{R}^N$. Let $\Phi = \{\varphi_j : j \in \mathbf{N}\}$ be the countable family of all functions $\varphi_j : \mathbf{R}^N \to [0,1]$ of the form $\varphi_j = p\chi_U$ ($p \in \mathbf{Q} \cap [0,1]$, $U \in \mathcal{U}$). For all $j \in \mathbf{N}$ and $q \in \mathbf{Q}^N$, define

$$E_{j,q} = \{x \in \Omega : f(x,q) \geq \varphi_j(q)\}.$$

Since $f(\cdot, q)$ is measurable then $E_{j,q}$ is measurable, and the set

$$E_j = \{x \in \Omega : f(x,q) \geq \varphi_j(q) \text{ for all } q \in \mathbf{Q}^N\}$$

is measurable since $E_j = \bigcap_q E_{j,q}$. Since $f(x,\cdot)$ is continuous, φ_j is l.s.c. and $\mathbf{Q}^N$ is dense in $\mathbf{R}^N$

$$E_j = \{x \in \Omega : f(x,q) \geq \varphi_j(q) \text{ for all } q \in \mathbf{R}^N\}.$$

Note that if $h : \mathbf{R}^N \to [0,1]$ is l.s.c. then (e.g. by Remark 1.2(v))

$$h = \sup\{\varphi \in \Phi : \varphi \leq h\}.$$

Hence, we have

$$\begin{aligned}
f(x,z) &= \sup\{\varphi(z) : \varphi \in \Phi, \ \varphi \leq f(x,\cdot)\} \\
&= \sup_j \varphi_j(z)\chi_{E_j}(x).
\end{aligned}$$

For all j we take a Borel set B_j with $B_j = E_j$ a.e. If we define

$$\tilde{f}(x,z) = \sup_j \varphi_j(z)\chi_{B_j}(x),$$

then $\tilde{f}$ is Borel and $f(x,\cdot) = \tilde{f}(x,\cdot)$ for a.e. x. $\qquad\square$

Remark 3.4 (a) If f is a Carathéodory integrand and

$$|f(x,s,A)| \leq c(1 + |s|^p + |A|^p) \tag{3.3}$$

for a.e. $x \in \Omega$, and for all $s \in \mathbf{R}^m$ and $A \in \mathbf{M}^{m \times n}$, then the functional F above is continuous with respect to the strong $\mathrm{W}^{1,p}$-convergence. To check this, it suffices, given a strongly converging sequence (u_h) in $\mathrm{W}^{1,p}(\Omega; \mathbf{R}^m)$, to apply Fatou's lemma to the sequences $(c(1 + |u_h|^p + |Du_h|^p) + f(x, u_h, Du_h))$, and $(c(1 + |u_h|^p + |Du_h|^p) - f(x, u_h, Du_h))$.

(b) If (f_j) is an increasing sequence of positive Borel functions, and f is the Borel function given by $f(x,s,A) = \sup_j f_j(x,s,A)$ for a.e. $x \in \Omega$ and for all $s \in \mathbf{R}^m$ and $A \in \mathbf{M}^{m \times n}$, then by Fatou's lemma

$$\int_\Omega f(x, u, Du)\, dx = \sup_j \int_\Omega f_j(x, u, Du)\, dx$$

for all $u \in W^{1,1}(\Omega; \mathbf{R}^m)$. This observation, together with (a), gives the lower semicontinuity of F above in the strong topology of $W^{1,p}(\Omega; \mathbf{R}^m)$ if f is a Borel integrand, l.s.c. in (s, A) and satisfying (3.3).

3.3 Rellich's Theorem and equivalent conditions for lower semicontinuity

We end the section by recalling the following imbedding theorem.

Theorem 3.5 (Rellich's Compactness Theorem) *Let Ω be a bounded open set with Lipschitz boundary. If a sequence (u_j) is bounded in $W^{1,p}(\Omega)$, then it is precompact in $L^q(\Omega)$ with respect to strong convergence for each $1 \leq q < p^*$ if $p < n$ (for all $1 \leq q < \infty$ if $p = n$, and for all q if $p > n$).*

This theorem will enable us to replace weak $W^{1,p}$-topologies by strong L^p-topologies in many problems, as explained in the following remark.

Remark 3.6 If Ω is a bounded open set with Lipschitz boundary, then, taking into account Theorem 3.5, Theorem 2.4 and Proposition 2.2(i), we see that a functional $F : W^{1,p}(\Omega) \to \overline{\mathbf{R}}$ is weakly l.s.c. in $W^{1,p}(\Omega)$ if and only if $F(u) \leq \liminf_j F(u_j)$ when $(u_j) \subset W^{1,p}(\Omega)$ satisfies one of the following conditions:

(a) $u_j \rightharpoonup u$ weakly in $L^p(\Omega)$ and $\sup_j \|Du_j\|_{L^p(\Omega; \mathbf{R}^n)} < \infty$;

(b) $u_j \to u$ strongly in $L^p(\Omega)$ and $\sup_j \|Du_j\|_{L^p(\Omega; \mathbf{R}^n)} < \infty$;

(c) $(1 \leq q < p^*)$ $u_j \to u$ in $L^q(\Omega)$ and $\sup_j \|Du_j\|_{L^p(\Omega; \mathbf{R}^n)} < \infty$.

In particular if F satisfies a growth condition of the form

$$c_1 \int_\Omega |Du|^p\, dx - c_2 \leq F(u), \tag{3.4}$$

then F is weakly l.s.c. in $W^{1,p}(\Omega)$ if and only if it is l.s.c. in $L^p(\Omega)$ (with respect to the strong convergence). The same remark applies to functionals in $W^{1,p}(\Omega; \mathbf{R}^m)$.

Remark 3.7 We will often use the fact that the set of continuous piecewise affine functions is dense with respect to the strong topology of $W^{1,p}(\Omega; \mathbf{R}^m)$ if Ω has Lipschitz boundary (or, equivalently, that it is dense with respect to the strong topology of $W^{1,p}_{\mathrm{loc}}(\Omega; \mathbf{R}^m)$ for a general Ω). This fact can be easily seen as a consequence of the density of $C^\infty(\overline{\Omega}; \mathbf{R}^m)$ functions in $W^{1,p}(\Omega; \mathbf{R}^m)$. A frequent application will be of the following type: let Ω have Lipschitz boundary, let G be a weakly l.s.c. functional on $W^{1,p}(\Omega; \mathbf{R}^m)$, and let $F(u) = \int_\Omega f(x, u, Du)\, dx$ with f a Carathéodory function satisfying the hypotheses of Remark 3.4(a). If $G(u) \leq F(u)$ for all continuous piecewise affine u, then $G \leq F$ on the whole

$W^{1,p}(\Omega; \mathbf{R}^m)$. In fact if $u \in W^{1,p}(\Omega; \mathbf{R}^m)$, we can choose u_j continuous piecewise affine and converging to u strongly in $W^{1,p}(\Omega; \mathbf{R}^m)$. We have then, by the lower semicontinuity of G,

$$G(u) \leq \liminf_j G(u_j) \leq \liminf_j F(u_j) = F(u),$$

the last equality being given by Remark 3.4(a). Note that by Rellich's Theorem the same remark applies if G is l.s.c. with respect to the $L^p(\Omega; \mathbf{R}^m)$ convergence.

3.4   Exercises

Exercise 3.1 Prove Poincaré's inequality (3.1) in the case $n = 1$ and $\Omega = (a, b)$ (use the Fundamental Theorem of Calculus and Hölder's inequality).

Exercise 3.2 Prove Poincaré's inequality (3.1) in the general case (use Fubini's Theorem and the previous exercise).

Exercise 3.3 Prove Poincaré's inequality (3.2) in the case $p > 1$.

Hint: clearly it suffices to consider the case $u_\Omega = 0$. Argue by contradiction, finding a sequence (u_j) such that $\int_\Omega |u_j|^p \, dx = 1$, $\int_\Omega |Du_j|^p \, dx \to 0$ and $\int_\Omega u_j \, dx = 0$. Deduce from Theorem 2.4 that (up to a subsequence) $u_j \rightharpoonup u$ weakly in $W^{1,p}(\Omega)$. From Proposition 2.2(ii) deduce that $Du = 0$ and hence that u is constant. Since by weak convergence $u_\Omega = 0$, deduce that $u = 0$. From Rellich's Theorem obtain that $\int_\Omega |u|^p \, dx = 1$, and hence a contradiction.

Exercise 3.4 Let $f : \Omega \times \mathbf{R}^m \times \mathbf{M}^{m \times n} \to \mathbf{R}$ be a Borel function such that $f(x, s, A) \geq c_1 |A|^p - c_2$ for all x, s and A. Prove that if $u \mapsto \int_U f(x, u, Du) \, dx$ is weakly l.s.c. with respect to the $W^{1,p}(U; \mathbf{R}^m)$ convergence for all open subsets $U \subset\subset \Omega$ with Lipschitz boundary, then $F(u) = \int_\Omega f(x, u, Du) \, dx$ is l.s.c. with respect to the $L^p(\Omega; \mathbf{R}^m)$ convergence (use Remark 3.6 and an approximation of Ω from the interior with smooth sets).

4

NECESSARY CONDITIONS FOR WEAK LOWER SEMICONTINUITY

We study properties of functions f for which the integral functional

$$\int_\Omega f(Du(x))\,dx$$

is weakly lower semicontinuous in some Sobolev space $W^{1,p}(\Omega;\mathbf{R}^m)$. This will be the case of functionals obtained by homogenization via Γ-convergence. In what follows $f : \mathbf{M}^{m\times n} \to [0,+\infty]$ is a Borel function.

4.1 General necessary conditions

Theorem 4.1 *If the integral functional*

$$\mathcal{F}(u) = \int_\Omega f(Du(x))\,dx$$

is (sequentially) weakly lower semicontinuous on $W^{1,p}(\Omega;\mathbf{R}^m)$ *(weakly* if* $p = \infty$*), then*

(i) *f is lower semicontinuous;*

(ii) *for every n-cube Y*

$$f\Big(\frac{1}{|Y|}\int_Y Dv(y)\,dy\Big) \le \frac{1}{|Y|}\int_Y f(Dv(y))\,dy \tag{4.1}$$

for all $v \in W^{1,p}_{\mathrm{loc}}(\mathbf{R}^n;\mathbf{R}^m)$ with Dv Y-periodic.

Proof If $A_j \to A$ in $\mathbf{M}^{m\times n}$, then we take $u_j = A_j x \to Ax$ to obtain $f(A) \le \liminf_j f(A_j)$. To prove the inequality in (4.1) we fix $v \in W^{1,p}_{\mathrm{loc}}(\mathbf{R}^n;\mathbf{R}^m)$ with Dv Y-periodic, and define

$$u_j(x) = \frac{1}{j}v(jx), \qquad u(x) = \Big(|Y|^{-1}\int_Y Dv(y)\,dy\Big)x.$$

By Example 2.10 $u_j \rightharpoonup u$ in $W^{1,p}(\Omega;\mathbf{R}^m)$ ($\rightharpoonup^*$ if $p = \infty$), so that

$$|\Omega|f\Big(\frac{1}{|Y|}\int_Y Dv(y)\,dy\Big) \le \liminf_j \int_\Omega f(Du_j)\,dx. \tag{4.2}$$

If $f(Dv(x)) \in \mathrm{L}^1(Y)$ then by Example 2.7

$$f(Du_j) \rightharpoonup \frac{1}{|Y|} \int_Y f(Dv(y)) \, dy$$

in $\mathrm{L}^1(\Omega)$ and the right-hand side of (4.2) converges to $|\Omega||Y|^{-1} \int_Y f(Dv(y)) \, dy$, so that (ii) follows. If $f(Dv(x)) \notin \mathrm{L}^1(Y)$ then the right-hand side of (4.2) is eventually infinite and (4.1) is verified all the same. $\qquad \square$

Condition (ii) of Theorem 4.1 can be rewritten as

$$f(A) = \min\Big\{ \frac{1}{|Y|} \int_Y f(D\varphi(x)) \, dx :$$

$$\varphi \in \mathrm{W}^{1,p}_{\mathrm{loc}}(\mathbf{R}^n; \mathbf{R}^m), D\varphi \ Y\text{-periodic} , \int_Y D\varphi \, dy = A|Y| \Big\}$$

$$= \min\Big\{ \frac{1}{|Y|} \int_Y f(A + D\varphi(x)) \, dx :$$

$$\varphi \in \mathrm{W}^{1,p}_{\mathrm{loc}}(\mathbf{R}^n; \mathbf{R}^m), \varphi \ Y\text{-periodic} \Big\}. \tag{4.3}$$

4.2 $\mathrm{W}^{1,p}$-quasiconvexity

Definition 4.2 *We say that a function $f : \mathbf{M}^{m \times n} \to [0, +\infty]$ is $\mathrm{W}^{1,p}$-quasiconvex at $A \in \mathbf{M}^{m \times n}$ if there exists a bounded open subset E of $\mathbf{R}^n$ with $|\partial E| = 0$ such that*

$$|E|f(A) \leq \int_E f(A + D\varphi(x)) \, dx \tag{4.4}$$

for every $\varphi \in \mathrm{W}^{1,p}_0(E; \mathbf{R}^m)$; that is, equivalently, such that the linear function Ax is a local minimum in $\mathrm{W}^{1,p}(E; \mathbf{R}^m)$:

$$f(A) = \min\Big\{ \frac{1}{|E|} \int_E f(A + D\varphi(x)) \, dx : \ \varphi \in \mathrm{W}^{1,p}_0(E; \mathbf{R}^m) \Big\}. \tag{4.5}$$

A function f will be called $\mathrm{W}^{1,p}$-quasiconvex if (4.4) holds for all $A \in \mathbf{M}^{m \times n}$.

Proposition 4.3 *If the integral functional $\mathcal{F}(u) = \int_\Omega f(Du(x)) \, dx$ is (sequentially) weakly lower semicontinuous on $\mathrm{W}^{1,p}(\Omega; \mathbf{R}^m)$ (weakly* if $p = \infty$), then f is $\mathrm{W}^{1,p}$-quasiconvex.*

Proof Let E be an n-cube in $\mathbf{R}^n$, let $A \in \mathbf{M}^{m \times n}$ and let $\varphi \in \mathrm{W}^{1,p}_0(E; \mathbf{R}^m)$, extended by periodicity to the whole $\mathbf{R}^n$. Define $v(x) = Ax + \varphi(x)$. By Theorem 4.1(ii) we have (4.4). $\qquad \square$

If f is $\mathrm{W}^{1,p}$-quasiconvex then property (4.4) extends to all bounded open subsets E with $|\partial E| = 0$. Before proving this fact we recall the following version of the Vitali Covering Theorem, whose proof we include for completeness since the statement differs slightly from the usual one.

Theorem 4.4 (Vitali Covering Theorem) *Let Ω be a bounded open subset of $\mathbf{R}^n$, and let $\mathcal{U}$ be a family of closed subsets of Ω. If $M > 1$ exists such that for each $U \in \mathcal{U}$, $B(x, r) \subseteq U \subseteq B(x, Mr)$ for some $x \in \Omega$, $r > 0$, and if for a.e. $x \in \Omega$ $\inf\{\operatorname{diam} U :\ x \in U \in \mathcal{U}\} = 0$, then there exists a disjoint (finite or) countable subfamily (U_j) of $\mathcal{U}$ such that $|\Omega \setminus \bigcup_j U_j| = 0$.*

Proof We denote by $\mathcal{F}$ the family of all closed balls $C(x, r) = \overline{B}(x, r)$ for which there exists $U \in \mathcal{U}$ such that $B(x, r) \subseteq U \subseteq B(x, Mr)$, and by $\mathcal{F}'$ the family of all closed balls $C(x, Mr)$ such that $C(x, r) \in \mathcal{F}$.

We will prove the theorem in three steps.

Step 1: there exists a (finite or) countable family $\mathcal{G}$ of disjoint balls of $\mathcal{F}'$ such that

$$\bigcup_{C(x,Mr)\in\mathcal{F}'} C(x, Mr) \subset \bigcup_{C(x,Mr)\in\mathcal{G}} C(x, 5Mr). \tag{4.6}$$

Let $D = \sup\{\operatorname{diam} C :\ C \in \mathcal{F}'\}$. For all $j = 1, 2, \ldots$ let $\mathcal{I}_j = \{C \in \mathcal{F}' :\ 2^{-j}D < \operatorname{diam} C \le 2^{-j+1}D\}$, and define $\mathcal{H}_j \subset \mathcal{F}'$ as follows.

(a) $\mathcal{H}_1$ is any maximal disjoint collection of balls in $\mathcal{I}_1$.

(b) Assuming $\mathcal{H}_1$, $\mathcal{H}_2$, ..., $\mathcal{H}_{k-1}$ have been selected, we choose $\mathcal{H}_k$ to be any maximal disjoint subcollection of $\{C \in \mathcal{I}_k : C \cap C' = \emptyset \text{ for all } C' \in \bigcup_{j=1}^{k-1} \mathcal{H}_j\}$.

Finally, we define $\mathcal{G} = \bigcup_{j=1}^{\infty} \mathcal{H}_j$. Clearly, $\mathcal{G}$ is a collection of disjoint balls and $\mathcal{G} \subset \mathcal{F}'$. In order to prove (4.6), fix $C(x, Mr) \in \mathcal{F}'$. Then, there exists an index k such that $C(x, Mr) \in \mathcal{I}_k$. By the maximality of $\mathcal{H}_k$, there exists a ball $C(x', Mr') \in \bigcup_{j=1}^{k} \mathcal{H}_j$ with $C(x, Mr) \cap C(x', Mr') \neq \emptyset$. But $Mr \le 2^{-j}D$ and $Mr' \ge 2^{-j-1}D$, so that $r \le 2r'$. Hence $C(x, Mr) \subset C(x', 5Mr')$, which gives (4.6).

Step 2: Fix θ so that $1 - (5M)^{-n} < \theta < 1$; there exists a finite collection $\{U_i\}_{i=1}^{N_1}$ of disjoint sets in $\mathcal{U}$ such that

$$\left|\Omega \setminus \bigcup_{i=1}^{N_1} U_i\right| \le \theta|\Omega|. \tag{4.7}$$

By Step 1, there exists a countable disjoint family $\mathcal{G}_1 \subset \mathcal{F}'$ such that

$$\Omega \subset \bigcup_{C(x,Mr)\in\mathcal{G}_1} C(x, 5Mr).$$

Thus

$$|\Omega| \le \sum_{C(x,Mr)\in\mathcal{G}_1} |C(x, 5Mr)|$$

$$= 5^n M^n \sum_{C(x,Mr)\in\mathcal{G}_1} |C(x, r)| = 5^n M^n \left|\bigcup_{C(x,Mr)\in\mathcal{G}_1} C(x, r)\right|.$$

Hence

$$\left| \Omega \setminus \bigcup_{C(x,Mr)\in\mathcal{G}_1} C(x,r) \right| \leq \left(1 - \frac{1}{5^n M^n}\right)|\Omega|.$$

Since $\mathcal{G}_1$ is countable, there exist balls $\{C(x_i, Mr_i)\}_{i=1}^{N_1}$ in $\mathcal{G}_1$ such that

$$\left| \Omega \setminus \bigcup_{i=1}^{N_1} C(x_i, r_i) \right| \leq \theta|\Omega|$$

and hence, by the definition of $\mathcal{F}$, for every $i = 1, \ldots, N_1$, there exists U_i, with $C(x_i, r_i) \subseteq U_i \subseteq C(x_i, Mr_i)$, satisfying (4.7).

Step 3: conclusion.

We construct the sequence (U_j) by induction. First, define $U_1, \ldots, U_{N_1}$ as in the previous step. Then, supposing that $U_1, \ldots U_{N_k}$ disjoint sets in $\mathcal{U}$ have been chosen such that

$$\left| \Omega \setminus \bigcup_{i=1}^{N_k} U_i \right| \leq \theta^k|\Omega|,$$

define

$$\Omega_{k+1} = \Omega \setminus \bigcup_{i=1}^{N_k} U_i, \qquad \mathcal{U}_{k+1} = \{U \in \mathcal{U} : U \subset \Omega_{k+1}\}.$$

Applying Step 2 with Ω_{k+1} and $\mathcal{U}_{k+1}$ in the place of Ω and $\mathcal{U}$, we obtain a finite collection $U_{N_k+1}, \ldots, U_{N_{k+1}}$ of disjoint sets of $\mathcal{U}_{k+1}$ such that

$$\left| \Omega_{k+1} \setminus \bigcup_{i=N_k}^{N_{k+1}} U_i \right| \leq \theta|\Omega_{k+1}|,$$

i.e.

$$\left| \Omega \setminus \bigcup_{i=1}^{N_{k+1}} U_i \right| \leq \theta^{k+1}|\Omega|.$$

Letting $k \to +\infty$ we deduce that (U_i) satisfies the thesis of the theorem. $\qquad \square$

The following proposition shows that the definition of $W^{1,p}$-quasiconvexity does not depend on the choice of the set E.

Proposition 4.5 *If f is $W^{1,p}$-quasiconvex at $A \in \mathbf{M}^{m\times n}$, then $|E'|f(A) \leq \int_{E'} f(A + D\varphi(x))\,dx$ for all bounded open sets E' with $|\partial E'| = 0$ and $\varphi \in W_0^{1,p}(E'; \mathbf{R}^m)$.*

Proof Fix a bounded open subset E as in Definition 4.2, and a bounded open set E' with $|\partial E'| = 0$. The family

$$\mathcal{U} = \{z + \varepsilon\overline{E'} : \ z \in \mathbf{R}^n, \ \varepsilon > 0, \ z + \varepsilon\overline{E'} \subset E\}$$

satisfies the hypotheses of Theorem 4.4. Hence there exists a (finite or) countable disjoint sequence $(z_j + \varepsilon_j\overline{E'})$ which covers E a.e. If $\varphi \in W_0^{1,p}(E'; \mathbf{R}^m)$, then we define $\psi \in W_0^{1,p}(E; \mathbf{R}^m)$ by

$$\psi(x) = \begin{cases} \varepsilon_j\varphi\left(\dfrac{x - z_j}{\varepsilon_j}\right) & \text{if } x \in z_j + \varepsilon_j E', \ j \in \mathbf{N} \\ 0 & \text{otherwise .} \end{cases}$$

By (4.4) we have

$$\begin{aligned} |E|f(A) &\leq \int_E f(A + D\psi(x))dx \\ &= \sum_j \int_{(z_j + \varepsilon_j E')} f\left(A + D\varphi\left(\frac{x - z_j}{\varepsilon_j}\right)\right) dx \\ &= \sum_j \left(\varepsilon_j^n|E'|\right)\frac{1}{|E'|} \int_{E'} f(A + D\varphi(y)) \, dy \\ &= \left(\sum_j |z_j + \varepsilon_j E'|\right) \frac{1}{|E'|} \int_{E'} f(A + D\varphi(y)) \, dy \\ &= \frac{|E|}{|E'|} \int_{E'} f(A + D\varphi(y)) \, dy, \end{aligned}$$

and the desired inequality follows. Note that in the first equality we use $|\partial E'| = 0$. $\qquad\square$

Remark 4.6 (i) If f takes only finite values, then the condition $|\partial E| = 0$ can be removed. In this case, in fact, if (4.4) holds for a non-empty bounded open set E, then given a second non-empty bounded open set E', we can consider $a \in \mathbf{R}^n$, $b > 0$ such that $a + bE' \subset E$. Given $\varphi \in W_0^{1,p}(E'; \mathbf{R}^m)$, we define

$$\psi(x) = \begin{cases} b\varphi\left(\dfrac{x - a}{b}\right) & \text{if } x \in a + bE' \\ 0 & \text{otherwise.} \end{cases}$$

We have $\psi \in W_0^{1,p}(E; \mathbf{R}^m)$ so that by (4.4)

$$\begin{aligned} |E|f(A) &\leq \int_E f(A + D\psi) \, dx \\ &= |E \setminus (a + bE')|f(A) + \int_{a+bE'} f\left(A + D\varphi\left(\frac{x - a}{b}\right)\right) dx \end{aligned}$$

$$= (|E| - b^n|E'|)f(A) + b^n \int_{E'} f(A + D\varphi) \, dx,$$

from which we see that (4.4) holds also for E'.

(ii) If $1 \le p \le q \le \infty$ then $W^{1,p}$-quasiconvexity implies $W^{1,q}$-quasiconvexity.

(iii) If $1 \le p < \infty$, f is continuous and satisfies the growth condition 'from above'

$$0 \le f(A) \le c(1 + |A|^p) \qquad \forall A \in \mathbf{M}^{m \times n}, \tag{4.8}$$

then f is $W^{1,p}$-quasiconvex if and only if it is $W^{1,\infty}$-quasiconvex. It suffices to remark that under this condition $\mathcal{F}$ is continuous for the strong $W^{1,p}$ convergence (Remark 3.4), and use the fact that C_0^∞ is dense in $W_0^{1,p}$ in the strong topology.

(iv) If $1 \le p < \infty$, and f satisfies the growth condition from below

$$c_1|A|^p - c_2 \le f(A) \qquad \forall A \in \mathbf{M}^{m \times n} \tag{4.9}$$

then f is $W^{1,p}$-quasiconvex if and only if it is $W^{1,1}$-quasiconvex. It suffices to take a Lipschitz domain E in Definition 4.2 and recall that $W_0^{1,1}(E; \mathbf{R}^m) \cap W^{1,p}(E; \mathbf{R}^m) = W_0^{1,p}(E; \mathbf{R}^m)$.

Remark 4.7 Condition (ii) of Theorem 4.1 is strictly stronger than $W^{1,p}$-quasiconvexity. Take for example $f : \mathbf{M}^{2 \times 2} \to \{0, +\infty\}$ defined by

$$f(A) = \begin{cases} 0 & \text{if } A = 0 \text{ or } A = e_1 \otimes e_1 \\ +\infty & \text{otherwise.} \end{cases}$$

Condition (4.1) is not satisfied by the function

$$v(x_1, x_2) = \begin{cases} (x_1 - \frac{1}{2}j, 0) & \text{if } j \in \mathbf{Z}, \, j \le x_1 < j + \frac{1}{2} \\[2mm] (\frac{1}{2}(j+1), 0) & \text{if } j \in \mathbf{Z}, \, j + \frac{1}{2} \le x_1 < j + 1. \end{cases}$$

In fact, the gradient Dv is the $(0,1)^2$-periodic piecewise constant function

$$Dv(x_1, x_2) = \begin{cases} e_1 \otimes e_1 & \text{if } j < x_1 < j + \frac{1}{2} \\ 0 & \text{if } j + \frac{1}{2} < x_1 < j + 1, \end{cases}$$

and

$$f\left(\int_Y Dv \, dy\right) = f\left(\frac{1}{2}e_1 \otimes e_1\right)$$

$$= +\infty > 0 = \frac{1}{2}f(e_1 \otimes e_1) + \frac{1}{2}f(0) = \int_Y f(Dv) \, dy.$$

On the other hand, if $A \in \mathbf{M}^{2 \times 2}$, $\varphi \in W_0^{1,1}(Y; \mathbf{R}^2)$, and $\int_Y f(A + D\varphi) \, dx < +\infty$, then either $D\varphi + A = 0$ or $D\varphi + A = e_1 \otimes e_1$. If $A \notin \{0, e_1 \otimes e_1\}$ then these conditions are void by the boundary condition; otherwise, we have $\varphi = 0$. In any case (4.4) is verified.

This example shows that in general not even $\mathrm{W}^{1,1}$-quasiconvexity is a sufficient condition for the weak* lower semicontinuity of $\mathcal{F}$ in $\mathrm{W}^{1,\infty}(\Omega; \mathbf{R}^m)$.

Example 4.8 Let $1 \leq p < 2$, and consider the function $f : \mathbf{M}^{2\times 2} \to [0, +\infty)$ given by

$$f(A) = |A|^p + |\det A| = |A|^p + |A_{11}A_{22} - A_{12}A_{21}|.$$

Let $E = B(0,1)$ be the open ball of centre 0 and radius 1 in $\mathbf{R}^2$. Note that the function $u(x) = \frac{x}{|x|}$ belongs to $\mathrm{W}^{1,p}(E; \mathbf{R}^2)$, and $\det Du = 0$ a.e. on E. If $t > 0$ is large enough then

$$|E|f(tI) = \left(t^p(\sqrt{2})^p + t^2\right)|E|$$
$$> t^p \int_E |Du|^p \, dx = \int_E |D(tu)|^p \, dx = \int_E f(D(tu)) \, dx.$$

Since $tu(x) - tx \in \mathrm{W}_0^{1,p}(E; \mathbf{R}^2)$ this implies that

$$|E|f(tI) > \inf\left\{ \int_E f(tI + D\varphi) \, dx : \varphi \in \mathrm{W}_0^{1,p}(E; \mathbf{R}^2) \right\};$$

hence, f is not $\mathrm{W}^{1,p}$-quasiconvex, and $\int_\Omega f(Du) \, dx$ is not weakly lower semicontinuous in $\mathrm{W}^{1,p}(\Omega; \mathbf{R}^2)$.

Proposition 4.9 *If* $f : \mathbf{M}^{m\times n} \to [0, +\infty)$ *is locally bounded, then a condition equivalent to* $\mathrm{W}^{1,\infty}$-*quasiconvexity is that for some non-empty bounded open set* Y *with* $|\partial Y| = 0$

$$|Y|f(A) \leq \liminf_j \int_Y f(A + Du_j) dy \tag{4.10}$$

for all sequences (u_j) *weakly* converging to* 0 *in* $\mathrm{W}^{1,\infty}(Y; \mathbf{R}^m)$ *and for all* $A \in \mathbf{M}^{m\times n}$.

Proof If property (4.10) holds then $\mathrm{W}^{1,\infty}$-quasiconvexity follows from the same arguments of Theorem 4.1 and Proposition 4.3. Conversely, if f is $\mathrm{W}^{1,\infty}$-quasiconvex, and $u_j \rightharpoonup^* 0$ in $\mathrm{W}^{1,\infty}(Y; \mathbf{R}^m)$, then we can take a sequence of positive numbers ε_j, and consider $\varphi_j(x) = 1 \wedge (\varepsilon_j^{-1}\mathrm{dist}\,(x, \partial Y))$. Since $v_j = \varphi_j u_j \in \mathrm{W}_0^{1,\infty}(Y; \mathbf{R}^m)$ we get

$$|Y|f(A) \leq \int_Y f(A + Dv_j) dy$$
$$= \int_Y f(A + \varphi_j Du_j + u_j D\varphi_j) \, dy$$
$$\leq \int_{\{\varphi_j = 1\}} f(A + Du_j) \, dy + \int_{\{\varphi_j < 1\}} f(A + \varphi_j Du_j + u_j D\varphi_j) \, dy$$

$$\leq \int_Y f(A + Du_j)\, dy + \int_{\{\varphi_j < 1\}} f(A + \varphi_j Du_j + u_j D\varphi_j)\, dy.$$

We can choose (ε_j) to converge to 0 and such that $|A + \varphi_j Du_j + u_j D\varphi_j| < c$ on Y, so that (4.10) follows, taking into account that f is bounded on bounded sets, and that $|\{\varphi_j < 1\}| \to 0$. $\qquad\square$

4.3   Rank-1-convexity

Definition 4.10 *A function* $f : \mathbf{M}^{m \times n} \to \overline{\mathbf{R}}$ *is rank-1-convex if for all* $A \in \mathbf{M}^{m \times n}$, $a \in \mathbf{R}^m$, $b \in \mathbf{R}^n$ *the function* $t \mapsto f(A + ta \otimes b)$ *is convex on* $\mathbf{R}$, *i.e.*

$$f(A + ta \otimes b) \leq tf(A + a \otimes b) + (1 - t)f(A) \tag{4.11}$$

for all $A \in \mathbf{M}^{m \times n}$, $a \in \mathbf{R}^m$, $b \in \mathbf{R}^n$ *and* $t \in (0, 1)$ *for which the right-hand side makes sense. Equivalently,* f *is rank-1-convex if and only if for all* $A, B \in \mathbf{M}^{m \times n}$ *such that* $\mathrm{rank}\,(A - B) \leq 1$

$$f(tA + (1 - t)B) \leq tf(A) + (1 - t)f(B) \tag{4.12}$$

for all $t \in (0, 1)$ *for which the right-hand side makes sense.*

Proposition 4.11 *If the integral functional*

$$\mathcal{F}(u) = \int_\Omega f(Du(x))\, dx$$

satisfies condition (ii) *of Theorem 4.1 for* $p = \infty$, *then* f *is rank-1-convex.*

Proof Let $A, B \in \mathbf{M}^{m \times n}$ such that $\mathrm{rank}(A - B) \leq 1$, and let $t \in (0, 1)$. Let $a \in \mathbf{R}^m$, $b \in \mathbf{R}^n$ be vectors such that $B - A = a \otimes b$. Consider the function $v \in \mathrm{W}^{1,\infty}_{\mathrm{loc}}(\mathbf{R}^n; \mathbf{R}^m)$ defined by

$$v(x) = \begin{cases} Ax + \langle b, x\rangle a - (1 - t)ja & \text{if } j \in \mathbf{Z},\ j \leq \langle b, x\rangle < j + t \\[2mm] Ax + (1 + j)ta & \text{if } j \in \mathbf{Z},\ j + t \leq \langle b, x\rangle < j + 1. \end{cases}$$

The gradient

$$Dv = \begin{cases} B & \text{if } j \in \mathbf{Z},\ j < \langle b, x\rangle < j + t \\[2mm] A & \text{if } j \in \mathbf{Z},\ j + t < \langle b, x\rangle < j + 1 \end{cases}$$

is Y-periodic, where Y is any n-cube with one face orthogonal to b and side-length 1. We have $\int_Y Dv\, dy = |Y|(tB + (1 - t)A)$, so that

$$f(tB + (1 - t)A) = f\left(\int_Y Dv\, dy\right)$$

$$\leq \int_Y f(Dv)\,dy = tf(B) + (1-t)f(A)$$

by condition (ii) of Theorem 4.1. $\qquad\square$

Corollary 4.12 *If the functional $\mathcal{F}$ is (sequentially) weakly* lower semicontinuous on $W^{1,\infty}(\Omega; \mathbf{R}^m)$ then f is rank-1-convex.*

Remark 4.13 (i) The example in Remark 4.7 shows a lower semicontinuous function f which is not rank-1-convex but is $W^{1,1}$-quasiconvex. If f is locally bounded, then rank-1-convexity follows from $W^{1,\infty}$-quasiconvexity. In fact, by Proposition 4.9, we can use the same argument as in Theorem 4.1(ii), with the function v in the proof of Proposition 4.11.

(ii) If $m = 1$ or $n = 1$ then rank-1-convexity is the same as *convexity*.

(iii) If f is rank-1-convex, and $0 \leq f(A) \leq c(1 + |A|^p)$, then f satisfies a local Lipschitz condition

$$|f(A) - f(B)| \leq c(1 + |A|^{p-1} + |B|^{p-1})|A - B| \tag{4.13}$$

for all $A, B \in \mathbf{M}^{m \times n}$ (see Exercise 4.1).

4.4 Exercises

Exercise 4.1 Prove the statement of Remark 4.13(iii).

Hint: use the fact that this inequality holds for convex functions (see the next chapter), and that we can write

$$f(B) - f(A) = \sum_{i=1}^{m}(f(A_i) - f(A_{i-1}))$$

with $|A_i| \leq |A| + |B|$ for all $i = 1, \ldots, m$, $A_0 = A$, $A_m = B$, $|A_i - A_{i-1}| \leq |A - B|$ and $\mathrm{rank}(A_i - A_{i-1}) \leq 1$ (e.g. $A_i = A + \sum_{k=1}^{i} e_k \otimes c_k$ for $i = 1, \ldots, m$, where $\sum_{k=1}^{m} e_k \otimes c_k = B - A$).

Exercise 4.2 Prove that if $f : \mathbf{M}^{m \times n} \to \mathbf{R}$ is rank-1-convex and bounded, then it is constant.

Hint: note that this is true for convex functions, and hence also for rank-1-convex functions, arguing componentwise.

5

SUFFICIENT CONDITIONS FOR WEAK LOWER SEMICONTINUITY

5.1 Convexity

First we consider the case of *convex* $f : \mathbf{R}^N \to \overline{\mathbf{R}}$, that is,

$$f(tu + (1 - t)v) \leq tf(u) + (1 - t)f(v)$$

for all $u, v \in \mathbf{R}^N$ and $t \in (0, 1)$ for which the right-hand side makes sense. If this condition holds then we easily deduce the following properties of f:

(i) if $f : \mathbf{R}^N \to \mathbf{R}$ then we have

$$f(v) \leq f(u) - \frac{f(v + t(u - v)) - f(v)}{t}$$

for all $u, v \in \mathbf{R}^N$ and $t \in (0, 1)$, and thus if f is differentiable then

$$f(v) \leq f(u) + \langle Df(v), v - u \rangle$$

for all $u, v \in \mathbf{R}^N$;

(ii) if ψ is as in Definition 1.5 and convex, then $T_\lambda^\psi f$ is convex for all $\lambda \geq 0$;

(iii) if $f : \mathbf{R}^N \to (-\infty, +\infty]$ is l.s.c. then there exists an increasing sequence (f_j) of convex smooth functions with f_j Lipschitz with Lipschitz constant j pointwise converging to f (see Exercise 1.5);

(iv) if $f : \mathbf{R}^N \to (-\infty, +\infty]$ is l.s.c. then *Jensen's inequality* holds: if μ is a finite positive measure on Ω, then

$$f\left(\frac{1}{\mu(\Omega)} \int_\Omega u \, d\mu\right) \leq \frac{1}{\mu(\Omega)} \int_\Omega f(u) \, d\mu$$

for all $u \in \mathrm{L}^1_\mu(\Omega)$;

(v) if $|f(u)| \leq c(1 + |u|^p)$ for all $u \in \mathbf{R}^N$, then

$$|f(u) - f(v)| \leq c(1 + |u|^{p-1} + |v|^{p-1})|u - v|$$

for all $u, v \in \mathbf{R}^N$. Clearly, it suffices to check this property when $N = 1$. In this case it is well known that the difference quotient of a convex function is non-decreasing. Thus we have, if $u < v < z = v + |v| + 1$,

$$\frac{f(u) - f(v)}{u - v} \leq \frac{f(z) - f(v)}{z - v} = \frac{f(z) - f(v)}{|v| + 1}$$

$$\leq \frac{c(1 + |v|^p)}{|v| + 1} \leq c(1 + |v|^{p-1}) \leq c(1 + |v|^{p-1} + |u|^{p-1}).$$

In the same way we get

$$\frac{f(u) - f(v)}{u - v} \geq -c(1 + |v|^{p-1} + |u|^{p-1}).$$

If $N > 1$ then the inequality is proven by arguing componentwise.

Theorem 5.1 *If* $f : \mathbf{R}^N \to [0, +\infty]$ *is a convex and l.s.c. function, then the functional*

$$\mathcal{F}(u) = \int_\Omega f(u) \, dx$$

is sequentially l.s.c. on $\mathrm{L}^1(\Omega; \mathbf{R}^N)$ *with respect to convergence in* $\mathcal{D}'(\Omega; \mathbf{R}^N)$.

Proof We can suppose by approximation that f is smooth and Lipschitz continuous with Lipschitz constant λ (see Remark 3.4(b), and use observation (iii) above). Let $u_j \to u$ in $\mathcal{D}'(\Omega; \mathbf{R}^N)$. Recall that we have

$$f(u(x)) \leq f(u_j(x)) + \langle Df(u(x)), u(x) - u_j(x) \rangle \tag{5.1}$$

for all $x \in \Omega$. First, we suppose $u \in C^\infty(\Omega; \mathbf{R}^N)$. If we take $\varphi \in C_0^\infty(\Omega)$ with $0 \leq \varphi \leq 1$, then we have from (5.1)

$$\int_\Omega \varphi f(u) \, dx \leq \int_\Omega \varphi f(u_j) \, dx + \int_\Omega \varphi \langle Df(u), u - u_j \rangle \, dx$$

$$\leq \int_\Omega f(u_j) \, dx + \int_\Omega \langle \varphi Df(u), u - u_j \rangle \, dx. \tag{5.2}$$

Since $\varphi Df(u) \in C_0^\infty(\Omega; \mathbf{R}^N)$

$$\lim_j \int_\Omega \langle \varphi Df(u), u - u_j \rangle \, dx = 0;$$

hence

$$\int_\Omega \varphi f(u) \, dx \leq \liminf_j \mathcal{F}(u_j)$$

and $\mathcal{F}(u) \leq \liminf_j \mathcal{F}(u_j)$.

In the general case we choose a sequence (v_k) in $C^\infty(\Omega; \mathbf{R}^N)$ with $v_k \to u$ in $\mathrm{L}^1(\Omega; \mathbf{R}^N)$. We have

$$\mathcal{F}(u) = \lim_k \mathcal{F}(v_k) \tag{5.3}$$

since

$$|\mathcal{F}(u) - \mathcal{F}(v_k)| \le \lambda \int_\Omega |u - v_k|\,dx \to 0$$

as $k \to \infty$. Now, for each k the sequence $(v_k + u_j - u)$ converges to the smooth function v_k in $\mathcal{D}'(\Omega; \mathbf{R}^N)$ as $j \to \infty$ so that, for what we have proven above,

$$\begin{aligned}
\mathcal{F}(v_k) &\le \liminf_j \mathcal{F}(v_k + u_j - u) \\
&\le \liminf_j \mathcal{F}(u_j) + \limsup_j |\mathcal{F}(v_k + u_j - u) - \mathcal{F}(u_j)| \\
&\le \liminf_j \mathcal{F}(u_j) + \lambda \int_\Omega |v_k - u|\,dx,
\end{aligned}$$

and, taking the limit as $k \to \infty$, we obtain $\mathcal{F}(u) \le \liminf_j \mathcal{F}(u_j)$ by (5.3). $\square$

Corollary 5.2 *Let $m = 1$ or $n = 1$, and let $f : \mathbf{M}^{m \times n} \to [0, +\infty]$ be a Borel function. The functional*

$$\mathcal{F}(u) = \int_\Omega f(Du)\,dx$$

is (sequentially) weakly l.s.c. in $\mathrm{W}^{1,1}(\Omega; \mathbf{R}^m)$ if and only if f is convex and l.s.c.

Proof The proof follows by applying Theorem 4.1, Corollary 4.12, and Remark 4.13(ii). $\square$

Corollary 5.3 *If $f : \mathbf{M}^{m \times n} \to \overline{\mathbf{R}}$ is convex and l.s.c., and there exists an increasing Borel function $\Phi : [0, +\infty) \to [0, +\infty]$ such that*

$$\lim_{t \to +\infty} \frac{\Phi(t)}{t} = +\infty \qquad\qquad f(A) \ge \Phi(|A|) - c$$

for all $A \in \mathbf{M}^{m \times n}$, then a solution to the minimum problem

$$\min\left\{ \int_\Omega f(Du)\,dx : \; u \in \mathrm{W}_0^{1,1}(\Omega; \mathbf{R}^m) \right\}$$

exists in $\mathrm{W}^{1,1}(\Omega; \mathbf{R}^m)$.

Proof By Theorem 3.1 and Theorem 2.19 the functional $\int_\Omega f(Du)\,dx$ is coercive on $\mathrm{W}_0^{1,1}(\Omega; \mathbf{R}^m)$. By Theorem 5.1 it is also l.s.c.; hence we can apply the direct methods of the calculus of variations. $\square$

Remark 5.4 If f is convex and l.s.c., and $f(A) \ge c_1|A|^p - c_2$ for all $A \in \mathbf{M}^{m \times n}$, then by Remark 3.6 the functional

$$\mathcal{F}(u) = \begin{cases} \displaystyle\int_\Omega f(Du)\,dx & \text{if } u \in \mathrm{W}^{1,p}(\Omega; \mathbf{R}^m) \\[2mm] +\infty & \text{otherwise} \end{cases}$$

is l.s.c. on $\mathrm{L}^p(\Omega; \mathbf{R}^m)$ with respect to the norm convergence.

Remark 5.5 Theorem 5.1 has been improved in many ways to obtain (sequential) lower semicontinuity theorems in the case of functionals

$$\int_\Omega f(x, u, Du)\, dx$$

with respect to the weak convergence in $W^{1,p}(\Omega; \mathbf{R}^m)$. In particular this is true if f is Carathéodory and also convex in the last variable.

5.2 Polyconvexity

Definition 5.6 *A function* $f : \mathbf{M}^{m \times n} \to (-\infty, +\infty]$ *is* polyconvex *if there exists a convex function* $g : \mathbf{R}^{\tau(n,m)} \to \overline{\mathbf{R}}$ *such that*

$$f(A) = g(M(A)) \qquad \text{for all } A \in \mathbf{M}^{m \times n}, \tag{5.4}$$

where $M(A)$ *represents the ordered vector of all the minors of order* $1, 2, \ldots, n \wedge m = \min\{n, m\}$, *and*

$$\tau(n, m) = \sum_{k=1}^{n \wedge m} \binom{m}{k} \binom{n}{k}. \tag{5.5}$$

Example 5.7 If $n = 1$ or $m = 1$, then we can take $M(A) = A$, and polyconvexity is the same as convexity. If $n = m = 2$ then we can take

$$M(A) = (A, \det A) = (A_{11}, A_{12}, A_{21}, A_{22}, A_{11}A_{22} - A_{12}A_{21}),$$

and f is polyconvex if and only if there exists a convex $g : \mathbf{R}^5 \to \overline{\mathbf{R}}$ such that

$$f(A) = g(A_{11}, A_{12}, A_{21}, A_{22}, A_{11}A_{22} - A_{12}A_{21}) = g(A, \det A).$$

For example the function $f(A) = |A|^p + |\det A|$ $(1 \le p < \infty)$ is polyconvex. Notice that this function is not $W^{1,p}$-quasiconvex if $1 \le p < 2$ (see Example 4.8).

It is possible to characterize polyconvex functions without invoking the existence of functions g as in the definition above, as explained in the following remark (for the proof of which we refer to Dacorogna (1989)).

Remark 5.8 A function $f : \mathbf{M}^{m \times n} \to [0, +\infty]$ is polyconvex if and only if $(\tau = \tau(n, m)$ as above)

$$f\Big(\sum_{j=1}^{\tau+1} t_j A_j\Big) \le \sum_{j=1}^{\tau+1} t_j f(A_j) \tag{5.6}$$

for all $A_j \in \mathbf{M}^{m \times n}$, $t_j \ge 0$ for $j = 1, \ldots, \tau + 1$ such that $\sum_{j=1}^{\tau+1} t_j = 1$ and

$$M\Big(\sum_{j=1}^{\tau+1} t_j A_j\Big) = \sum_{j=1}^{\tau+1} t_j M(A_j). \tag{5.7}$$

If $n = m = 2$, then this means that $f : \mathbf{M}^{2\times 2} \to [0, +\infty]$ is polyconvex if and only if

$$f\Big(\sum_{j=1}^{6} t_j A_j\Big) \le \sum_{j=1}^{6} t_j f(A_j)$$

for all $A_1, \ldots, A_6 \in \mathbf{M}^{2\times 2}$, $t_1, \ldots, t_6 \ge 0$ with $\sum_{j=1}^{6} t_j = 1$ such that

$$\det\Big(\sum_{j=1}^{6} t_j A_j\Big) = \sum_{j=1}^{6} t_j \det A_j.$$

Remark 5.9 If $f : \mathbf{M}^{m\times n} \to [0, +\infty]$ is a polyconvex function then a function $g : \mathbf{R}^{\tau(n,m)} \to \overline{\mathbf{R}}$ such that (5.4) holds can be obtained by setting

$$g(X) = \inf\Big\{ \sum_{j=1}^{\tau(n,m)+1} t_j f(A_j) : \sum_{j=1}^{\tau(n,m)+1} t_j M(A_j) = X, \ \sum_{j=1}^{\tau(n,m)+1} t_j = 1, t_j \ge 0 \Big\} \quad (5.8)$$

for all $X \in \mathbf{R}^{\tau(n,m)}$.

Theorem 5.10 *Let $f : \mathbf{M}^{m\times n} \to [0, +\infty]$ be a polyconvex function, such that a positive l.s.c. convex g exists satisfying (5.4). Then the functional*

$$\mathcal{F}(u) = \int_{\Omega} f(Du)\, dx$$

is (sequentially) weakly l.s.c. on $\mathbf{W}^{1,n\wedge m}(\Omega; \mathbf{R}^m)$.

Proof We prove the theorem only in the case $n = m = 2$. If $n \wedge m = 1$ then we are in the case of Corollary 5.2; if $n \vee m > 2$ the argument of the proof can be easily generalized (see Ball (1977) or Dacorogna (1989)).

First we observe that if $u = (u^1, u^2) \in \mathbf{W}^{1,2}$ then we can write

$$\det Du = \operatorname{div}(u^1 D_2 u^2, -u^1 D_1 u^2) \qquad (5.9)$$

in $\mathcal{D}'(\Omega)$. In fact if $u \in C^2(\Omega; \mathbf{R}^2)$ then the equality holds pointwise in Ω so that we have in particular

$$\int_{\Omega} \varphi \det Du\, dx = - \int_{\Omega} (u^1 D_2 u^2 D_1\varphi - u^1 D_1 u^2 D_2\varphi)\, dx \qquad (5.10)$$

for all $\varphi \in C_0^\infty(\Omega)$, and this equality extends to arbitrary $u \in \mathbf{W}^{1,2}$ by a density argument.

Now let $u_j \rightharpoonup u$ in $\mathbf{W}^{1,2}$. Note that $\det Du_j, \det Du \in \mathbf{L}^1(\Omega)$. Since

$$u_j^1 \to u^1 \text{ strongly in } \mathbf{L}^2_{\text{loc}}(\Omega)$$

$$Du_j^2 \rightharpoonup Du^2 \text{ weakly in } L^2(\Omega; \mathbf{R}^2),$$

fixed $\varphi \in C_0^\infty(\Omega)$, then from the equality

$$\int_\Omega \varphi \det Du_j \, dx = -\int_\Omega (u_j^1 D_2 u_j^2 D_1\varphi - u_j^1 D_1 u_j^2 D_2\varphi) \, dx$$

we deduce

$$\lim_j \int_\Omega \varphi \det Du_j \, dx = -\lim_j \int_\Omega (u_j^1 D_2 u_j^2 D_1\varphi - u_j^1 D_1 u_j^2 D_2\varphi) \, dx$$

$$= -\int_\Omega (u^1 D_2 u^2 D_1\varphi - u^1 D_1 u^2 D_2\varphi) \, dx$$

$$= \int_\Omega \varphi \det Du \, dx;$$

that is, $\det Du_j \to \det Du$ in $\mathcal{D}'(\Omega)$.

Since we have $(Du_j, \det Du_j) \to (Du, \det Du)$ in $\mathcal{D}'(\Omega; \mathbf{R}^5)$ we get

$$\mathcal{F}(u) = \int_\Omega g(Du, \det Du) \, dx$$

$$\leq \liminf_j \int_\Omega g(Du_j, \det Du_j) \, dx = \liminf_j \mathcal{F}(u_j)$$

by Theorem 5.1. $\qquad\square$

Remark 5.11 If $f : \mathbf{M}^{m\times n} \to [0, +\infty]$ is polyconvex and l.s.c., and

$$\lim_{|A|\to\infty} \frac{f(A)}{|A|} = +\infty,$$

then the function g defined by (5.8) gives a convex l.s.c. function as required by Theorem 5.10. In fact, let $X_k \to X$. By the lower semicontinuity and coerciveness assumptions on f, there exist two sequences (t_j^k) and (A_j^k) such that the minimum in (5.8) for $g(X_k)$ is attained. By a compactness argument we can suppose that $t_j^k \to t_j$ and $A_j^k \to A_j$ with $\sum_j t_j M(A_j) = X$, so that

$$g(X) \leq \sum_{j=1}^{\tau(n,m)+1} t_j f(A_j) \leq \sum_{j=1}^{\tau(n,m)+1} t_j \liminf_k f(A_j^k)$$

$$= \sum_{j=1}^{\tau(n,m)+1} \liminf_k t_j^k f(A_j^k) \leq \liminf_k \sum_{j=1}^{\tau(n,m)+1} t_j^k f(A_j^k)$$

$$= \liminf_k g(X_k),$$

so that g is l.s.c. It is also convex by construction.

Corollary 5.12 *If $f : \mathbf{M}^{m \times n} \to [0, +\infty]$ is polyconvex, l.s.c., and*

$$f(A) \geq c_1 |A|^p - c_2$$

for some $p \geq n \wedge m$, then the functional

$$\mathcal{F}(u) = \begin{cases} \displaystyle\int_\Omega f(Du)\,dx & u \in \mathrm{W}^{1,p}(\Omega; \mathbf{R}^m) \\ +\infty & u \in \mathrm{L}^p(\Omega; \mathbf{R}^m) \setminus \mathrm{W}^{1,p}(\Omega; \mathbf{R}^m) \end{cases}$$

is l.s.c. with respect to the norm convergence of $\mathrm{L}^p(\Omega; \mathbf{R}^m)$.

Example 5.13 We can give an existence result for 2-dimensional finite elasticity problems. If $f : \mathbf{M}^{2 \times 2} \to [0, +\infty]$ is polyconvex, l.s.c., and

$$f(A) \geq c_1 |A|^2 + c_2 |\det A|^{-\beta} - c_3 \quad \text{if } \det A > 0$$

$(\beta > 0)$, $f(A) = +\infty$ if $\det A \leq 0$, then we have the existence of solutions to the problem

$$\min\left\{ \int_\Omega f(Du)\,dx - \int_\Omega \langle h, u \rangle\,dx : \right.$$
$$\left. u - u_0 \in \mathrm{W}_0^{1,2}(\Omega; \mathbf{R}^2),\ \det Du > 0 \text{ a.e. on } \Omega \right\}$$

for $h \in \mathrm{L}^2(\Omega; \mathbf{R}^2)$ and $u_0 \in \mathrm{W}^{1,2}$. It suffices to apply the direct method of the calculus of variations and Corollary 5.12.

5.3 Quasiconvexity

Definition 5.14 *We say that $f : \mathbf{M}^{m \times n} \to \mathbf{R}$ is* quasiconvex *if f is continuous, and for all $A \in \mathbf{M}^{m \times n}$ and for every bounded open set E of $\mathbf{R}^n$*

$$|E| f(A) \leq \int_E f(A + D\varphi)\,dx$$

for every $\varphi \in C_0^\infty(E; \mathbf{R}^m)$.

Remark 5.15 Note that it suffices to check the definition of quasiconvexity for one fixed bounded open set E, as in Remark 4.6(i).

If $1 \leq p < \infty$, and f satisfies the growth condition from above $0 \leq f(A) \leq c(1 + |A|^p)$ for all $A \in \mathbf{M}^{m \times n}$, then f is quasiconvex $\iff f$ is $\mathrm{W}^{1,\infty}$-quasiconvex $\iff f$ is $\mathrm{W}^{1,p}$-quasiconvex $\implies f$ is rank-1-convex, and the local Lipschitz condition

$$|f(A) - f(B)| \leq c(1 + |A|^{p-1} + |B|^{p-1})|A - B| \tag{5.11}$$

holds for all $A, B \in \mathbf{M}^{m \times n}$ (see Remarks 4.6(iii) and 4.13(iii)).

Theorem 5.16 *If $1 \le p < \infty$, and $f : \mathbf{M}^{m \times n} \to \mathbf{R}$ is a quasiconvex function satisfying*

$$0 \le f(A) \le c(1 + |A|^p) \qquad \text{for all } A \in \mathbf{M}^{m \times n}, \tag{5.12}$$

then the functional

$$\mathcal{F}(u) = \int_{\Omega} f(Du)\, dx$$

is (sequentially) weakly l.s.c. on $\mathrm{W}^{1,p}(\Omega; \mathbf{R}^m)$.

Proof We divide the proof into two steps. In the first one we prove the theorem in the case of Ω a ball and u linear; in the second one we recover the general result by a covering and approximation argument.

Step 1. Let $A \in \mathbf{M}^{m \times n}$, and let $\Omega = B(x_0, r)$ be the open ball in $\mathbf{R}^n$ of centre x_0 and radius $r > 0$. We write B_s in place of $B(x_0, s)$ for $0 < s \le r$. Let (u_j) be a sequence converging weakly to the linear function Ax in $\mathrm{W}^{1,p}(B_r; \mathbf{R}^m)$. We suppose without loss of generality that

$$\lim_j \mathcal{F}(u_j) = \liminf_j \mathcal{F}(u_j) < +\infty.$$

Fix $0 < s < r$ and $0 < \varepsilon < r - s$, and define

$$\varphi(x) = \varphi^{s,\varepsilon}(x) = \left(\frac{1}{\varepsilon}\mathrm{dist}(x, B_s)\right) \wedge 1$$
$$v_j(x) = v_j^{s,\varepsilon}(x) = u_j(x) + \varphi(x)(Ax - u_j(x))$$

for $x \in B_r$. Note that $v_j = u_j$ on B_s and $v_j = Ax$ on $B_r \setminus B_{s+\varepsilon}$; in particular $v_j - Ax \in \mathrm{W}_0^{1,p}(B_r; \mathbf{R}^m)$ and, since f is $\mathrm{W}^{1,p}$-quasiconvex (Remark 5.15),

$$\mathcal{F}(Ax) = |B_r| f(A) \le \int_{B_r} f(Dv_j)\, dx. \tag{5.13}$$

We estimate the right-hand side of this inequality:

$$\int_{B_r} f(Dv_j)\, dx = \int_{B_s} f(Du_j)\, dx + \int_{B_r \setminus B_{s+\varepsilon}} f(A)\, dx$$
$$+ \int_{B_{s+\varepsilon} \setminus B_s} f(Du_j + \varphi(A - Du_j) + (Ax - u_j)D\varphi)\, dx$$
$$\le \int_{B_r} f(Du_j)\, dx + |B_r \setminus B_s| f(A)$$
$$+ c\int_{B_{s+\varepsilon} \setminus B_s} (1 + |Du_j + \varphi(A - Du_j) + (Ax - u_j)D\varphi|^p)\, dx$$
$$\le \int_{B_r} f(Du_j)\, dx + |B_r \setminus B_s| f(A)$$

$$+ c \int_{B_{s+\varepsilon} \setminus B_s} (1 + |Du_j|^p + |A|^p)\, dx$$

$$+ \frac{c}{\varepsilon^p} \int_{B_{s+\varepsilon} \setminus B_s} |Ax - u_j|^p\, dx$$

$$= \int_{B_r} f(Du_j)\, dx + |B_r \setminus B_s| f(A)$$

$$+ c\mu_j(B_{s+\varepsilon} \setminus B_s) + \frac{c}{\varepsilon^p} \int_{B_r} |Ax - u_j|^p\, dx,$$

having set $\mu_j(E) = \int_E (1 + |Du_j|^p + |A|^p)\, dx$. Passing to the limit as $j \to \infty$ we get

$$\liminf_j \mathcal{F}(v_j) \le \liminf_j \mathcal{F}(u_j)$$

$$+ |B_r \setminus B_s| f(A) + c \limsup_j \mu_j(\overline{B_{s+\varepsilon} \setminus B_s}) \qquad (5.14)$$

owing to the strong convergence of u_j to Ax in $L^p(\Omega; \mathbf{R}^m)$.

Note that (μ_j) is a bounded sequence of Borel measures on B_r. We can suppose then, passing possibly to a subsequence, that there exists a Borel measure μ such that μ_j converges weakly* to μ in the sense of measures. In particular μ is finite, and for all compact sets K $\limsup_j \mu_j(K) \le \mu(K)$, so that

$$\limsup_j \mu_j(\overline{B_{s+\varepsilon} \setminus B_s}) \le \mu(\overline{B_{s+\varepsilon} \setminus B_s}). \qquad (5.15)$$

As μ is finite we have $\mu(\partial B_s) = 0$ for every s except for at most a countable set of radii $s \in (0, r)$, and

$$\lim_{\varepsilon \to 0+} \mu(\overline{B_{s+\varepsilon} \setminus B_{s-\varepsilon}}) = 0 \qquad (5.16)$$

for all such s. If we choose s with these properties in (5.14), then passing to the limit first as $\varepsilon \to 0$ and subsequently as $s \to r$, we obtain

$$\mathcal{F}(Ax) \le \liminf_j \mathcal{F}(u_j)$$

by (5.13).

Step 2. Let Ω be any bounded open set, and let $u_j \rightharpoonup u$ in $W^{1,p}(\Omega; \mathbf{R}^m)$. Fix $\varepsilon > 0$, and define

$$\mathcal{U}^\varepsilon = \left\{ B = \overline{B(x,r)} \subseteq \Omega :\ \exists A \in \mathbf{M}^{m \times n},\ \int_B |Du - A|^p\, dx \le \varepsilon^p |B| \right\}. \qquad (5.17)$$

Note that if x_0 is a Lebesgue point for Du, then $B(x_0, r)$ belongs to $\mathcal{U}^\varepsilon$ for r sufficiently small, and we can take $A = Du(x_0)$ in (5.17). In particular $\mathcal{U}^\varepsilon$ satisfies the requirements of Theorem 4.4, so that there exists a disjoint sequence $(B_j) = (B(x_j, r_j)) \subset \mathcal{U}^\varepsilon$ such that $|\Omega \setminus \bigcup_j B_j| = 0$. Consider for every j a matrix

$A_j \in \mathbf{M}^{m \times n}$ such that $\int_{B_j} |Du(x) - A_j|^p \, dx \le \varepsilon^p |B_j|$, and define a function $\psi \in \mathrm{L}^p(\Omega; \mathbf{M}^{m \times n})$ by $\psi(x) = A_j$ on B_j. Note that we have

$$\int_\Omega |Du - \psi|^p \, dx = \sum_j \int_{B_j} |Du(x) - A_j|^p \, dx \le \varepsilon^p |\Omega|. \qquad (5.18)$$

We can write

$$\begin{aligned}
\int_{B_j} f(Du_k) \, dx = {} & \int_{B_j} (f(Du_k) - f(Du_k - Du + \psi)) \, dx \\
& + \int_{B_j} (f(Du_k - Du + A_j) - f(A_j)) \, dx \\
& + \int_{B_j} (f(A_j) - f(Du)) \, dx + \int_{B_j} f(Du) \, dx.
\end{aligned}$$

By the previous step we have

$$\liminf_k \int_{B_j} (f(Du_k - Du + A_j) - f(A_j)) \, dx \ge 0;$$

hence for all $N \in \mathbf{N}$

$$\begin{aligned}
& \liminf_k \sum_{j=1}^N \int_{B_j} f(Du_k) \, dx \\
\ge {} & \liminf_k \sum_{j=1}^N \int_{B_j} (f(Du_k) - f(Du_k - Du + \psi)) \, dx \\
& + \sum_{j=1}^N \liminf_k \int_{B_j} (f(Du_k - Du + A_j) - f(A_j)) \, dx \\
& + \sum_{j=1}^N \int_{B_j} (f(A_j) - f(Du)) \, dx + \sum_{j=1}^N \int_{B_j} f(Du) \, dx \\
\ge {} & -\limsup_k \int_\Omega |f(Du_k) - f(Du_k - Du + \psi)| \, dx \\
& - \int_\Omega |f(\psi) - f(Du)| \, dx + \sum_{j=1}^N \int_{B_j} f(Du) \, dx \\
\ge {} & \sum_{j=1}^N \int_{B_j} f(Du) \, dx - c\varepsilon. \qquad (5.19)
\end{aligned}$$

In the last inequality we have used (5.18) together with the estimates

$$\int_\Omega |f(Du_k) - f(Du_k - Du + \psi)| \, dx$$

$$\le c\Big(\int_\Omega (1 + |Du_k|^{p-1} + |Du_k - Du + \psi|^{p-1})^{p/(p-1)} \, dx\Big)^{(p-1)/p}$$

$$\times \Big(\int_\Omega |Du - \psi|^p \, dx\Big)^{1/p},$$

$$\int_\Omega |f(Du) - f(\psi)| \, dx$$

$$\le c\Big(\int_\Omega (1 + |Du|^{p-1} + |\psi|^{p-1})^{p/(p-1)} \, dx\Big)^{(p-1)/p} \Big(\int_\Omega |Du - \psi|^p \, dx\Big)^{1/p},$$

which follow from the Lipschitz condition (5.11) and Hölder's inequality. Finally, from (5.19) we have

$$-c\varepsilon + \sum_{j=1}^{N} \int_{B_j} f(Du) \, dx \le \liminf_k \sum_{j=1}^{N} \int_{B_j} f(Du_k) \, dx$$

$$\le \liminf_k \int_\Omega f(Du_k) \, dx,$$

whence we obtain the thesis letting $N \to \infty$, and then $\varepsilon \to 0$. $\square$

Remark 5.17 In the case $p = \infty$, the functional $\mathcal{F}$ is weakly* l.s.c. on the space $\mathrm{W}^{1,\infty}(\Omega; \mathbf{R}^m)$ under the only condition that $f : \mathbf{M}^{m \times n} \to \mathbf{R}$ be quasiconvex (see Exercise 5.4).

5.4   Exercises

Exercise 5.1 Prove Jensen's inequality.

    Hint: if f is differentiable then integrate the inequality

$$f\Big(\frac{1}{\mu(\Omega)} \int_\Omega u \, d\mu\Big) \le f(u(x)) + \Big\langle Df\Big(\frac{1}{\mu(\Omega)} \int_\Omega u \, d\mu\Big), \Big(\frac{1}{\mu(\Omega)} \int_\Omega u \, d\mu\Big) - u(x)\Big\rangle.$$

In the general case use an approximation procedure.

Exercise 5.2 (Banach-space Jensen's inequality) Let X be a Banach space, and let $f : X \to [0, +\infty]$ be convex and l.s.c. Let $(E, \mathcal{E}, \mu)$ be a measure space with $\mu \ge 0$ and $\mu(E) = 1$. Then

$$f\Big(\int_E u \, d\mu\Big) \le \int_E f(u) \, d\mu$$

for every $u \in \mathrm{L}^1_\mu(E; X)$.

    Solution: let X' be the dual space of X, and let $\langle \cdot, \cdot \rangle$ denote the duality pairing between X' and X. Since f is convex and l.s.c., by the Hahn Banach Theorem for every $x \in X$ and for every $t < f(x)$ there exist $x' \in X'$ and $a \in \mathbf{R}$ such that

$$t < \langle x', x \rangle + a \qquad \text{and} \qquad \langle x', y \rangle + a \leq f(y) \quad \text{for every } y \in X. \tag{5.20}$$

In particular this holds for $x = \int_E u \, d\mu$ and $t < f(\int_E u \, d\mu)$. On the other hand, from (5.20) we get

$$\langle x', x \rangle + a = \int_E \left(\langle x', u(s) \rangle + a \right) d\mu \leq \int_E f(u) \, d\mu, \tag{5.21}$$

so (5.20) and (5.21) imply $t < \int_E f(u) \, d\mu$ for every $t < f(\int_E u \, d\mu)$ and conclude the proof.

Exercise 5.3 Prove that if f is convex and finite, then it is locally Lipschitz. Deduce that a quasiconvex function is locally Lipschitz.

Exercise 5.4 Prove the statement of Remark 5.17.

Hint: follow the proof of Theorem 5.16, using in Step 1 the method described in Proposition 4.9. In the second step replace p by 1, and use the local Lipschitz continuity of f in (5.19).

We recall that a function $A : \mathbf{R}^n \to \mathbf{R}^n$ is said to be *monotone* if

$$\langle A(\xi_1) - A(\xi_2), \xi_1 - \xi_2 \rangle \geq 0$$

for every ξ_1, $\xi_2 \in \mathbf{R}^n$; a monotone function A is called *maximal monotone* if $\langle y - A(\xi), \eta - \xi \rangle \geq 0$ for every $\xi \in \mathbf{R}^n$ implies that $y = A(\eta)$.

Exercise 5.5 Let $f : \mathbf{R}^n \to \mathbf{R}$ be convex and differentiable. Prove that $Df : \mathbf{R}^n \to \mathbf{R}^n$ is maximal monotone.

Solution: by the property (i) in Section 5.1, we obtain immediately the monotonicity of Df. For its maximal monotonicity we must show that $\langle y - Df(\zeta), \eta - \zeta \rangle \geq 0$ for every $\zeta \in \mathbf{R}^n$ implies that $y = Df(\eta)$. Thus, given η and y with $y \neq Df(\eta)$ we must prove that there exists $\zeta \in \mathbf{R}^n$ such that $\langle y - Df(\zeta), \eta - \zeta \rangle < 0$. Replacing f by the l.s.c. convex function $g(x) = f(x + \eta) - \langle x, y \rangle$ if necessary, we can reduce the argument to $\eta = 0$ and $y = 0$. We assume therefore that $0 \neq Df(0)$, and we show that there exists $\zeta \in \mathbf{R}^n$ such that $\langle Df(\zeta), \zeta \rangle < 0$. Notice that the l.s.c. coercive strictly convex function $f(x) + \frac{1}{2}|x|^2$ admits a unique minimum point $\zeta \in \mathbf{R}^n$ satisfying $Df(\zeta) + \zeta = 0$. If $\zeta = 0$, then $Df(\zeta) = 0$, contrary to $Df(0) \neq 0$. If $\zeta \neq 0$, then $\zeta = -Df(\zeta)$, and we have $|\zeta|^2 = \langle \zeta, \zeta \rangle = \langle -Df(\zeta), \zeta \rangle > 0$ and $\langle Df(\zeta), \zeta \rangle < 0$, concluding the proof.

6

THE STRUCTURE OF QUASICONVEX FUNCTIONS

From Remark 5.17 we deduce that quasiconvexity is a sufficient condition for the lower semicontinuity of integral functionals with respect to the weak* convergence in $W^{1,\infty}$; hence quasiconvex functions are rank-1-convex by Corollary 4.12. More generally we will show that for continuous f we have

$$f \text{ polyconvex} \implies f \text{ quasiconvex} \implies f \text{ rank-1-convex},$$

and these implications cannot be reversed in general.

6.1 Quasiconvexity of polyconvex functions

Remark 6.1 If $A \in \mathbf{M}^{m \times n}$ and $\varphi \in C_0^\infty(\Omega; \mathbf{R}^m)$, then

$$\int_\Omega M(A + D\varphi) \, dx = M(A)|\Omega|. \tag{6.1}$$

In the case $n = m = 2$ this means that

$$\int_\Omega D\varphi \, dx = 0, \quad \text{and} \quad \int_\Omega \det(A + D\varphi) \, dx = |\Omega| \det A.$$

The first equality is obvious, while the second one is easily checked by writing $v(x) = Ax + \varphi(x)$, and recalling that $\det Dv = \operatorname{div}(v^1 D_2 v^2, -v^1 D_1 v^2)$ so that $\int_\Omega \det Dv \, dx$ depends only on the values of Dv at the boundary of Ω. In the case $n \vee m > 2$ one can proceed likewise, writing each minor as a divergence.

Proposition 6.2 *If* $f : \mathbf{M}^{m \times n} \to \mathbf{R}$ *is continuous and polyconvex, then it is quasiconvex.*

Proof Let $g : \mathbf{R}^{\tau(n,m)} \to \mathbf{R}$ be a convex function such that $f(A) = g(M(A))$. If E is a bounded open set, then we have by Jensen's inequality

$$
\begin{aligned}
\int_E f(A + D\varphi) \, dx &= \int_E g(M(A + D\varphi)) \, dx \\
&\geq |E| \, g\Big(\frac{1}{|E|} \int_E M(A + D\varphi) \, dx\Big) \\
&= |E| \, g(M(A)) = |E| \, f(A)
\end{aligned}
$$

for all $A \in \mathbf{M}^{m \times n}$ and $\varphi \in C_0^\infty(E; \mathbf{R}^m)$. $\qquad\square$

6.2 Quasiconvexification

In order to exhibit a quasiconvex function which is not polyconvex we need a way to construct non-trivial quasiconvex functions. This will be done by 'quasiconvexification'.

Definition 6.3 *Let $f : \mathbf{M}^{m \times n} \to \mathbf{R}$ be a Borel function, and let E be a bounded open subset of $\mathbf{R}^n$ with $|\partial E| = 0$. We define the* quasiconvexification Qf *of f by*

$$Qf(A) = \inf\Big\{\frac{1}{|E|} \int_E f(A + D\varphi)\, dx : \varphi \in C_0^\infty(E; \mathbf{R}^m)\Big\} \tag{6.2}$$

for all $A \in \mathbf{M}^{m \times n}$.

Remark 6.4 The definition of Qf does not depend on E. The proof of this statement can be performed in a similar way as that of Proposition 4.5. In fact, let E and E' be two bounded open subsets of $\mathbf{R}^n$ with $|\partial E| = |\partial E'| = 0$, and let Qf and $Q'f$ be the corresponding quasiconvexifications of f given by (6.2). With fixed $\varepsilon > 0$ we can find $x_1, \ldots, x_N \in \mathbf{R}^n$ and $\rho_1, \ldots, \rho_N > 0$ such that $E_j = x_j + \rho_j E \subset E'$, $E_j \cap E_i = \emptyset$ if $i \neq j$, and

$$\Big|E' \setminus \bigcup_{j=1}^{N} E_j\Big| \leq \varepsilon |E'|.$$

With fixed $A \in \mathbf{M}^{m \times n}$, we choose $\varphi \in C_0^\infty(E; \mathbf{R}^m)$ such that

$$\int_E f(A + D\varphi)\, dx \leq |E|(Qf(A) + \varepsilon),$$

and we define $\psi \in C_0^\infty(E'; \mathbf{R}^m)$ by

$$\psi(x) = \begin{cases} \rho_j \varphi\Big(\dfrac{x - x_j}{\rho_j}\Big) & \text{if } x \in E_j \\ 0 & \text{otherwise.} \end{cases}$$

Note that

$$\int_{E_j} f(A + D\psi)\, dx = \rho_j^n \int_E f(A + D\varphi)\, dx \leq |E_j|(Qf(A) + \varepsilon);$$

hence

$$|E'|Q'f(A) \leq \int_{E'} f(A + D\psi)\, dx$$
$$= f(A)\Big|E' \setminus \bigcup_{j=1}^{N} E_j\Big| + \sum_{j=1}^{N} \int_{E_j} f(A + D\psi)\, dx$$

$$\leq \varepsilon |E'| f(A) + \sum_{j=1}^{N} |E_j| (Qf(A) + \varepsilon)$$

$$\leq \varepsilon |E'| (f(A) + 1) + |E'| Qf(A).$$

By the arbitrariness of ε we get $Q'f(A) \leq Qf(A)$. In the same way we prove the opposite inequality.

Proposition 6.5 *If $f : \mathbf{M}^{m \times n} \to \mathbf{R}$ is a locally bounded Borel function, then Qf is a continuous function.*

Proof We use the notation $B_s = B(0, s)$. It is clear that by a translation argument it suffices to prove that Qf is continuous at 0. Let $A_j \to 0$. We set

$$r_j = 1 - \left(\frac{2|A_j|}{1 - 2|A_j|} \right). \tag{6.3}$$

Let $\varphi_j \in C_0^\infty(B_{r_j}; \mathbf{R}^m)$ such that

$$\int_{B_{r_j}} f(D\varphi_j)\, dx \leq |B_{r_j}| \left(Qf(0) + \frac{1}{j} \right).$$

We define $\zeta_j \in C_0^\infty(B_1; \mathbf{R}^m)$ by

$$\zeta_j(x) = \begin{cases} \varphi_j(x) - A_j x & \text{if } x \in B_{r_j} \\[2mm] -\psi_j(x) A_j x & \text{if } x \in B_1 \setminus B_{r_j}, \end{cases}$$

where $\psi_j \in C_0^\infty(B_1)$, $0 \leq \psi_j \leq 1$, $\psi_j \equiv 1$ on B_{r_j} and $|D\psi_j| \leq \frac{2}{1-r_j}$. We obtain the estimate

$$
\begin{aligned}
|B_1| Qf(A_j) &\leq \int_{B_1} f(A_j + D\zeta_j)\, dx \\
&= \int_{B_1 \setminus B_{r_j}} f(A_j - A_j x D\psi_j - A_j \psi_j)\, dx + \int_{B_{r_j}} f(D\varphi_j)\, dx \\
&\leq |B_1 \setminus B_{r_j}| \sup\left\{ f(X) : |X| \leq \left(\frac{2}{1-r_j} + 2 \right)|A_j| \right\} \\
&\quad + |B_{r_j}| \left(Qf(0) + \frac{1}{j} \right).
\end{aligned}
$$

Taking into account that by (6.3)

$$\left(\frac{2}{1-r_j} + 2 \right)|A_j| = 1 \qquad \text{and} \qquad r_j \to 1,$$

we get

$$\limsup_j Qf(A_j) \leq Qf(0).$$

Vice versa, let now $\varphi_j \in C_0^\infty(B_{r_j}; \mathbf{R}^m)$ satisfy

$$\int_{B_{r_j}} f(A_j + D\varphi_j)\, dx \leq |B_{r_j}| \left(Qf(A_j) + \frac{1}{j} \right).$$

We set

$$\zeta_j(x) = \begin{cases} \varphi_j(x) + A_j x & \text{if } x \in B_{r_j} \\[2mm] \psi_j(x) A_j x & \text{if } x \in B_1 \setminus B_{r_j}, \end{cases}$$

where $\psi_j \in C_0^\infty(B_1)$ is as above. We obtain

$$\begin{aligned}
|B_1| Qf(0) &\leq \int_{B_1} f(D\zeta_j)\, dx \\
&= \int_{B_1 \setminus B_{r_j}} f(A_j x D\psi_j + A_j \psi_j)\, dx \\
&\quad + \int_{B_{r_j}} f(A_j + D\varphi_j)\, dx \\
&\leq |B_1 \setminus B_{r_j}| \sup\left\{ f(X) : |X| \leq \left(\frac{2}{1 - r_j} + 1 \right) |A_j| \right\} \\
&\quad + |B_{r_j}| \left(Qf(A_j) + \frac{1}{j} \right) \\
&\leq |B_1 \setminus B_{r_j}| \sup\{ f(X) : |X| \leq 1 \} + |B_{r_j}| \left(Qf(A_j) + \frac{1}{j} \right);
\end{aligned}$$

hence $Qf(0) \leq \liminf_j Qf(A_j)$. $\qquad\square$

Proposition 6.6 *If $f : \mathbf{M}^{m\times n} \to \mathbf{R}$ is a locally bounded Borel function, then Qf satisfies*

$$|\Omega| Qf(A) \leq \int_\Omega Qf(A + D\varphi)\, dx \tag{6.4}$$

for all $A \in \mathbf{M}^{m\times n}$ and for all $\varphi \in \mathrm{W}^{1,\infty}(\Omega; \mathbf{R}^m)$ piecewise affine with compact support in Ω.

Proof By a translation argument it suffices to show (6.4) for $A = 0$. Let $\varphi(x) = \sum_{j=1}^N (z_j + A_j x)\chi_{E_j}$ be a $\mathrm{W}^{1,\infty}$ piecewise affine function with compact support in Ω.

Fix $\varepsilon > 0$. Let $\varphi_j \in C_0^\infty(E_j; \mathbf{R}^m)$ be such that

$$\int_{E_j} f(A_j + D\varphi_j)\, dx \leq |E_j|(Qf(A_j) + \varepsilon),$$

and let $\psi \in \mathrm{W}_0^{1,\infty}(\Omega; \mathbf{R}^m)$ be defined by

$$\psi(x) = \begin{cases} z_j + A_j x + \varphi_j(x) & x \in E_j \\ 0 & x \in \Omega \setminus \bigcup_j E_j. \end{cases}$$

We slightly modify ψ to smooth it out near $\bigcup_j \partial E_j$: for $\eta > 0$ let

$$U_\eta = \left\{ x \in \Omega : \ \text{dist}\left(x, \bigcup_j \partial E_j\right) < \eta \right\};$$

if η is small enough, so that

$$U_\eta \cap \bigcup_j \text{spt}\,\varphi_j = \emptyset,$$

then we choose $\psi_\eta \in C_0^\infty(\Omega; \mathbf{R}^m)$ such that

$$\text{spt}\,(\psi - \psi_\eta) \subset U_\eta \qquad \text{and} \qquad \|D\psi_\eta\|_{L^\infty(\Omega;\mathbf{R}^m)} \le \|D\psi\|_{L^\infty(\Omega;\mathbf{R}^m)}$$

(such a function can be easily obtained by taking ψ_η equal to a convolution of ψ on U_η). Then

$$\lim_{\eta \to 0+} \int_{E_j} f(D\psi_\eta)\,dx = \int_{E_j} f(D\psi)\,dx$$

for all $j = 1, \dots, N$.

Define

$$\Omega_\eta = U_\eta \cup \bigcup_{j=1}^N E_j.$$

Since $\psi_\eta \in C_0^\infty(\Omega_\eta; \mathbf{R}^m)$, we have

$$|\Omega_\eta|\, Qf(0) \le \int_{\Omega_\eta} f(D\psi_\eta)\,dx$$

$$\le \sum_{j=1}^N \int_{E_j} f(D\psi_\eta)\,dx + \left|\Omega_\eta \setminus \bigcup_{j=1}^N E_j\right| \sup\{f(A) : \ |A| \le \|D\psi\|_\infty\}.$$

Hence,

$$|\Omega|\, Qf(0) = \lim_{\eta \to 0+} \left(|\Omega \setminus \Omega_\eta|\, Qf(0) + |\Omega_\eta|\, Qf(0) \right)$$

$$\le \left|\Omega \setminus \bigcup_{j=1}^N E_j\right| Qf(0) + \lim_{\eta \to 0+} \sum_{j=1}^N \int_{E_j} f(D\psi_\eta)\,dx$$

$$= \left|\Omega \setminus \bigcup_{j=1}^N E_j\right| Qf(0) + \sum_{j=1}^N \int_{E_j} f(D\psi)\,dx$$

$$\leq \int_\Omega Qf(D\varphi)\,dx + \varepsilon|\Omega|.$$

By the arbitrariness of $\varepsilon > 0$ inequality (6.4) is proven. $\square$

Proposition 6.7 *If $f : \mathbf{M}^{m\times n} \to \mathbf{R}$ is a locally bounded Borel function, then Qf is quasiconvex.*

Proof By Proposition 6.5 Qf is continuous. It suffices then to remark that inequality (6.4) holds for all $C_0^\infty(\Omega; \mathbf{R}^m)$ by approximation in $\mathbf{W}^{1,\infty}(\Omega; \mathbf{R}^m)$ with piecewise affine functions with compact support in Ω. $\square$

Remark 6.8 If $f : \mathbf{M}^{m\times n} \to \mathbf{R}$ is a locally bounded Borel function, then

$$Qf = \sup\{g : \mathbf{M}^{m\times n} \to \mathbf{R} : \ g \text{ quasiconvex}, g \leq f\}; \tag{6.5}$$

that is, Qf is the *quasiconvex envelope* of f. In fact, if we denote by h the right-hand side of (6.5), then $Qf \leq h$ since Qf is quasiconvex and $Qf \leq f$. On the other hand, if g is quasiconvex and $g \leq f$, then $g = Qg \leq Qf$, so that $h \leq Qf$.

6.3 Example of a quasiconvex non-polyconvex function

Before giving an example of a quasiconvex non-polyconvex function we state the following remark about growth conditions satisfied by polyconvex functions.

Remark 6.9 If $f : \mathbf{M}^{m\times n} \to [0, +\infty)$ is a continuous polyconvex function satisfying the growth condition $f(A) \leq c(1+|A|^p)$ for some $p < 2$ then f is convex.

It suffices to prove this fact for $m = n = 2$. In this case, if $f(A) = g(A, \det A)$ with g convex then we can write

$$g(A, r) = \sup\{a_j + \langle A_j, A\rangle + b_j r : \ j \in \mathbf{N}\}$$

for a suitable choice of $a_j, b_j \in \mathbf{R}$, $A_j \in \mathbf{M}^{2\times 2}$. If $b_j = 0$ for all j then $f(A) = g(A, 0)$ and f is convex. Otherwise, we may suppose $b_1 \neq 0$. If $b_1 > 0$ then we can estimate $f(tI)$ and get

$$f(tI) = g(tI, t^2) \geq a_1 + t\langle A_1, I\rangle + t^2 b_1,$$

thus contradicting the growth estimate above for t large. If $b_1 < 0$ then the same conclusion follows by considering the matrices $t(e_1 \otimes e_1 - e_2 \otimes e_2)$.

Proposition 6.10 *If $n \geq 2$, $m \geq 2$ then there exists a function $f : \mathbf{M}^{m\times n} \to [0, +\infty)$ that is quasiconvex but not polyconvex.*

Proof It suffices to consider the case $m = n = 2$. We prove that the function $f = Qg$ satisfies the thesis of the proposition, where $g : \mathbf{M}^{2\times 2} \to [0, +\infty)$ is defined by

$$g(A) = \begin{cases} 1 + |A|^p & \text{if } A \notin \{I, -I\} \\ 0 & \text{if } A \in \{I, -I\} \end{cases}$$

with $1 < p < 2$. Clearly $0 \leq Qg(A) \leq 1 + |A|^p$; hence by Remark 6.9 Qg is polyconvex if and only if it is convex.

We have trivially $Qg(I) = Qg(-I) = 0$. We prove now that $Qg(0) > 0$. This will show that Qg is not convex.

Were $Qg(0) = 0$, then we could find a sequence $(\varphi_j) \in C_0^\infty(\Omega; \mathbf{R}^2)$ such that

$$\lim_j \int_\Omega g(D\varphi_j)\, dx = 0.$$

Since we have

$$g(A) \geq c(|A_{11} - A_{22}|^p + |A_{12} + A_{21}|^p),$$

we also obtain

$$\lim_j \int_\Omega (|D_1\varphi_j^1 - D_2\varphi_j^2|^p + |D_1\varphi_j^2 + D_2\varphi_j^1|^p)\, dx = 0.$$

Since we can write

$$\Delta\varphi_j^1 = \operatorname{div}(D_1\varphi_j^1 - D_2\varphi_j^2, D_1\varphi_j^2 + D_2\varphi_j^1)$$
$$\Delta\varphi_j^2 = \operatorname{div}(D_1\varphi_j^2 + D_2\varphi_j^1, -D_1\varphi_j^1 + D_2\varphi_j^2)$$

and we have the estimate (see Appendix C)

$$\|D\varphi_j^i\|_{\mathrm{L}^p(\Omega;\mathbf{R}^2)}^p \leq c\|\Delta\varphi_j^i\|_{\mathrm{W}^{-1,p}(\Omega)}^p$$
$$\leq c\int_\Omega (|D_1\varphi_j^1 - D_2\varphi_j^2|^p + |D_1\varphi_j^2 + D_2\varphi_j^1|^p)\, dx$$

(where $\mathrm{W}^{-1,p}(\Omega)$ is the dual space of $\mathrm{W}_0^{1,p'}(\Omega)$), we get $\lim_j \int_\Omega |D\varphi_j|^p\, dx = 0$, so that

$$\lim_j \int_\Omega g(D\varphi_j)\, dx = |\Omega| > 0,$$

contradicting the hypothesis. $\qquad\qquad\qquad\qquad\qquad\qquad\qquad\qquad\square$

6.4   Example of a rank-1-convex non-quasiconvex function

Finally we show that rank-1-convexity does not imply quasiconvexity.

Remark 6.11 A continuous function $f : \mathbf{M}^{m \times n} \to \mathbf{R}$ is quasiconvex if and only if

$$f(A)|Y| \leq \int_Y f(A + D\varphi(y))\, dy \tag{6.6}$$

for every n-cube Y and $\varphi \in C^\infty(\mathbf{R}^n; \mathbf{R}^m)$ Y-periodic. In fact, (6.6) follows from Theorem 4.1 by Remark 5.17, while from (6.6) we deduce that f is quasiconvex taking $\varphi \in C_0^\infty(\Omega; \mathbf{R}^m)$, Y any n-cube containing Ω, and extending φ first to 0 in $Y \setminus \Omega$, and then to the whole $\mathbf{R}^n$ by Y-periodicity. Note that (6.6) holds for all Y-periodic $\varphi \in \mathrm{W}_{\mathrm{loc}}^{1,p}(\mathbf{R}^n; \mathbf{R}^m)$ if $|f(B)| \leq c(1 + |B|^p)$ for all $B \in \mathbf{M}^{m \times n}$ as in Remark 3.4(a).

Proposition 6.12 *If $n \geq 2$, $m \geq 3$ then there exists a function $f : \mathbf{M}^{m \times n} \to [0, +\infty)$ that is rank-1-convex but not quasiconvex.*

Proof It suffices to consider the case $n = 2$ and $m = 3$.

We first define a rank-1-convex function g on the linear subspace of $\mathbf{M}^{3 \times 2}$

$$L = \{ r(e_1 \otimes e_1) + s(e_2 \otimes e_2) + te_3 \otimes (e_1 + e_2) : \ r, s, t \in \mathbf{R} \}$$

by

$$g\big(r(e_1 \otimes e_1) + s(e_2 \otimes e_2) + te_3 \otimes (e_1 + e_2)\big) = rst.$$

The function g is linear on rank-1 lines in L, and so it is rank-1-convex. It is not quasiconvex on L: we can take $w : \mathbf{R}^2 \to \mathbf{R}^3$ given by

$$w(x_1, x_2) = (\cos x_1, -\sin x_2, \cos(x_1 + x_2)),$$

whose gradient is

$$Dw(x_1, x_2) = -\sin x_1 (e_1 \otimes e_1) - \cos x_2 (e_2 \otimes e_2) - \sin(x_1 + x_2) e_3 \otimes (e_1 + e_2).$$

We have $Dw \in L$, and

$$g(Dw(x_1, x_2)) = -(\sin x_1)^2 (\cos x_2)^2 - \sin x_1 \sin x_2 \cos x_1 \cos x_2,$$

so that, taking $Y = [0, 2\pi]^2$,

$$\int_Y g(Dw)\, dy = -\int_Y (\sin x_1)^2 (\cos x_2)^2\, dx_1\, dx_2 = -\pi^2 < 0 = g(0),$$

and (6.6) is violated with $A = 0$.

We have to extend the function g to a rank-1-convex function defined on the whole $\mathbf{M}^{3 \times 2}$. Since $|Dw| \leq 2$, we have

$$\int_Y \left(g(Dw) + \frac{1}{100}(|Dw|^2 + |Dw|^4) \right) dy \leq -\pi^2 + \frac{4}{5}\pi^2 < 0, \qquad (6.7)$$

so that the function $g(A) + \frac{1}{100}(|A|^2 + |A|^4)$ is still rank-1-convex and non-quasiconvex on L. Let $P : \mathbf{M}^{3 \times 2} \to L$ be the orthogonal projection on L. We want to show that there exists $k \in \mathbf{N}$ such that the function $f : \mathbf{M}^{3 \times 2} \to \mathbf{R}$ defined by

$$f(A) = g(P(A)) + \frac{1}{100}(|A|^2 + |A|^4) + k|A - P(A)|^2$$

is rank-1-convex.

Since f is differentiable, we must show that for all $A, X \in \mathbf{M}^{3 \times 2}$ with $\operatorname{rank} X = 1$ and $|X| = 1$ we have

$$\frac{d^2}{dt^2}f(A+tX)\Big|_{t=0} \geq 0. \tag{6.8}$$

For every $A, X \in \mathbf{M}^{3\times 2}$

$$\frac{d^2}{dt^2}f(A+tX)_{|t=0} = \frac{d^2}{dt^2}g(P(A)+tP(X))\Big|_{t=0} + \frac{1}{50}|X|^2 + \frac{1}{25}|A|^2|X|^2$$
$$+ \frac{2}{25}\langle A, X\rangle^2 + 2k|X - P(X)|^2, \tag{6.9}$$

where $\langle A, X\rangle = \frac{1}{2}\frac{d}{dt}|A+tX|^2\Big|_{t=0}$. The function $X \mapsto g(P(X))$ is a homogeneous polynomial of degree 3; hence

$$\frac{d^2}{dt^2}g(P(A)+tP(X))\Big|_{t=0} \geq -c|A||X|^2$$

for all A, X. In particular, from (6.9) we get

$$\frac{d^2}{dt^2}f(A+tX)\Big|_{t=0} \geq \left(-c|A| + \frac{1}{25}|A|^2\right)|X|^2,$$

so that (6.8) is satisfied if $|A| \geq 25c$. From (6.9) we also have

$$\frac{d^2}{dt^2}f(A+tX)\Big|_{t=0} \geq \frac{d^2}{dt^2}g(P(A)+tP(X))\Big|_{t=0}$$
$$+ \frac{1}{50}|X|^2 + 2k|X - P(X)|^2. \tag{6.10}$$

Let us call $F_k(A, X)$ the right-hand side of (6.10). We show that for some $k \in \mathbf{N}$ we have $F_k(A, X) > \frac{1}{100}$ on the set

$$\mathcal{K} = \{(A, X) : |A| \leq 25c, |X| = 1, \operatorname{rank} X = 1\}.$$

We argue by contradiction: if no such $k \in \mathbf{N}$ exists then we can find $(A_j, X_j) \in \mathcal{K}$ such that

$$F_j(A_j, X_j) \leq \frac{1}{100}$$

for all $j \in \mathbf{N}$. Since $\mathcal{K}$ is compact we may assume that $(A_j, X_j) \to (\overline{A}, \overline{X}) \in \mathcal{K}$. From (6.10) it follows immediately that $\overline{X} \in L$, and that

$$\frac{d^2}{dt^2}g(P(\overline{A})+t\overline{X})\Big|_{t=0} \leq -\frac{1}{100}$$

(recall that $|\overline{X}| = 1$) contradicting the fact that g is rank-1-convex on L. $\qquad\square$

Remark 6.13 (Open problem) It is not known whether or not rank-1-convexity implies quasiconvexity in the case $n \geq 2$, $m = 2$.

Part II

Γ-convergence

7

A NAÏVE INTRODUCTION TO Γ-CONVERGENCE

7.1 Definition and basic properties

Now we face the problem of finding a suitable notion of 'convergence' (which we shall call Γ-convergence) with the following feature: *if (X, d) is a metric space, if (f_j) is a sequence of* equi-mildly coercive *functions on X (i.e. there exists a compact set K such that* $\inf_X f_j = \inf_K f_j$ *for all $j \in \mathbf{N}$), and if f_j Γ-converge' to a function f_∞, then*

$$\exists \min_X f_\infty = \lim_j \inf_X f_j.$$

This property is clearly the fundamental requirement for a good variational convergence. The equality above can be seen as a 'Weierstrass theorem along the sequence (f_j)', where f_∞ replaces $\overline{f}$ in Theorem 1.9.

We follow the proof of Theorem 1.9 to find what we must ask of our convergence: take a sequence $(x_j) \subset K$ such that

$$\liminf_j f_j(x_j) = \liminf_j \inf_X f_j;$$

without loss of generality we can suppose that $x_j \to x_\infty \in K$. We then have

$$\inf_X f_\infty \le f_\infty(x_\infty) \ldots \liminf_j f_j(x_j) = \liminf_j \inf_X f_j, \tag{7.1}$$

and the first requirement is that we can substitute the ellipsis by '$\le$'; the second one is that

$$\limsup_j \inf_X f_j \le \inf_X f_\infty, \tag{7.2}$$

so that all inequalities in (7.1) and (7.2) become equalities.

Thus, we are led to the following definition.

Definition 7.1 *We say that a sequence $f_j : X \to \overline{\mathbf{R}}$ Γ(d)-converges to $f_\infty :$ $X \to \overline{\mathbf{R}}$ if for all $x \in X$ we have*

(i) (lim inf inequality) *for every sequence (x_j) converging to x*

$$f_\infty(x) \le \liminf_j f_j(x_j); \tag{7.3}$$

(ii) (existence of a recovery sequence) *there exists a sequence (x_j) converging to x such that*

$$f_\infty(x) \ge \limsup_j f_j(x_j), \tag{7.4}$$

or, equivalently by (7.3),

$$f_\infty(x) = \lim_j f_j(x_j). \tag{7.5}$$

The function f_∞ is called the $\Gamma(d)$-limit of (f_j), and we write $f_\infty = \Gamma(d)\text{-}\lim_j f_j$. If no confusion is likely, we can drop the dependence on the metric d, and write simply $\Gamma\text{-}\lim_j$, Γ-convergence, and so on.

We immediately obtain the required result as follows.

Theorem 7.2 *Let (X, d) be a metric space, let (f_j) be a sequence of equi-mildly coercive functions on X, and let $f_\infty = \Gamma(d)\text{-}\lim_j f_j$; then*

$$\exists \min_X f_\infty = \lim_j \inf_X f_j. \tag{7.6}$$

Moreover, if (x_j) is a converging sequence such that $\lim_j f_j(x_j) = \lim_j \inf_X f_j$, then its limit is a minimum point for f_∞.

Proof Take a precompact sequence $(\tilde{x}_j)$ in X such that

$$\liminf_j f_j(\tilde{x}_j) = \liminf_j \inf_X f_j.$$

There exists a subsequence $(\tilde{x}_{j_k})$ converging to some x_∞, and such that

$$\lim_k f_{j_k}(\tilde{x}_{j_k}) = \liminf_j \inf_X f_j.$$

Using (7.3) with $x = x_\infty$, and $x_j = \tilde{x}_j$ if $j \in \{j_k : k \in \mathbf{N}\}$, $x_j = x_\infty$ if $j \neq j_k$ for all $k \in \mathbf{N}$, we obtain

$$\inf_X f_\infty \leq f_\infty(x_\infty)$$
$$\leq \liminf_j f_j(x_j) \leq \lim_k f_{j_k}(x_{j_k}) = \liminf_j \inf_X f_j.$$

From (7.4) we have that for all $x \in X$

$$\limsup_j \inf_X f_j \leq f_\infty(x);$$

hence

$$\limsup_j \inf_X f_j \leq \inf_X f_\infty,$$

and (7.6) is proven. If $x_j \to x_\infty$, and $\lim_j f_j(x_j) = \lim_j \inf_X f_j$, then also the last statement of Theorem 7.2 is proven. $\qquad\square$

From the definition of Γ-convergence we immediately obtain the properties contained in the following remark.

Remark 7.3 If $f_\infty = \Gamma(d)\text{-}\lim_j f_j$ then the following properties hold, whose proof is an easy consequence of the definition of Γ-convergence.

(i) f_∞ is d-lower semicontinuous. In fact, if $x^k \to x$, then by (7.5) for each $k \in \mathbf{N}$ we can find a sequence $(x_j^k)_j$ with $x_j^k \to x^k$ and $\lim_j f_j(x_j^k) = f_\infty(x^k)$. We set $\sigma_0 = 0$,

$$\sigma_i = \inf\left\{k > \sigma_{i-1} : \ d(x_j^i, x^i) \leq \frac{1}{i}, \ |f_\infty(x^i) - f_j(x_j^i)| \leq \frac{1}{i} \ \forall j \geq k\right\},$$

and we define $x_j = x_j^k$ for $\sigma_k \leq j < \sigma_{k+1}$. We have $x_j \to x$, and by (7.3)

$$f_\infty(x) \leq \liminf_j f_j(x_j) = \liminf_k f_\infty(x^k).$$

(ii) By (7.3) and (7.5)

$$\begin{aligned}
f_\infty(x) &= \min\{\liminf_j f_j(x_j) : \ x_j \to x\} \\
&= \min\{\limsup_j f_j(x_j) : \ x_j \to x\}.
\end{aligned} \tag{7.7}$$

(iii) From (7.3) and (7.5) it follows that for every increasing sequence of integers (j_k) $\ f_\infty = \Gamma(d)\text{-}\lim_k f_{j_k}$.

(iv) If (x_j) is a precompact sequence such that $\lim_j f_j(x_j) = \lim_j \inf_X f_j$, then each limit of a converging subsequence of (x_j) is a minimum point for f_∞. In fact, if (x_{j_k}) is a converging subsequence, it suffices to follow the proof of Theorem 7.2 for the sequence (f_{j_k}) recalling Remark 7.3(iii) above.

(v) If g is a continuous function then $f_\infty + g = \Gamma(d)\text{-}\lim_j(f_j + g)$.

We can state some simple but important cases when the $\Gamma(d)$-limit does exist, and it is easily computed.

Remark 7.4 (i) If $f_j = f$ for all $j \in \mathbf{N}$, then

$$\Gamma(d)\text{-}\lim_j f_j = \overline{f}; \tag{7.8}$$

more generally, this holds if $f_j \to f$ uniformly;

(ii) if $f_j \to f$ decreasingly then $\Gamma(d)\text{-}\lim_j f_j = \overline{f}$;

(iii) if $f_j \leq f_{j+1}$ for all $j \in \mathbf{N}$, then

$$\Gamma(d)\text{-}\lim_j f_j = \sup_j \overline{f}_j = \lim_j \overline{f}_j; \tag{7.9}$$

in particular if f_j is l.s.c. for every $j \in \mathbf{N}$, then

$$\Gamma(d)\text{-}\lim_j f_j = \lim_j f_j. \tag{7.10}$$

7.2 Lower and upper Γ-limits

Proposition 7.5 *Let $f_j : X \to \overline{\mathbf{R}}$, and let $x \in X$, $\lambda \in \overline{\mathbf{R}}$. The following statements are equivalent to each other:*

(i) *for all sequences (x_j) such that $x_j \to x$ we have*

$$\lambda \leq \liminf_j f_j(x_j), \tag{7.11}$$

and there exists $x_j \to x$ such that

$$\lambda = \liminf_j f_j(x_j); \tag{7.12}$$

(ii) $\lambda = \min\{\liminf_j f_j(x_j) : \ x_j \to x\}$;

(iii) $\lambda = \inf\{\liminf_j f_j(x_j) : \ x_j \to x\}$;

(iv) $\lambda = \sup\limits_{U \in \mathcal{N}(x)} \liminf_j \inf\limits_{y \in U} f_j(y) = \sup\limits_{U \in \mathcal{N}(x)} \sup\limits_{k \in \mathbf{N}} \inf\limits_{j \geq k} \inf\limits_{y \in U} f_j(y).$

Proof Trivially (i) $\iff$ (ii) $\implies$ (iii). Moreover, (iii) $\implies$ (i): clearly (7.11) holds. If $\lambda = +\infty$ then (7.12) is trivially true. We suppose $\lambda \in \mathbf{R}$; for every $k \in \mathbf{N}$ let $x_j^k \to x$ and

$$\liminf_j f_j(x_j^k) \leq \lambda + \frac{1}{k}.$$

If we define $\sigma_0 = 0$,

$$\sigma_k = \min\Big\{ h > \sigma_{k-1} : \ f_h(x_h^k) \leq \lambda + \frac{2}{k}, d(x_j^k, x) \leq \frac{1}{k} \ \forall j \geq h \Big\},$$

and $x_j = x_j^k$ for $\sigma_k \leq j < \sigma_{k+1}$, then $\liminf_j f_j(x_j) \leq \liminf_k f_{\sigma_k}(x_{\sigma_k}) \leq \lambda$ so that (x_j) satisfies (7.12). The same proof fits the case $\lambda = -\infty$.

Now we prove that (iii) $\iff$ (iv): let λ be defined by (iii), and denote

$$\lambda' = \sup\limits_{U \in \mathcal{N}(x)} \liminf_j \inf\limits_{y \in U} f_j(y) = \sup\limits_{k \in \mathbf{N}} \liminf_j \inf\limits_{d(x,y)<1/k} f_j(y).$$

Let $x_j \to x$; with fixed $U \in \mathcal{N}(x)$ we have $x_j \in U$ definitely, so that $\inf_{y \in U} f_j(y) \leq f_j(x_j)$. This means that $\liminf_j \inf_{y \in U} f_j(y) \leq \liminf_j f_j(x_j)$ for every $U \in \mathcal{N}(x)$; that is, $\lambda' \leq \lambda$. It remains to prove that $\lambda \leq \lambda'$. If $\lambda' = +\infty$ there is nothing to prove. Suppose $\lambda' \in \mathbf{R}$. For all $k \in \mathbf{N}$ we have

$$\liminf_j \inf\Big\{ f_j(y) : \ d(x,y) < \frac{1}{k} \Big\} < \lambda' + \frac{1}{k},$$

so that there exists an increasing sequence of integers (σ_k) such that

$$\inf\Big\{ f_{\sigma_k}(y) : \ d(x,y) < \frac{1}{k} \Big\} < \lambda' + \frac{1}{k}$$

for every $k \in \mathbf{N}$. We can find y_k such that $d(x, y_k) < \frac{1}{k}$ and $f_{\sigma_k}(y_k) < \lambda' + \frac{1}{k}$. If we set $x_j = y_k$ if $j = \sigma_k$, and $x_j = x$ if $j \neq \sigma_k$ for every $k \in \mathbf{N}$, then

$$\lambda \leq \liminf_j f_j(x_j) \leq \liminf_k f_{\sigma_k}(y_k) \leq \lambda'.$$

If $\lambda' = -\infty$, we remark that for all $k \in \mathbf{N}$

$$\liminf_j \inf\left\{ f_j(y) : \ d(x,y) < \frac{1}{k} \right\} < -k,$$

and we proceed likewise. $\qquad\square$

Proposition 7.6 *Let* $f_j : X \to \overline{\mathbf{R}}$, *and let* $x \in X$, $\lambda \in \overline{\mathbf{R}}$. *The following statements are equivalent to each other:*

(i) *for all sequences* (x_j) *such that* $x_j \to x$ *we have*

$$\lambda \leq \limsup_j f_j(x_j), \tag{7.13}$$

and there exists $x_j \to x$ *such that*

$$\lambda = \limsup_j f_j(x_j); \tag{7.14}$$

(ii) $\lambda = \min\{\limsup_j f_j(x_j) : \ x_j \to x\}$;

(iii) $\lambda = \inf\{\limsup_j f_j(x_j) : \ x_j \to x\}$;

(iv) $\lambda = \displaystyle\sup_{U \in \mathcal{N}(x)} \limsup_j \inf_{y \in U} f_j(y) = \sup_{U \in \mathcal{N}(x)} \inf_{k \in \mathbf{N}} \sup_{j \geq k} \inf_{y \in U} f_j(y)$.

Proof As above, we have trivially (i) $\Longleftrightarrow$ (ii) $\Longrightarrow$ (iii). Moreover, (iii) $\Longrightarrow$ (i): clearly (7.13) holds. If $\lambda = +\infty$ then (7.14) is trivially satisfied. Let $\lambda = -\infty$; for every $k \in \mathbf{N}$ let $x_j^k \to x$ and

$$\limsup_j f_j(x_j^k) \leq -k.$$

If we define $\sigma_0 = 0$,

$$\sigma_k = \min\left\{ h > \sigma_{k-1} : \ f_j(x_j^k) \leq 1 - k, \ d(x_j^k, x) \leq \frac{1}{k} \ \forall j \geq h \right\},$$

and $x_j = x_j^k$ for $\sigma_k \leq j < \sigma_{k+1}$, then $\limsup_j f_j(x_j) = -\infty$ so that (x_j) satisfies (7.14). In the same way we proceed if $\lambda \in \mathbf{R}$.

Now we prove that (iii) $\Longleftrightarrow$ (iv): let λ be defined by (iii), and denote

$$\lambda' = \sup_{U \in \mathcal{N}(x)} \limsup_j \inf_{y \in U} f_j(y) = \sup_{k \in \mathbf{N}} \limsup_j \inf_{d(x,y)<1/k} f_j(y).$$

Let $x_j \to x$; with fixed $U \in \mathcal{N}(x)$ we have $x_j \in U$ definitely, so that $\inf_{y \in U} f_j(y) \leq f_j(x_j)$. This means that $\limsup_j \inf_{y \in U} f_j(y) \leq \limsup_j f_j(x_j)$ for every $U \in \mathcal{N}(x)$; that is, $\lambda' \leq \lambda$. It remains to prove that $\lambda \leq \lambda'$. If $\lambda' = +\infty$ there is nothing to prove. Suppose $\lambda' = -\infty$. For all $k \in \mathbf{N}$ we have

$$\limsup_j \inf\left\{ f_j(y) : \ d(x,y) < \frac{1}{k} \right\} < -k,$$

so that there exists an increasing sequence of integers (σ_k) such that

$$\inf\left\{ f_j(y) : \ d(x,y) < \frac{1}{k} \right\} < -k$$

for all $j \geq \sigma_k$; for such j we can find $y_j^k \in X$ with $d(x, y_j^k) < \frac{1}{k}$ and $f_j(y_j^k) < -k$. If we set $x_j = x$ if $j < \sigma_1$, and $x_j = y_j^k$ for $\sigma_k \leq j < \sigma_{k+1}$, then $x_j \to x$ in X and $\lim_j f_j(x_j) = -\infty$, and $\lambda = -\infty$. If $\lambda' \in \mathbf{R}$ we proceed likewise. $\square$

Definition 7.7 *Let $f_j : X \to \overline{\mathbf{R}}$, and $x \in X$. We say that $\lambda \in \overline{\mathbf{R}}$ is the $\Gamma(d)$-lower limit of the sequence (f_j) at x, and we write*

$$\lambda = \Gamma(d)\text{-}\liminf_j f_j(x) \tag{7.15}$$

if any of the equivalent conditions of Proposition 7.5 is satisfied. We say that $\lambda \in \overline{\mathbf{R}}$ is the $\Gamma(d)$-upper limit of the sequence (f_j) at x, and we write

$$\lambda = \Gamma(d)\text{-}\limsup_j f_j(x) \tag{7.16}$$

if any of the equivalent conditions of Proposition 7.6 is satisfied. If we have

$$\Gamma(d)\text{-}\liminf_j f_j(x) = \lambda = \Gamma(d)\text{-}\limsup_j f_j(x), \tag{7.17}$$

then we write

$$\lambda = \Gamma(d)\text{-}\lim_j f_j(x), \tag{7.18}$$

and we say that λ is the $\Gamma(d)$-limit of the sequence (f_j) at x. Again, if the dependence on the metric d is clear, we may drop it from the notation.

Remark 7.8 By condition (iii) of Proposition 7.5 and condition (iii) of Proposition 7.6 the $\Gamma(d)$-lower limit and the $\Gamma(d)$-upper limit exist at every $x \in X$. Moreover, as in Remark 7.3(i), they define l.s.c. functions on X.

Definition 7.7 is coherent with Definition 7.1, and we can say that a sequence (f_j) $\Gamma(d)$-converges to f_∞ if and only if the $\Gamma(d)$-limit exists, and we have $\lambda = f_\infty(x)$ in (7.18), at all $x \in X$.

7.3   Further properties. Compactness

Proposition 7.9 *Let (X, d) be a separable metric space, and for all $j \in \mathbf{N}$ let $f_j : X \to \overline{\mathbf{R}}$ be a function. Then there is an increasing sequence of integers (j_k) such that the $\Gamma(d)$-$\lim_k f_{j_k}(x)$ exists for all $x \in X$.*

Proof Let (U_k) be a countable base of open sets for the topology of X. Since $\overline{\mathbf{R}}$ is compact, there exists an increasing sequence of integers $(\sigma_j^0)_j$ along which the limit

$$\lim_j \inf_{y \in U_0} f_{\sigma_j^0}(y)$$

exists, and for all $k \geq 1$ we define $(\sigma_j^k)_j$ as any subsequence of $(\sigma_j^{k-1})_j$ along which the limit

$$\lim_j \inf_{y \in U_k} f_{\sigma_j^k}(y)$$

exists. The 'diagonal' sequence $j_k = \sigma_k^k$, being a subsequence of all $(\sigma_j^l)_j$, has the property that the limit

$$\lim_k \inf_{y \in U_l} f_{j_k}(y)$$

exists for all $l \in \mathbf{N}$. In particular we have

$$\liminf_k \inf_{y \in U_l} f_{j_k}(y) = \limsup_k \inf_{y \in U_l} f_{j_k}(y)$$

for all $l \in \mathbf{N}$, and we deduce (7.17) by property (iv) of Proposition 7.5 and by property (iv) of Proposition 7.6. $\qquad\square$

Remark 7.10 If (X, d) is not separable, then Proposition 7.9 fails. As an example, we can take $X = \{-1, 1\}^{\mathbf{N}}$ equipped with the discrete topology. X is metrizable, and Γ-convergence on X is equivalent to pointwise convergence. We take the sequence $f_j : X \to \{-1, 1\}$ defined by $f_j(\mathbf{x}) = x_j$ if $\mathbf{x} = (x_0, x_1, \ldots)$. If (f_{j_k}) is a subsequence of (f_j) and we define $\mathbf{x}$ by $x_{j_k} = (-1)^k$, and $x_j = 1$ if $j \notin \{j_k : k \in \mathbf{N}\}$, then the limit $\lim_k f_{j_k}(\mathbf{x})$ does not exist. Hence no subsequence of (f_j) Γ-converges.

Proposition 7.11 *We have $\Gamma(d)$-$\lim_j f_j = f_\infty$ if and only if for every subsequence (f_{j_k}) there exists a further subsequence which $\Gamma(d)$-converges to f_∞.*

Proof Clearly if f_j $\Gamma(d)$-converges to f_∞, then every subsequence of f_j $\Gamma(d)$-converges to the same limit (Remark 7.3(iii)).

For every increasing sequence of integers (j_k) we have

$$\Gamma(d)\text{-}\liminf_j f_j \le \Gamma(d)\text{-}\liminf_k f_{j_k}$$
$$\le \Gamma(d)\text{-}\limsup_k f_{j_k} \le \Gamma(d)\text{-}\limsup_j f_j.$$

Hence if $\Gamma(d)$-$\lim_k f_{j_k}(x) = f_\infty(x)$ but the $\Gamma(d)$-$\lim_j f_j(x)$ does not exist we have either

$$f_\infty(x) < \Gamma(d)\text{-}\limsup_j f_j(x)$$

or

$$f_\infty(x) > \Gamma(d)\text{-}\liminf_j f_j(x).$$

In the first case we have

$$f_\infty(x) < \sup_{U \in \mathcal{N}(x)} \limsup_j \inf_{y \in U} f_j(y),$$

so that there exists $U \in \mathcal{N}(x)$ such that

$$f_\infty(x) < \limsup_j \inf_{y \in U} f_j(y).$$

This means that a subsequence (f_{j_k}) of (f_j) exists along which

$$f_\infty(x) < \liminf_k \inf_{y \in U} f_{j_k}(y),$$

so that $f_\infty(x) < \Gamma(d)\text{-}\liminf_k f_{j_k}(x)$ leads to a contradiction. In the second case a sequence x_j converging to x exists such that $\liminf_j f_j(x_j) < f_\infty(x)$. We can extract a subsequence (x_{j_k}) of (x_j) such that $\lim_k f_j(x_{j_k}) < f_\infty(x)$. This means that $\Gamma(d)\text{-}\limsup_k f_{j_k}(x) < f_\infty(x)$, thus giving a contradiction. $\qquad\square$

Remark 7.12 If (X, d) is a topological vector space, then it is easy to check, by Definition 7.1, that if $f_\infty = \Gamma\text{-}\lim_j f_j$ then we have:

(i) if every f_j is positively homogeneous of degree α (i.e. if $f_j(tx) = t^\alpha f_j(x)$ for all $x \in X$ and $t > 0$), then so is f_∞;

(ii) if every f_j is a positive quadratic form, then so is f_∞ (note that f is a positive quadratic form if and only if f is positive, positively homogeneous of degree 2, and $f(x + y) + f(x - y) = 2f(x) + 2f(y)$; for a proof we refer to Dal Maso (1993) p. 128);

(iii) if every f_j is convex then f_∞ is convex.

Proposition 7.13 *We have*

$$\Gamma(d)\text{-}\liminf_j f_j = \Gamma(d)\text{-}\liminf_j \overline{f}_j \tag{7.19}$$

$$\Gamma(d)\text{-}\limsup_j f_j = \Gamma(d)\text{-}\limsup_j \overline{f}_j. \tag{7.20}$$

Proof We have

$$
\begin{aligned}
\Gamma(d)\text{-}\liminf_j \overline{f}_j(x) &= \sup_{U \in \mathcal{N}(x)} \liminf_j \inf_{y \in U} \sup_{V \in \mathcal{N}(y)} \inf_{z \in V} f_j(z) \\
&= \sup_{U \in \mathcal{N}(x)} \liminf_j \inf_{y \in U} \sup_{V \in \mathcal{N}(y)} \inf_{z \in V \cap U} f_j(z) \\
&\geq \sup_{U \in \mathcal{N}(x)} \liminf_j \inf_{z \in U} f_j(z) \\
&= \Gamma(d)\text{-}\liminf_j f_j(x),
\end{aligned}
$$

while the opposite inequality is trivial. The same proof works for the $\Gamma(d)$-upper limit. $\qquad\square$

7.4   Exercises

Exercise 7.1 Prove the statements of Remark 7.4 and of Remark 7.12 (they are all easy consequences of the definition of Γ-limit).

Exercise 7.2 Show that in general $\Gamma\text{-}\lim_j(f_j + g_j) \neq \Gamma\text{-}\lim_j f_j + \Gamma\text{-}\lim_j g_j$, even if g_j is l.s.c. and independent of j (recall Exercise 1.1).

$$8$$

THE INDIRECT METHODS OF Γ-CONVERGENCE

8.1 Γ-limits and Yosida transforms

Proposition 1.7 admits a counterpart for Γ-limits, which sometimes permits their computation via the solution of Euler equations.

Proposition 8.1 *Let $f_j : X \to [0, +\infty]$ be a sequence of functions, and let $\psi : [0, +\infty) \to [0, +\infty)$ be a strictly increasing continuous function with $\psi(0) = 0$. We have*

$$\Gamma(d)\text{-}\liminf_j f_j(x) = \sup_{\lambda \geq 0} \liminf_j T_\lambda^\psi f_j(x), \tag{8.1}$$

$$\Gamma(d)\text{-}\limsup_j f_j(x) = \sup_{\lambda \geq 0} \limsup_j T_\lambda^\psi f_j(x) \tag{8.2}$$

for all $x \in X$.

Proof We only prove (8.1). With fixed $x \in X$, let

$$r = \Gamma(d)\text{-}\liminf_j f_j(x) = \sup_{\varepsilon \to 0} \liminf_j \inf_{d(x,y)<\varepsilon} f_j(y),$$

$$s = \sup_{\lambda \geq 0} \liminf_j T_\lambda^\psi f_j(x).$$

Let $t < r$. There exists $\varepsilon > 0$ such that

$$t < \liminf_j \inf_{d(x,y)<\varepsilon} f_j(y);$$

hence there exists $k \in \mathbf{N}$ such that

$$t < \inf_{d(x,y)<\varepsilon} f_j(y)$$

for all $j \geq k$. For all $\lambda > \frac{t}{\psi(\varepsilon)}$ we get

$$t < \inf_{y \in X} \{f_j(y) + \lambda\psi(d(x,y))\} = T_\lambda^\psi f_j(x)$$

for all $j \geq k$. Hence

$$t < \liminf_j T_\lambda^\psi f_j(x) \leq s.$$

By the arbitrariness of $t < r$ we get $r \leq s$.

We now prove that $s \leq r$. If $\lambda \geq 0$ and $\varepsilon > 0$, then

$$T_\lambda^\psi f_j(x) \leq \inf_{d(x,y)<\varepsilon} \{f_j(y) + \lambda\psi(d(x,y))\}$$

$$\leq \inf_{d(x,y)<\varepsilon} f_j(y) + \lambda\psi(\varepsilon)$$

and

$$\liminf_j T_\lambda^\psi f_j(x) \leq \liminf_j \inf_{d(x,y)<\varepsilon} f_j(y) + \lambda\psi(\varepsilon)$$

$$\leq r + \lambda\psi(\varepsilon).$$

Letting $\varepsilon \to 0$ we obtain

$$\liminf_j T_\lambda^\psi f_j(x) \leq r,$$

and finally, taking the supremum for $\lambda \geq 0$, $s \leq r$. $\qquad\square$

8.2 An example: Γ-limits of quadratic functionals

Example 8.2 We can use Proposition 8.1 to prove the Γ-convergence of a sequence of functionals via the study of the limit behaviour of the corresponding Euler equations. As an example we can consider the case of quadratic functionals of the form

$$F_\varepsilon(u) = \int_\Omega \sum_{i,j=1}^n a_{ij}\left(\frac{x}{\varepsilon}\right) D_i u D_j u \, dx, \qquad u \in W_0^{1,2}(\Omega) \tag{8.3}$$

where (a_{ij}) is a symmetric matrix of 1-periodic measurable functions satisfying

$$c_1|\xi|^2 \leq \sum_{i,j=1}^n a_{ij}(x)\xi_i\xi_j \leq c_2|\xi|^2 \qquad \forall x \in \mathbf{R}^n, \xi \in \mathbf{R}^n. \tag{8.4}$$

The following result can be proven on the convergence of the solutions of boundary value problems for the Euler operators of the functionals F_ε: *there exist real numbers q_{ij} such that for all $\lambda \geq 0$ and $g \in L^2(\Omega)$ the solutions u_ε of the problems*

$$\begin{cases} -\sum_{i,j=1}^n D_j\left(a_{ij}\left(\frac{x}{\varepsilon}\right) D_i u_\varepsilon\right) + \lambda u_\varepsilon = g \\ u_\varepsilon \in W_0^{1,2}(\Omega) \end{cases} \tag{8.5}$$

converge weakly in $W^{1,2}(\Omega)$ as $\varepsilon \to 0$ to the solution u of the problem

$$\begin{cases} -\sum_{i,j=1}^n q_{ij} D_j D_i u + \lambda u = g \\ u \in W_0^{1,2}(\Omega) \end{cases} \tag{8.6}$$

(see e.g. Bensoussan *et al.* (1978)).

Let $v \in W_0^{1,2}(\Omega)$; we can consider the minimum problems

$$T_\lambda^2 F_\varepsilon(v) = \min_{u \in W_0^{1,2}(\Omega)} \left(F_\varepsilon(u) + \lambda \int_\Omega |u - v|^2 \, dx \right) \tag{8.7}$$

for $\lambda \geq 0$. This problem admits a unique solution, which satisfies the corresponding Euler equation

$$\begin{cases} - \displaystyle\sum_{i,j=1}^n D_j \left(a_{ij} \left(\frac{x}{\varepsilon} \right) D_i u_\varepsilon \right) + \lambda u_\varepsilon = \lambda v \\[2mm] u_\varepsilon \in W_0^{1,2}(\Omega). \end{cases} \tag{8.8}$$

Substituting (8.8) in (8.7) we obtain

$$T_\lambda^2 F_\varepsilon(v) = \lambda \int_\Omega v(v - u_\varepsilon) \, dx. \tag{8.9}$$

Taking $g = \lambda v$ in (8.5) and (8.6) we obtain that the functions u_ε weakly converge in $W^{1,2}(\Omega)$ as $\varepsilon \to 0$ to the solution u of the problem

$$\begin{cases} - \displaystyle\sum_{i,j=1}^n q_{ij} D_j D_i u + \lambda u = \lambda v \\[2mm] u \in W_0^{1,2}(\Omega), \end{cases} \tag{8.10}$$

so that by (8.9) we get

$$\lim_{\varepsilon \to 0+} T_\lambda^2 F_\varepsilon(v) = \lambda \int_\Omega v(v - u) \, dx = T_\lambda^2 F_\infty(v),$$

where

$$F_\infty(u) = \int_\Omega \sum_{i,j=1}^n q_{ij} D_i u D_j u \, dx, \qquad u \in W_0^{1,2}(\Omega). \tag{8.11}$$

In particular we can apply Proposition 8.1 with $X = W_0^{1,2}(\Omega)$ equipped with the $L^2(\Omega)$ distance, obtaining that for every sequence (ε_j) of positive numbers converging to 0 we have

$$\Gamma(L^2(\Omega))\text{-}\liminf_j F_{\varepsilon_j}(u) = \Gamma(L^2(\Omega))\text{-}\limsup_j F_{\varepsilon_j}(u)$$
$$= \sup_{\lambda \geq 0} T_\lambda^2 F_\infty(u) \tag{8.12}$$

for every $u \in W_0^{1,2}(\Omega)$. Since $f(\xi) = \sum_{i,j=1}^n q_{ij} \xi_i \xi_j$ is a convex function the functional F_∞ is lower semicontinuous with respect to the $L^2(\Omega)$ convergence. Hence, by Proposition 1.7 we get the existence of the Γ-limit

$$\Gamma(\mathrm{L}^2(\Omega))\text{-}\lim_j F_{\varepsilon_j}(u) = F_\infty(u) \tag{8.13}$$

for all $u \in \mathrm{W}_0^{1,2}(\Omega)$ and for all sequences (ε_j) converging to 0.

Note that from (8.13) we may obtain the Γ-convergence of sequences of functionals for which the Euler equations cannot be written. For example, by Remark 7.3(v), we can add to the functionals F_ε a term of the form

$$G(u) = \int_\Omega g(x, u(x))\, dx,$$

with g any Carathéodory function (possibly not smooth) satisfying

$$0 \le g(x, s) \le c(1 + |s|^2)$$

for all $x \in \Omega$ and $s \in \mathbf{R}$, which can be proven to be continuous with respect to the L^2 convergence as in Remark 3.4(a).

9

DIRECT METHODS. AN INTEGRAL REPRESENTATION RESULT

9.1 Localization

The direct methods of Γ-convergence for integral functionals consist in proving general abstract compactness results that assure the existence of Γ-converging sequences, and then in recovering enough information on the structure of the Γ-limits as to obtain a representation in a suitable form. For this purpose, it is useful to consider at the same time the dependence of the Γ-limits as set functions and as integrals. In order to highlight this point of view, usually called the 'localization method', we will consider variational functionals of the form

$$F(u, U) = \int_U f(x, Du(x))\, dx, \tag{9.1}$$

with $u \in W^{1,p}(\Omega; \mathbf{R}^m)$ and $U \in \mathcal{A}(\Omega)$ (the family of all open subsets of Ω). We now discuss some properties of abstract functionals F that guarantee equality (9.1) for some f.

9.2 Integral representation on Sobolev spaces

It is important to single out the properties which assure that functionals defined on pairs function-set can be written in an integral form, as in the following theorem.

Theorem 9.1 *Let Ω be a bounded open subset of $\mathbf{R}^n$, and let $1 \leq p < \infty$. Let $F : W^{1,p}(\Omega; \mathbf{R}^m) \times \mathcal{A}(\Omega) \to [0, +\infty)$ be a functional satisfying the following conditions:*

(i) (locality) F is local, i.e. $F(u, U) = F(v, U)$ if $u = v$ a.e. on $U \in \mathcal{A}(\Omega)$;

(ii) (measure property) for all $u \in W^{1,p}(\Omega; \mathbf{R}^m)$ the set function $F(u, \cdot)$ is the restriction of a Borel measure to $\mathcal{A}(\Omega)$;

(iii) (growth condition) there exists $c > 0$ and $a \in L^1(\Omega)$ such that

$$F(u, U) \leq c \int_U (a(x) + |Du|^p)\, dx$$

for all $u \in W^{1,p}(\Omega; \mathbf{R}^m)$ and $U \in \mathcal{A}(\Omega)$;

(iv) (translation invariance in u) $F(u + z, U) = F(u, U)$ for all $z \in \mathbf{R}^m$, $u \in W^{1,p}(\Omega; \mathbf{R}^m)$ and $U \in \mathcal{A}(\Omega)$;

(v) (lower semicontinuity) *for all $U \in \mathcal{A}(\Omega)$ $F(\cdot, U)$ is sequentially lower semicontinuous with respect to the weak convergence in $\mathrm{W}^{1,p}(\Omega; \mathbf{R}^m)$.*

Then there exists a Carathéodory function $f : \Omega \times \mathbf{M}^{m \times n} \to [0, +\infty)$ satisfying the growth condition

$$0 \leq f(x, A) \leq c(a(x) + |A|^p) \tag{9.2}$$

for all $x \in \Omega$ and $A \in \mathbf{M}^{m \times n}$, such that

$$F(u, U) = \int_U f(x, Du(x))\, dx \tag{9.3}$$

for all $u \in \mathrm{W}^{1,p}(\Omega; \mathbf{R}^m)$ and $U \in \mathcal{A}(\Omega)$.

Proof *Step 1: definition of f.*

Fix $A \in \mathbf{M}^{m \times n}$; by (ii) $F(Ax, \cdot)$ can be extended to a Borel measure on Ω which, by (iii), is absolutely continuous with respect to the Lebesgue measure. Hence there exists a density function $g_A \in \mathrm{L}^1(\Omega)$ such that

$$F(Ax, U) = \int_U g_A(x)\, dx$$

for all $U \in \mathcal{A}(\Omega)$. We set

$$f(x, A) = g_A(x)$$

for all $x \in \Omega$ and $A \in \mathbf{M}^{m \times n}$. Note that by condition (iii), with fixed $A \in \mathbf{M}^{m \times n}$

$$0 \leq f(x, A) \leq c(a(x) + |A|^p)$$

for a.e. $x \in \Omega$.

Step 2: integral representation on piecewise affine functions.

Let $U \in \mathcal{A}(\Omega)$ and let $u \in \mathrm{W}^{1,p}(\Omega; \mathbf{R}^m)$ be piecewise affine on U, i.e. we can write

$$u_{|U} = \sum_{j=1}^{N} \chi_{U_j}(A_j x + z_j),$$

where the sets U_j are disjoint open sets with $|U \setminus \bigcup_{j=1}^{N} U_j| = 0$, $A_j \in \mathbf{M}^{m \times n}$, $z_j \in \mathbf{R}^m$ for $j = 1, \ldots, N$. By (i), (ii), (iv) and Step 1

$$F(u, U) = \sum_{j=1}^{N} F(u, U_j) = \sum_{j=1}^{N} F(A_j x + z_j, U_j)$$

$$= \sum_{j=1}^{N} F(A_j x, U_j) = \sum_{j=1}^{N} \int_{U_j} f(x, A_j)\, dx$$

$$= \sum_{j=1}^{N} \int_{U_j} f(x, Du)\, dx = \int_U f(x, Du)\, dx;$$

that is, the representation (9.3).

Step 3: rank-1-convexity of f.

We want to show that, with fixed $A, B \in \mathbf{M}^{m \times n}$ such that $\mathrm{rank}(B - A) = 1$, and $t \in (0, 1)$

$$f(y, tB + (1 - t)A) \le tf(y, B) + (1 - t)f(y, A)$$

for all $y \in \Omega$. By definition we can take

$$f(y, A) = \limsup_{\rho \to 0+} \frac{F(Ax, B(y, \rho))}{|B(y, \rho)|}$$

for all $y \in \Omega$ and $A \in \mathbf{M}^{m \times n}$. Hence it will suffice to show that if $B(y, \rho) \subset \Omega$ then

$$F((tB + (1 - t)A)x, B(y, \rho)) \le tF(Bx, B(y, \rho)) + (1 - t)F(Ax, B(y, \rho)). \quad (9.4)$$

Let $a \in \mathbf{R}^m$, $b \in \mathbf{R}^n$ be vectors such that $B - A = a \otimes b$. Consider as in Proposition 4.11 the function $v \in W^{1,\infty}_{\mathrm{loc}}(\mathbf{R}^n; \mathbf{R}^m)$ defined by

$$v(x) = \begin{cases} Ax + \langle b, x \rangle a - (1 - t)ja & \text{if } j \in \mathbf{Z}, \ j \le \langle b, x \rangle < j + t \\[2mm] Ax + (1 + j)ta & \text{if } j \in \mathbf{Z}, \ j + t \le \langle b, x \rangle < j + 1. \end{cases}$$

We also define the sets

$$E_A = \{x \in \mathbf{R}^n : \exists j \in \mathbf{Z} : \ j + t \le \langle b, x \rangle < j + 1\},$$
$$E_B = \{x \in \mathbf{R}^n : \exists j \in \mathbf{Z} : \ j \le \langle b, x \rangle < j + t\}.$$

If we set $u_j(x) = \frac{1}{j}v(jx)$, we have (see Examples 2.7 and 2.10)

$$u_j \rightharpoonup^* (tB + (1 - t)A)x \qquad \text{weakly* in } W^{1,\infty}(\Omega; \mathbf{R}^m)$$
$$\chi_{\frac{1}{j}E_A} \rightharpoonup^* 1 - t \qquad \text{weakly* in } L^\infty(\Omega)$$
$$\chi_{\frac{1}{j}E_B} \rightharpoonup^* t \qquad \text{weakly* in } L^\infty(\Omega);$$

moreover, $Du_j = A$ on $\frac{1}{j}E_A$ and $Du_j = B$ on $\frac{1}{j}E_B$. Hence by (v) and (iv) we obtain

$$F((tB + (1 - t)A)x, B(y, \rho))$$
$$\le \liminf_j F(u_j, B(y, \rho))$$
$$= \liminf_j \left(F\left(Ax, \frac{1}{j}E_A \cap B(y, \rho)\right) + F\left(Bx, \frac{1}{j}E_B \cap B(y, \rho)\right) \right)$$
$$= \liminf_j \left(\int_{B(y,\rho)} \chi_{\frac{1}{j}E_A} g_A(x)\, dx + \int_{B(y,\rho)} \chi_{\frac{1}{j}E_B} g_B(x)\, dx \right)$$

$$= t \int_{B(y,\rho)} g_B(x)\, dx + (1-t) \int_{B(y,\rho)} g_A(x)\, dx$$
$$= tF(Bx, B(y,\rho)) + (1-t)F(Ax, B(y,\rho)),$$

proving (9.4), and finally the rank-1-convexity of $f(y, \cdot)$, taking the $\limsup$ as $\rho \to 0+$. By Remark 4.13(iii) we have that $f(y, \cdot)$ is locally Lipschitz, and hence f is a Carathéodory function.

Step 4: an inequality by continuity.

As a consequence of Step 3 and (9.2) the functional $u \mapsto \int_U f(x, Du)\, dx$ is continuous with respect to the strong convergence of $\mathrm{W}^{1,p}(U; \mathbf{R}^m)$. If $U \subset\subset \Omega$ we can find a sequence $u_j \in \mathrm{W}^{1,p}(\Omega; \mathbf{R}^m)$ converging strongly to u in $\mathrm{W}^{1,p}(\Omega; \mathbf{R}^m)$, and such that their restrictions to U are piecewise affine. Then

$$F(u, U) \le \liminf_j F(u_j, U)$$
$$= \lim_j \int_U f(x, Du_j)\, dx = \int_U f(x, Du)\, dx$$

by (v) and Step 3.

Step 5: equality by translation.

Let $u \in \mathrm{W}^{1,p}(\Omega; \mathbf{R}^m)$; we consider the functional $G : \mathrm{W}^{1,p}(\Omega; \mathbf{R}^m) \times \mathcal{A}(\Omega) \to [0, +\infty)$ defined by

$$G(v, U) = F(u + v, U)$$

that satisfies all hypotheses (i)–(v) of Theorem 9.1 (with different c and a in condition (iii)). Hence, by Steps 1–4 above, there exists a Carathéodory function $\psi : \Omega \times \mathbf{M}^{m \times n} \to [0, +\infty)$ such that

$$G(v, U) \le \int_U \psi(x, Dv)\, dx$$

for all $v \in \mathrm{W}^{1,p}(\Omega; \mathbf{R}^m)$ and for all $U \subset\subset \Omega$ open sets, with equality for v piecewise affine on U. Let us take an open set $U \subset\subset \Omega$, and u_j piecewise affine on U converging strongly to u in $\mathrm{W}^{1,p}(\Omega; \mathbf{R}^m)$ as in Step 4. We have

$$\int_U \psi(x, 0)\, dx = G(0, U) = F(u, U)$$
$$\le \int_U f(x, Du)\, dx = \lim_j \int_U f(x, Du_j)\, dx$$
$$= \lim_j F(u_j, U) = \lim_j G(u_j - u, U)$$
$$\le \lim_j \int_U \psi(x, Du_j - Du)\, dx = \int_U \psi(x, 0)\, dx;$$

hence all inequalities are in fact equalities, and in particular

$$F(u, U) = \int_U f(x, Du) \, dx$$

for all $U \subset\subset \Omega$.

Step 6: integral representation.

By (ii) the integral representation obtained in Step 5 holds for all open subsets U of Ω. $\qquad\square$

9.3 Integral representation of homogeneous functionals

The following result, which is a corollary to Theorem 9.1, characterizes a class of integral functionals with integrand independent of the space variable.

Proposition 9.2 *Let* $F : W^{1,p}(\Omega; \mathbf{R}^m) \times \mathcal{A}(\Omega) \to [0, +\infty)$. *There exists a quasiconvex* $f : \mathbf{M}^{m \times n} \to [0, +\infty)$ *satisfying*

$$0 \le f(A) \le c(1 + |A|^p) \qquad \forall A \in \mathbf{M}^{m \times n} \tag{9.5}$$

such that the functional F can be represented by

$$F(u, U) = \int_U f(Du) \, dx \tag{9.6}$$

if and only if conditions (i)–(v) *of Theorem 9.1 hold and in addition*

(vi)(translation invariance in x)

$$F(Ax, B(y, \rho)) = F(Ax, B(z, \rho))$$

for all $A \in \mathbf{M}^{m \times n}$, $y, z \in \Omega$, *and* $\rho > 0$ *such that* $B(y, \rho) \cup B(z, \rho) \subset \Omega$.

Proof If (i)–(vi) hold, then we obtain the representation (9.3), and

$$f(y, A) = \limsup_{\rho \to 0+} \frac{F(Ax, B(y, \rho))}{|B(y, \rho)|}$$
$$= \limsup_{\rho \to 0+} \frac{F(Ax, B(z, \rho))}{|B(z, \rho)|} = f(z, A)$$

for all $y, z \in \Omega$. Hence f is independent of the first variable, and we can take the function a in (iii) to be constant. Since quasiconvexity is a necessary and sufficient condition for the sequential lower semicontinuity of $F(\cdot, \Omega)$ with respect to the weak convergence in $W^{1,p}(\Omega; \mathbf{R}^m)$ (see Theorem 5.16, Proposition 4.3 and Remark 5.15), then f is quasiconvex. The converse implication is obvious. $\qquad\square$

10

INCREASING SET FUNCTIONS

10.1 Increasing set functions

Definition 10.1 *A set function* $\alpha : \mathcal{A}(\Omega) \to [0, +\infty]$ *is called an* increasing set function *if* $\alpha(\emptyset) = 0$ *and* $\alpha(V) \leq \alpha(U)$ *if* $V \subset U$. *An increasing set function is said to be* subadditive *if*

$$\alpha(U \cup V) \leq \alpha(U) + \alpha(V) \tag{10.1}$$

for all $U, V \in \mathcal{A}(\Omega)$; α *is said to be* superadditive *if*

$$\alpha(U \cup V) \geq \alpha(U) + \alpha(V) \tag{10.2}$$

for all $U, V \in \mathcal{A}(\Omega)$ *with* $U \cap V = \emptyset$; α *is said to be* inner regular *if*

$$\alpha(U) = \sup\{\alpha(V) : \ V \in \mathcal{A}(\Omega), \ V \subset\subset U\} \tag{10.3}$$

for all $U \in \mathcal{A}(\Omega)$.

10.2 A characterization of measures as set functions

It will be useful to characterize measures as increasing set functions enjoying special properties which are often satisfied by Γ-limits. The following criterion is due to De Giorgi and Letta (1977).

Theorem 10.2 (Measure property criterion) *Let* $\alpha : \mathcal{A}(\Omega) \to [0, +\infty]$ *be an increasing set function. The following statements are equivalent:*

(i) α *is the restriction to* $\mathcal{A}(\Omega)$ *of a Borel measure on* Ω;

(ii) α *is subadditive, superadditive and inner regular;*

(iii) *the set function*

$$\beta(E) = \inf\{\alpha(U) : \ U \in \mathcal{A}(\Omega), E \subseteq U\} \tag{10.4}$$

defines a Borel measure on Ω.

Proof Clearly β is an extension of α; hence (iii) $\Rightarrow$ (i) $\Rightarrow$ (ii). It remains to prove that (ii) $\Rightarrow$ (iii).

Since β is defined as a Carathéodory outer measure it will suffice to show that: (a) β is countably subadditive (on all subsets of Ω), i.e.

$$\beta(U) \le \sum_{j=0}^{\infty} \beta(U_j) \qquad \text{if} \qquad U \subset \bigcup_{j=0}^{\infty} U_j;$$

(b) open sets are measurable.

We first show that α is countably subadditive on $\mathcal{A}(\Omega)$: let $U \in \mathcal{A}(\Omega)$, and let (U_j) be a sequence in $\mathcal{A}(\Omega)$ such that $U \subseteq \bigcup_j U_j$. If $t < \alpha(U)$ then by the inner regularity of α there exists $V \in \mathcal{A}(\Omega)$ such that $V \subset\subset U \subseteq \bigcup_j U_j$ and $\alpha(V) > t$. Since $\overline{V}$ is compact we can suppose then that $\overline{V} \subseteq \bigcup_{j=0}^{N} U_j$ so that, by the finite additivity of α,

$$t < \alpha(V) \le \sum_{j=0}^{N} \alpha(U_j) \le \sum_{j=0}^{\infty} \alpha(U_j);$$

hence, by the arbitrariness of $t < \alpha(U)$ we get $\alpha(U) \le \sum_{j=0}^{\infty} \alpha(U_j)$. From the countable subadditivity of α we immediately obtain the countable subadditivity of β on all subsets of Ω: let $E \subset \Omega$, and let (E_j) be a sequence of subsets of Ω such that $E \subseteq \bigcup_j E_j$. Fix $\varepsilon > 0$; for all $j \in \mathbf{N}$ let $U_j \in \mathcal{A}(\Omega)$ be such that $E_j \subseteq U_j$ and

$$\alpha(U_j) \le \beta(E_j) + 2^{-j}\varepsilon.$$

We obtain

$$\beta(E) \le \alpha\Big(\bigcup_{j=0}^{\infty} U_j\Big) \le \sum_{j=0}^{\infty} \alpha(U_j) \le \sum_{j=0}^{\infty} \beta(E_j) + 2\varepsilon,$$

and the desired inequality follows by letting $\varepsilon \to 0$.

It remains to show that open sets are measurable, i.e. that for every $U \in \mathcal{A}(\Omega)$ we have

$$\beta(E \cap U) + \beta(E \setminus U) \le \beta(E) \tag{10.5}$$

for all subsets E of Ω, the opposite inequality coming from the subadditivity of β. We argue by contradiction: suppose that $U \in \mathcal{A}(\Omega)$ and a subset E of Ω exists such that

$$\beta(E) < \beta(E \cap U) + \beta(E \setminus U).$$

By definition of β we can find $V \in \mathcal{A}(\Omega)$ with $E \subseteq V$, and

$$\alpha(V) < \alpha(V \cap U) + \beta(V \setminus U).$$

By the inner regularity of α there exists $W \in \mathcal{A}(\Omega)$ with $W \subset\subset V \cap U$ and

$$\begin{aligned}
\alpha(V) &< \alpha(W) + \beta(V \setminus U) \\
&\le \alpha(W) + \alpha(V \setminus \overline{W}).
\end{aligned} \tag{10.6}$$

By the superadditivity of α we have

$$\alpha(W) + \alpha(V \setminus \overline{W}) \le \alpha(V),$$

which together with (10.6) yields the desired contradiction. $\qquad\square$

10.3 Increasing set functions and compactness of Γ-limits

The properties of increasing set functions will be used to obtain an integral representation for the Γ-limits as in Chapter 9, and other results as in the following theorem.

Theorem 10.3 (Compactness) *Let* $(F_\varepsilon) : \mathrm{L}^p(\Omega; \mathbf{R}^m) \times \mathcal{A}(\Omega) \to [0, +\infty]$ $(\varepsilon > 0)$ *be a family of functionals. Suppose that for every sequence* (ε_k) *of positive real numbers converging to* 0 *and for every* $u \in \mathrm{W}^{1,p}(\Omega; \mathbf{R}^m)$

$$\alpha'(U) = \Gamma(\mathrm{L}^p)\text{-}\liminf_k F_{\varepsilon_k}(u, U)$$

$$\alpha''(U) = \Gamma(\mathrm{L}^p)\text{-}\limsup_k F_{\varepsilon_k}(u, U)$$

(the Γ-limits are performed with respect to the $\mathrm{L}^p(U; \mathbf{R}^m)$ *convergence) define inner regular increasing set functions. Then for every sequence* (ε_j) *of positive real numbers converging to* 0 *there exists a subsequence* (ε_{j_k}) *such that the Γ-limit*

$$F(u, U) = \Gamma(\mathrm{L}^p)\text{-}\lim_k F_{\varepsilon_{j_k}}(u, U)$$

exists for all $U \in \mathcal{A}(\Omega)$ *and* $u \in \mathrm{W}^{1,p}(\Omega; \mathbf{R}^m)$.

Proof Consider the countable family $\mathcal{R} = (R_j)$ of all finite unions of open rectangles of Ω with rational vertices. Fix a sequence (ε_j) converging to 0; by Proposition 7.9 and a diagonal procedure we can find a subsequence (ε_{j_k}) such that the Γ-limit

$$F(u, R) = \Gamma(\mathrm{L}^p)\text{-}\lim_k F_{\varepsilon_{j_k}}(u, R)$$

exists for all $R \in \mathcal{R}$ and $u \in \mathrm{W}^{1,p}(\Omega; \mathbf{R}^m)$. Let $U \in \mathcal{A}(\Omega)$ and $u \in \mathrm{W}^{1,p}(\Omega; \mathbf{R}^m)$. Since by hypothesis the Γ-lim inf and Γ-lim sup define inner regular increasing set functions, we have

$$
\begin{aligned}
\Gamma(\mathrm{L}^p)\text{-}\liminf_k F_{\varepsilon_{j_k}}(u, U) &= \alpha'(U) \\
&= \sup\{\alpha'(V) :\ V \in \mathcal{A}(\Omega),\ V \subset\subset U\} \\
&= \sup\{\alpha'(R) :\ R \in \mathcal{R},\ R \subset\subset U\} \\
&= \sup\{\alpha''(R) :\ R \in \mathcal{R},\ R \subset\subset U\} \\
&= \sup\{\alpha''(V) :\ V \in \mathcal{A}(\Omega),\ V \subset\subset U\} \\
&= \alpha''(U) = \Gamma(\mathrm{L}^p)\text{-}\limsup_k F_{\varepsilon_{j_k}}(u, U);
\end{aligned}
$$

that is, the thesis. Note that we have used the fact that if $V \subset\subset U \subseteq \Omega$ then there exists $R \in \mathcal{R}$ such that $V \subset\subset R \subset\subset U$, and that by construction $\alpha'(R) = \alpha''(R)$ for all $R \in \mathcal{R}$. $\qquad\square$

11

THE FUNDAMENTAL ESTIMATE

11.1 Fundamental estimates

It is convenient to highlight some properties from which it is possible to guarantee that the Γ-limits define inner regular increasing set functions. In this section $p \geq 1$.

Definition 11.1 *Let $U, U' \in \mathcal{A}(\Omega)$ satisfy $U' \subset\subset U$. We say that a function φ is a* cut-off function *between U' and U if $\varphi \in C_0^\infty(U)$, $0 \leq \varphi \leq 1$, and $\varphi \equiv 1$ on U'*

Definition 11.2 *Let $F : L^p(\Omega; \mathbf{R}^m) \times \mathcal{A}(\Omega) \to [0, +\infty]$ be a functional. We say that F satisfies the L^p-fundamental estimate if for every $U, U', V \in \mathcal{A}(\Omega)$ with $U' \subset\subset U$ and $\sigma > 0$ there exists $M_\sigma > 0$ such that for all $u, v \in L^p(\Omega; \mathbf{R}^m)$ there exists a cut-off function φ between U' and U such that*

$$F(\varphi u + (1 - \varphi)v, U' \cup V) \leq (1 + \sigma)(F(u, U) + F(v, V))$$
$$+ M_\sigma \int_{(U \cap V) \setminus U'} |u - v|^p \, dx + \sigma. \quad (11.1)$$

We say that a family $\mathcal{F}$ of functionals defined on $L^p(\Omega; \mathbf{R}^m) \times \mathcal{A}(\Omega)$ with values in $[0, +\infty]$ satisfies uniformly *the L^p-fundamental estimate if every $F \in \mathcal{F}$ satisfies the fundamental estimate, and the constant M_σ can be chosen uniformly on $\mathcal{F}$. If $\mathcal{F} = (F_\varepsilon)_{(\varepsilon > 0)}$, then we say that (F_ε) satisfies the L^p-fundamental estimate as $\varepsilon \to 0$ if for every $U, U', V \in \mathcal{A}(\Omega)$ with $U' \subset\subset U$ and $\sigma > 0$ there exist $M_\sigma > 0$ and $\varepsilon_\sigma > 0$ such that for all $u, v \in L^p(\Omega; \mathbf{R}^m)$ and $\varepsilon < \varepsilon_\sigma$ there exists a cut-off function φ between U' and U such that*

$$F_\varepsilon(\varphi u + (1 - \varphi)v, U' \cup V) \leq (1 + \sigma)(F_\varepsilon(u, U) + F_\varepsilon(v, V))$$
$$+ M_\sigma \int_{(U \cap V) \setminus U'} |u - v|^p \, dx + \sigma. \quad (11.2)$$

The definition of fundamental estimate extends to sequences (F_j) with obvious changes.

Remark 11.3 The definitions above will suffice to treat integral functionals defined on Sobolev spaces. All the reasoning in the sequel can be carried over in the same way if in place of (11.1) we require that

$$F(\varphi u + (1 - \varphi)v, U' \cup V) \leq (1 + \sigma)(F(u, U) + F(v, V))$$

$$+ M_\sigma\, \omega\Big(\|u - v\|_{L^p((U\cap V)\setminus U';\mathbf{R}^m)}\Big) + \sigma,$$

where $\omega : \mathbf{R} \to [0, +\infty)$ is a continuous function with $\omega(0) = 0$. Note that in this case we can also take $p = \infty$.

The definition of the L^p-fundamental estimate can be generalized to cover more general functionals requiring that for every $U, U', V \in \mathcal{A}(\Omega)$ with $U' \subset\subset U$ and $\sigma > 0$ there exists $M_\sigma > 0$ such that for all $u \in L^p(U; \mathbf{R}^m)$ and $v \in L^p(V; \mathbf{R}^m)$ there exists a function $w \in L^p(U' \cup V; \mathbf{R}^m)$ such that $w = u$ in U', $w = v$ in $V \setminus U$,

$$\int_{(U\cap V)\setminus U'} (|u - w|^p + |v - w|^p)\, dx \leq 2 \int_{(U\cap V)\setminus U'} |u - v|^p\, dx,$$

and

$$F(w, U' \cup V) \leq (1 + \sigma)(F(u, U) + F(v, V))$$
$$+ M_\sigma \int_{(U\cap V)\setminus U'} |u - v|^p\, dx + \sigma. \qquad (11.3)$$

In some cases it is not possible to choose w in the form $\varphi u + (1 - \varphi)v$ as in Definition 11.2. As an example we can consider the functional $F : L^1(\Omega) \times \mathcal{A}(\Omega) \to [0, +\infty]$ defined by

$$F(u, U) = \begin{cases} \displaystyle\int_{\partial E \cap U} f(x)d\Sigma & \text{if } u = \chi_E,\ E \subset \Omega \text{ with piecewise } C^1 \text{ boundary} \\ +\infty & \text{otherwise,} \end{cases}$$

where f is a fixed bounded Borel function, and the integral with respect to $d\Sigma$ stands for the surface integration on ∂E. It is not difficult to prove that F satisfies (11.3), but it trivially does not satisfy (11.1).

Example 11.4 Consider the functional $F : L^p(\Omega; \mathbf{R}^m) \times \mathcal{A}(\Omega) \to [0, +\infty]$ defined by

$$F(u, U) = \begin{cases} \displaystyle\int_U f(x, Du(x))dx & \text{if } u \in W^{1,1}(\Omega; \mathbf{R}^m) \\ +\infty & \text{otherwise,} \end{cases} \qquad (11.4)$$

where $f : \Omega \times \mathbf{M}^{m\times n} \to [0, +\infty)$ is a Borel function, convex in the second variable, such that there exists $C > 0$ such that

$$0 \leq f(x, A) \leq C(1 + |A|^p), \qquad f(x, 2A) \leq C(1 + f(x, A)) \qquad (11.5)$$

for all $x \in \Omega$ and $A \in \mathbf{M}^{m\times n}$. In order to prove (11.1) it suffices to consider $u, v \in W^{1,1}(\Omega; \mathbf{R}^m)$, and fix $U, U', V \in \mathcal{A}(\Omega)$ with $U' \subset\subset U$. We set $\delta = \text{dist}\,(U', \partial U)$, and we fix $N \in \mathbf{N}$, $N > 0$. With fixed $k \in \{1, \ldots, N\}$ let φ be a cut-off function between $\{x \in U : N\mathrm{dist}\,(x, U') < \delta(k-1)\}$ and $\{x \in U : N\mathrm{dist}\,(x, U') < \delta k\}$ with $|D\varphi|\delta \le 2N$. We define for all $k = 1, \dots, N$

$$C_k = \{x \in U : \delta(k-1) \le N\mathrm{dist}\,(x, U') < \delta k\}.$$

We then have

$$F(\varphi u + (1 - \varphi)v, U' \cup V) \le F(u, U) + F(v, V)$$
$$+ \int_{C_k \cap V} f(x, \varphi Du + (1 - \varphi)Dv + (u - v)D\varphi)\, dx$$
$$= F(u, U) + F(v, V)$$
$$+ \int_{C_k \cap V} f\left(x, 2\left(\frac{1}{2}(\varphi Du + (1 - \varphi)Dv) + \frac{1}{2}(u - v)D\varphi\right)\right) dx$$
$$\le F(u, U) + F(v, V)$$
$$+ C\int_{C_k \cap V} \left(1 + f\left(x, \frac{1}{2}(\varphi Du + (1 - \varphi)Dv) + \frac{1}{2}(u - v)D\varphi\right)\right) dx$$
$$\le F(u, U) + F(v, V)$$
$$+ C\int_{C_k \cap V} \left(1 + \frac{1}{2}f(x, \varphi Du + (1 - \varphi)Dv) + \frac{1}{2}f(x, (u - v)D\varphi)\right) dx$$
$$\le F(u, U) + F(v, V)$$
$$+ C\int_{C_k \cap V} \left(1 + \frac{1}{2}\varphi f(x, Du) + \frac{1}{2}(1 - \varphi)f(x, Dv) + \frac{1}{2}f(x, (u - v)D\varphi)\right) dx$$
$$\le F(u, U) + F(v, V)$$
$$+ C\int_{C_k \cap V} (1 + f(x, Du) + f(x, Dv) + C(1 + |D\varphi|^p|u - v|^p))\, dx.$$

Since

$$\sum_{k=1}^{N} \int_{C_k \cap V} (1 + C + f(x, Du) + f(x, Dv))\, dx$$
$$\le \int_{(U \cap V) \setminus U'} (1 + C + f(x, Du) + f(x, Dv))\, dx$$
$$\le (1 + C)|U \cap V| + F(u, U) + F(v, V)$$

we can choose k such that

$$\int_{C_k \cap V} (1 + C + f(x, Du) + f(x, Dv))\, dx \le \frac{1}{N}\Big((1 + C)|U \cap V| + F(u, U) + F(v, V)\Big).$$

We then have

$$F(\varphi u + (1 - \varphi)v, U' \cup V) \le \left(1 + \frac{C}{N}\right)(F(u, U) + F(v, V))$$

$$+ \frac{C}{N}(1+C)|U \cap V| + C^2\left(\frac{2N}{\delta}\right)^p \int_{C_k \cap V} |u-v|^p \, dx$$

$$\leq \left(1+\frac{C}{N}\right)(F(u,U)+F(v,V))$$

$$+ \frac{C}{N}(1+C)|U \cap V| + C^2\left(\frac{2N}{\delta}\right)^p \int_{(U \cap V)\setminus U'} |u-v|^p \, dx.$$

If we choose

$$N = N_\sigma = \left[\max\left\{\frac{C}{\sigma}, \frac{C}{\sigma}(1+C)|U \cap V|\right\}\right] + 1,$$

($[t]$ denotes the integer part of t) and

$$M_\sigma = C^2\left(\frac{2N}{\delta}\right)^p,$$

then (11.1) is satisfied.

11.2 Subadditivity of Γ-limits

From the fundamental estimate we can derive some inequalities for the Γ-limits.

Proposition 11.5 *Let (F_ε) be a family of functionals defined on $L^p(\Omega; \mathbf{R}^m) \times \mathcal{A}(\Omega)$ with values in $[0, +\infty]$ satisfying the L^p-fundamental estimate as $\varepsilon \to 0$, and let (ε_j) be a sequence of positive real numbers converging to 0. If for every $u \in L^p(\Omega; \mathbf{R}^m)$ and $U \in \mathcal{A}(\Omega)$ we denote*

$$F'(u,U) = \Gamma(L^p)\text{-}\liminf_j F_{\varepsilon_j}(u,U) \tag{11.6}$$

$$F''(u,U) = \Gamma(L^p)\text{-}\limsup_j F_{\varepsilon_j}(u,U), \tag{11.7}$$

then we have

$$F'(u, U' \cup V) \leq F'(u,U) + F''(u,V) \tag{11.8}$$

$$F''(u, U' \cup V) \leq F''(u,U) + F''(u,V) \tag{11.9}$$

for all $u \in L^p(\Omega; \mathbf{R}^m)$ and $U, U', V \in \mathcal{A}(\Omega)$ with $U' \subset\subset U$.

Proof We prove only (11.8), the proof of (11.9) being completely analogous.

By Propositions 7.5 and 7.6 there exist two sequences (u_j) and (v_j) of functions converging to u in $L^p(\Omega; \mathbf{R}^m)$, such that

$$F'(u,U) = \liminf_j F_{\varepsilon_j}(u_j, U),$$

$$F''(u,V) = \limsup_j F_{\varepsilon_j}(v_j, V).$$

By the L^p-fundamental estimate as $\varepsilon \to 0$, applied to the functions u_j and v_j, with fixed $\sigma > 0$ we can find $M_\sigma, \varepsilon_\sigma > 0$ such that for all $\varepsilon_j < \varepsilon_\sigma$ there exists a sequence $w_j = \varphi_j u_j + (1 - \varphi_j) v_j$, where φ_j are suitable cut-off functions between U' and U, such that

$$F_{\varepsilon_j}(w_j, U' \cup V) \leq (1 + \sigma)(F_{\varepsilon_j}(u_j, U) + F_{\varepsilon_j}(v_j, V)) + M_\sigma \int_{U \cap V} |u_j - v_j|^p \, dx + \sigma.$$

Since $w_j \to u$ in $\mathrm{L}^p(\Omega; \mathbf{R}^m)$, and $\int_{U \cap V} |u_j - v_j|^p \, dx \to 0$, we have

$$\begin{aligned}
F'(u, U' \cup V) &\leq \liminf_j F_{\varepsilon_j}(w_j, U' \cup V) \\
&\leq (1 + \sigma)(\liminf_j F_{\varepsilon_j}(u_j, U) + \limsup_j F_{\varepsilon_j}(v_j, V)) + \sigma \\
&= (1 + \sigma)(F'(u, U) + F''(u, V)) + \sigma.
\end{aligned}$$

By the arbitrariness of σ we have the thesis. $\qquad\square$

From the previous proposition we obtain some inner regularity results, provided that a growth estimate is satisfied.

Proposition 11.6 *Let* (F_ε) *be as in Proposition* 11.5*, and let* F' *and* F'' *be defined by* (11.6) *and* (11.7)*, respectively. Let* $q \geq 1$*, and let* $u \in \mathrm{W}^{1,q}(\Omega; \mathbf{R}^m) \cap \mathrm{L}^p(\Omega; \mathbf{R}^m)$*; if* $F'(u, \cdot)$ *and* $F''(u, \cdot)$ *are increasing set functions and*

$$F''(u, U) \leq c \int_U (1 + |Du|^q) \, dx \tag{11.10}$$

for all $U \in \mathcal{A}(\Omega)$*, then* $F'(u, \cdot)$ *and* $F''(u, \cdot)$ *are inner regular increasing set functions. Moreover,* $F''(u, \cdot)$ *is subadditive.*

Proof Fix $W \in \mathcal{A}(\Omega)$. Let $K \subset W$ be a compact subset of W. Choose $U', U \in \mathcal{A}(\Omega)$ such that
$$K \subset U' \subset\subset U \subset\subset W,$$
and define $V = W \setminus K$. By Proposition 11.5 we have

$$\begin{aligned}
F'(u, W) &\leq F'(u, U' \cup V) \leq F'(u, U) + F''(u, V) \\
&= F'(u, U) + F''(u, W \setminus K)
\end{aligned}$$

and

$$\begin{aligned}
F''(u, W) &\leq F''(u, U' \cup V) \leq F''(u, U) + F''(u, V) \\
&= F''(u, U) + F''(u, W \setminus K).
\end{aligned}$$

If $u \in \mathrm{W}^{1,q}(\Omega; \mathbf{R}^m)$, by (11.10) we have

$$F'(u, W) \leq \sup\{F'(u, U) : \ U \subset\subset W\} + c \int_{W \setminus K} (1 + |Du|^q) \, dx,$$

and

$$F''(u, W) \leq \sup\{F''(u, U) : \ U \subset\subset W\} + c \int_{W \setminus K} (1 + |Du|^q) \, dx.$$

The last term of these two inequalities can be taken arbitrarily small as $|W \setminus K| \to 0$. The opposite inequalities

$$F'(u, W) \geq \sup\{F'(u, U) : \ U \subset\subset W\}$$

and

$$F''(u, W) \geq \sup\{F''(u, U) : \ U \subset\subset W\}$$

are trivial since we suppose that $F'(u, \cdot)$ and $F''(u, \cdot)$ are increasing set functions.

The subadditivity of $F''(u, \cdot)$ follows from Proposition 11.5 and from its inner regularity. In fact, let $U, V \in \mathcal{A}(\Omega)$, and $u \in \mathrm{W}^{1,q}(\Omega; \mathbf{R}^m) \cap \mathrm{L}^p(\Omega; \mathbf{R}^m)$. Let $U' \subset\subset U$, $U \in \mathcal{A}(\Omega)$, and let $W \in \mathcal{A}(\Omega)$ be such that $W \subseteq U' \cup V$. Then we have by (11.9)

$$F''(u, W) \leq F''(u, U' \cup V) \leq F''(u, V) + F''(u, U).$$

Taking the supremum for $W \in \mathcal{A}(\Omega)$, $W \subset\subset U \cup V$, by the inner regularity of $F''(u, \cdot)$ we have

$$F''(u, U \cup V) \leq F''(u, V) + F''(u, U),$$

and the thesis. $\qquad\qquad\qquad\qquad\qquad\qquad\qquad\qquad\qquad\qquad\qquad\quad \square$

11.3 Γ-limits and boundary values

Further properties can be obtained if we suppose that estimates of the form (11.10) are satisfied uniformly.

Proposition 11.7 *Let (F_ε) be a family of functionals defined on $\mathrm{W}^{1,p}(\Omega; \mathbf{R}^m) \times \mathcal{A}(\Omega)$ with values in $[0, +\infty]$ satisfying the L^p-fundamental estimate as $\varepsilon \to 0$ (we regard these functionals as extended to $+\infty$ on $\mathrm{L}^p(\Omega; \mathbf{R}^m) \setminus \mathrm{W}^{1,p}(\Omega; \mathbf{R}^m)$), and let (ε_j) be a sequence of positive real numbers converging to 0. Let*

$$F_{\varepsilon_j}(u, U) \leq c \int_U (1 + |Du|^p) \, dx \qquad\qquad (11.11)$$

hold for all $U \in \mathcal{A}(\Omega)$ and $u \in \mathrm{W}^{1,p}(\Omega; \mathbf{R}^m)$. If we take $\phi \in \mathrm{W}^{1,p}(\Omega; \mathbf{R}^m)$, and we define $G_{\varepsilon_j}^\phi : \mathrm{L}^p(\Omega; \mathbf{R}^m) \to [0, +\infty]$

$$G_{\varepsilon_j}^\phi(u) = \begin{cases} F_{\varepsilon_j}(u, \Omega) & \text{if } u - \phi \in \mathrm{W}_0^{1,p}(\Omega; \mathbf{R}^m) \\[2ex] +\infty & \text{otherwise,} \end{cases}$$

then we have

$$F'(u, \Omega) = \Gamma(\mathrm{L}^p)\text{-}\liminf_j F_{\varepsilon_j}(u, \Omega)$$

$$= \Gamma(\mathrm{L}^p)\text{-}\liminf_j G^\phi_{\varepsilon_j}(u) \tag{11.12}$$

$$F''(u,\Omega) = \Gamma(\mathrm{L}^p)\text{-}\limsup_j F_{\varepsilon_j}(u,\Omega)$$
$$= \Gamma(\mathrm{L}^p)\text{-}\limsup_j G^\phi_{\varepsilon_j}(u) \tag{11.13}$$

for all $u \in \mathrm{W}^{1,p}(\Omega;\mathbf{R}^m)$ such that $u - \phi \in \mathrm{W}^{1,p}_0(\Omega;\mathbf{R}^m)$.

Proof We prove only equality (11.12), the proof of (11.13) being completely analogous.

Since $F_{\varepsilon_j}(u,\Omega) \leq G^\phi_{\varepsilon_j}(u)$ for all $u \in \mathrm{L}^p(\Omega;\mathbf{R}^m)$, we have trivially

$$F'(u,\Omega) = \Gamma(\mathrm{L}^p)\text{-}\liminf_j F_{\varepsilon_j}(u,\Omega)$$
$$\leq \Gamma(\mathrm{L}^p)\text{-}\liminf_j G^\phi_{\varepsilon_j}(u).$$

We have to prove that if $w - \phi \in \mathrm{W}^{1,p}_0(\Omega;\mathbf{R}^m)$ then

$$F'(w,\Omega) \geq \Gamma(\mathrm{L}^p)\text{-}\liminf_j G^\phi_{\varepsilon_j}(w),$$

i.e. that for all $\sigma > 0$ there exists a sequence $w_j \to w$ in $\mathrm{L}^p(\Omega;\mathbf{R}^m)$ such that $w_j - \phi \in \mathrm{W}^{1,p}_0(\Omega;\mathbf{R}^m)$, and

$$2\sigma + (1+\sigma)F'(w,\Omega) \geq \liminf_j F_{\varepsilon_j}(w_j,\Omega). \tag{11.14}$$

Let (u_j) be a sequence in $\mathrm{W}^{1,p}(\Omega;\mathbf{R}^m)$ converging to w in $\mathrm{L}^p(\Omega;\mathbf{R}^m)$ such that $F'(w,\Omega) = \liminf_j F_{\varepsilon_j}(u_j,\Omega)$. As in the previous proposition we can fix a compact K of Ω and $U, U' \in \mathcal{A}(\Omega)$ such that $K \subseteq U' \subset\subset U \subset\subset \Omega$. With fixed $\sigma > 0$, by the L^p-fundamental estimate as $\varepsilon \to 0$ applied to $v = w$ on $V = \Omega \setminus K$ and $u = u_j$ on U, we can find a function $w_j = \varphi_j u_j + (1 - \varphi_j)w$, with φ_j suitable cut-off functions between U' and U, such that

$$F_{\varepsilon_j}(w_j,\Omega) \leq (1+\sigma)(F_{\varepsilon_j}(u_j,\Omega) + F_{\varepsilon_j}(w,\Omega \setminus K))$$
$$+ M_\sigma \int_\Omega |w - u_j|^p\, dx + \sigma$$
$$\leq (1+\sigma)\Big(F_{\varepsilon_j}(u_j,\Omega) + c\int_{\Omega\setminus K}(1 + |Dw|^p)\, dx\Big)$$
$$+ M_\sigma \int_\Omega |w - u_j|^p\, dx + \sigma.$$

We have $w_j = w$ on $\Omega \setminus U$, so that $w_j - \phi \in \mathrm{W}^{1,p}_0(\Omega;\mathbf{R}^m)$; moreover,

$$\liminf_j F_{\varepsilon_j}(w_j,\Omega) \leq (1+\sigma)\Big(F'(w,\Omega) + c\int_{\Omega\setminus K}(1 + |Dw|^p)\, dx\Big) + \sigma.$$

It now suffices to choose K such that $(1+\sigma)c\int_{\Omega\setminus K}(1 + |Dw|^p)\, dx \leq \sigma$ to obtain (11.14) and to conclude the proof. $\qquad\square$

Remark 11.8 The statement of Proposition 11.7 can be rephrased by saying that if (11.11) holds, together with the L^p-fundamental estimate as $\varepsilon \to 0$, then minimizing sequences for the Γ-limits can be taken with the same boundary values as their limit.

Remark 11.9 Note that (11.11) implies (11.10) with $q = p$. Note also that the conclusions of Proposition 11.7 may not hold if we assume only (11.10) with $p = q$. We can take for example $n = 2$ and $F_\varepsilon(u, U) = \int_U f(\frac{x}{\varepsilon}, Du)\, dx$, $u \in W^{1,2}(U)$, where

$$
f(x_1, x_2, A) = \begin{cases} 0 & \text{if } A = 0 \\[2mm] +\infty & \text{if } A \neq 0 \text{ and } (x_1 - [x_1])^2 + x_2^2 < \frac{1}{9} \\[2mm] |A|^2 & \text{otherwise;} \end{cases}
$$

that is, we must have u constant on each ball of centre $\varepsilon(k, 0)$ and radius $\varepsilon/3$, $k \in \mathbf{Z}$, intersecting U, in order to have $F(u, U) < \infty$. It can be immediately seen that if $\varepsilon_j \to 0$ then $\Gamma(L^2(U))\text{-}\lim_j F_{\varepsilon_j}(u, U) = \int_U |Du|^2\, dx$ for all U bounded open subsets of $\mathbf{R}^2$ and $u \in W^{1,2}(U)$, but Proposition 11.7 does not hold taking, for example, $\Omega = [0, 1]^2$ and $\phi(x, y) = x$, since $G^\phi_{\varepsilon_j}(u) = +\infty$ if $u - \phi \in W^{1,2}_0(\Omega)$.

11.4   Exercises

Exercise 11.1 For all $j \in \mathbf{N}$, let $f_j : \Omega \times \mathbf{R}^n \to [0, +\infty)$ be a Borel function satisfying $f_j(x, z) \leq c(1 + |z|^p)$, and let $v \in W^{1,p}_0(\Omega)$ with $v > 0$ on Ω. Let $u, w_j \in W^{1,p}(\Omega)$, and let $w_j \to 0$ in $L^p(\Omega)$ be such that the limit $\lim_j \int_\Omega f_j(x, Du + Dw_j)\, dx$ exists. Define $v_j = (-v) \vee (w_j \wedge v)$, so that $v_j \in W^{1,p}_0(\Omega)$. Prove that

$$
\limsup_j \int_\Omega f_j(x, Du + Dv_j)\, dx \leq \lim_j \int_\Omega f_j(x, Du + Dw_j)\, dx.
$$

Deduce an alternative proof for Proposition 11.7 in the scalar case $m = 1$.

Hint: define $E_j = \{v_j \neq w_j\}$. Note that

$$
\int_\Omega f_j(x, Du + Dv_j)\, dx \leq \int_{\Omega \setminus E_j} f_j(x, Du + Dw_j)\, dx
$$
$$
+ c \int_{E_j} (1 + |Du + Dv|^p)\, dx
$$

and $|E_j| \to 0$.

12

INTEGRAL FUNCTIONALS WITH STANDARD GROWTH CONDITIONS

12.1 Standard growth conditions

In this chapter we consider the Γ-limits of integral functionals

$$\int_\Omega f(x, u, Du)\, dx, \tag{12.1}$$

where $f : \Omega \times \mathbf{R}^m \times \mathbf{M}^{m \times n} \to [0, +\infty)$ is a Borel function satisfying a growth condition of the form

$$\alpha|A|^p \le f(x, s, A) \le \beta(1 + |A|^p) \tag{12.2}$$

for all $x \in \Omega$, $s \in \mathbf{R}^m$ and $A \in \mathbf{M}^{m \times n}$, $p \ge 1$. Conditions of this type are usually called *standard growth conditions of order p*.

Definition 12.1 *Let $\alpha, \beta > 0$ and $p \ge 1$. A functional $F : \mathrm{W}^{1,p}(\Omega; \mathbf{R}^m) \times \mathcal{A}(\Omega) \to [0, +\infty)$ belongs to the class $\mathcal{F}(\alpha, \beta, p)$ if there exists a Borel function $f : \Omega \times \mathbf{R}^m \times \mathbf{M}^{m \times n} \to [0, +\infty)$ satisfying (12.2) such that we can write*

$$F(u, U) = \int_U f(x, u, Du)\, dx \tag{12.3}$$

for all $u \in \mathrm{W}^{1,p}(\Omega; \mathbf{R}^m)$ and $U \in \mathcal{A}(\Omega)$.

12.2 Fundamental estimate

Proposition 12.2 *The family $\mathcal{F}(\alpha, \beta, p)$ satisfies the L^p-fundamental estimate uniformly (each functional is extended to $+\infty$ on $\mathrm{L}^p(\Omega; \mathbf{R}^m) \setminus \mathrm{W}^{1,p}(\Omega; \mathbf{R}^m)$).*

Proof Let $F \in \mathcal{F}(\alpha, \beta, p)$. Let $U, U', V \in \mathcal{A}(\Omega)$, with $U' \subset\subset U$, and define $\delta = \mathrm{dist}\,(U', \partial U)$. Take $0 < \eta < \delta$ and $0 < r < \delta - \eta$. Let φ be a cut-off function between $\{x \in U : \mathrm{dist}\,(x, U') < r\}$ and $\{x \in U : \mathrm{dist}\,(x, U') < r + \eta\}$, with $|D\varphi| \le 2/\eta$. If $u, v \in \mathrm{W}^{1,p}(\Omega; \mathbf{R}^m)$ then we have, setting $V_r^\eta = \{x \in V : r < \mathrm{dist}\,(x, U') < r + \eta\}$,

$$F(\varphi u + (1 - \varphi)v, U' \cup V)$$
$$= \int_{U' \cup V} f(x, \varphi u + (1 - \varphi)v, \varphi Du + (1 - \varphi)Dv + (u - v)D\varphi)\, dx$$

$$= \int_{\{x\in V:\ \mathrm{dist}\,(x,U')\geq r+\eta\}} f(x,v,Dv)\,dx$$

$$+ \int_{\{x\in U'\cup V:\ \mathrm{dist}\,(x,U')\leq r\}} f(x,u,Du)\,dx$$

$$+ \int_{V_r^\eta} f(x,\varphi u + (1-\varphi)v, \varphi Du + (1-\varphi)Dv + (u-v)D\varphi)\,dx$$

$$\leq F(v,V) + F(u,U)$$

$$+ \beta \int_{V_r^\eta} (1 + |\varphi Du + (1-\varphi)Dv + (u-v)D\varphi|^p)\,dx$$

$$\leq F(v,V) + F(u,U)$$

$$+ \beta 4^{p-1} \int_{V_r^\eta} (1 + |Du|^p + |Dv|^p + |D\varphi|^p|u-v|^p)\,dx$$

$$\leq F(v,V) + F(u,U)$$

$$+ \beta 4^{p-1} \int_{V_r^\eta} \left(1 + |Du|^p + |Dv|^p + \frac{2^p}{\eta^p}|u-v|^p\right)dx$$

$$\leq F(v,V) + F(u,U)$$

$$+ \beta 4^{p-1} \int_{V_r^\eta} (1 + |Du|^p + |Dv|^p)\,dx$$

$$+ \beta \frac{2^{3p-2}}{\eta^p} \int_{(U\cap V)\setminus U'} |u-v|^p\,dx.$$

Let

$$\mu(E) = \beta 4^{p-1} \int_E (1 + |Du|^p + |Dv|^p)\,dx;$$

note that by (12.2)

$$\mu(U\cap V) \leq \beta 4^{p-1}|U\cap V| + 4^{p-1}\frac{\beta}{\alpha}(F(u,U) + F(v,V)).$$

Moreover, for every $N = 1,2,\ldots$

$$\mu(U\cap V) \geq \sum_{k=1}^N \mu\left(\left\{x\in V:\ \delta\frac{k-1}{N} < \mathrm{dist}\,(x,U') < \delta\frac{k}{N}\right\}\right)$$

Hence for every $N = 1,2,\ldots$ there exists $k \in \{1,\ldots,N\}$ such that

$$\mu\left(\left\{x\in V:\ \delta\frac{k-1}{N} < \mathrm{dist}\,(x,U') < \delta\frac{k}{N}\right\}\right) \leq \beta 4^{p-1}\frac{1}{N}|U\cap V|$$

$$+ 4^{p-1}\frac{1}{N}\frac{\beta}{\alpha}(F(u,U) + F(v,V)).$$

The L^p-fundamental estimate (11.1) is proven by taking, with fixed $\sigma > 0$,

$$N \geq \frac{1}{\sigma} \max\left\{\beta 4^{p-1}|U \cap V|, 4^{p-1}\frac{\beta}{\alpha}\right\}, \quad \eta = \frac{\delta}{N} \quad \text{and} \quad r = \frac{k-1}{N}\delta.$$

In this case

$$M_\sigma = \beta \frac{2^{3p-2}N^p}{\delta^p}$$

depends only on $U, U', V, \alpha, \beta, p$ and hence it can be chosen uniformly on the family $\mathcal{F}(\alpha, \beta, p)$. $\qquad\square$

12.3 Compactness for the Γ-limits

Proposition 12.3 *Let (F_ε) be a family in $\mathcal{F}(\alpha, \beta, p)$. Then for every sequence (ε_j) of positive real numbers converging to 0 there exists a further subsequence (ε_{j_k}) such that the Γ-limit*

$$F(u, U) = \Gamma(\mathrm{L}^p)\text{-}\lim_k F_{\varepsilon_{j_k}}(u, U)$$

exists for all $u \in \mathrm{W}^{1,p}(\Omega; \mathbf{R}^m)$ and $U \in \mathcal{A}(\Omega)$, and $F(u, \cdot)$ is the restriction of a Borel measure to $\mathcal{A}(\Omega)$.

Proof We extend our functionals equal to $+\infty$ on $\mathrm{L}^p(\Omega; \mathbf{R}^m) \setminus \mathrm{W}^{1,p}(\Omega; \mathbf{R}^m)$. By (12.2) we obtain that Proposition 11.6 holds with $q = p$. Hence, we can apply Theorem 10.3 and obtain the existence of the Γ-limit for every $U \in \mathcal{A}(\Omega)$ and $u \in \mathrm{W}^{1,p}(\Omega; \mathbf{R}^m)$. We check the measure property through the criterion given by Theorem 10.2: the superadditivity is obvious, while the inner regularity is given by Proposition 11.6; we deduce the subadditivity of the set function $F(u, \cdot)$ from Proposition 11.6. $\qquad\square$

Remark 12.4 The same conclusions of Proposition 12.3 follow if we suppose in the place of (12.2) that

$$g(x, A) \leq f(x, s, A) \leq c(1 + g(x, A)),$$

with $g(x, \cdot)$ convex, $g(x, 2A) \leq c(1 + g(x, A))$, and $g(x, A) \leq c(1 + |A|^p)$ for all $x \in \Omega$, $s \in \mathbf{R}^m$ and $A \in \mathbf{M}^{m \times n}$, taking into account Example 11.4.

Theorem 12.5 *Let (f_ε) be a family of Borel functions with $f_\varepsilon : \Omega \times \mathbf{M}^{m \times n} \to [0, +\infty)$ satisfying the estimate*

$$\alpha|A|^p \leq f_\varepsilon(x, A) \leq \beta(1 + |A|^p) \tag{12.4}$$

for all $x \in \Omega$ and $A \in \mathbf{M}^{m \times n}$, and let

$$F_\varepsilon(u, U) = \int_U f_\varepsilon(x, Du)\, dx \tag{12.5}$$

if $u \in \mathrm{W}^{1,p}(\Omega; \mathbf{R}^m)$. Then, for every sequence (ε_j) of positive real numbers converging to 0 there exists a subsequence (ε_{j_k}) and a Carathéodory function

$\varphi : \Omega \times \mathbf{M}^{m \times n} \to [0, +\infty)$ *satisfying the same growth estimate as* f_ε *such that, if we define*

$$F_0(u, U) = \int_U \varphi(x, Du)\, dx \tag{12.6}$$

for $u \in \mathrm{W}^{1,p}(\Omega; \mathbf{R}^m)$, *we have*

$$F_0(u, U) = \Gamma(\mathrm{L}^p)\text{-}\lim_k F_{\varepsilon_{j_k}}(u, U) \tag{12.7}$$

for all $u \in \mathrm{W}^{1,p}(\Omega; \mathbf{R}^m)$ *and* $U \in \mathcal{A}(\Omega)$.

Proof The functionals, extended to $+\infty$ on $\mathrm{L}^p(\Omega; \mathbf{R}^m) \backslash \mathrm{W}^{1,p}(\Omega; \mathbf{R}^m)$, belong to the class $\mathcal{F}(\alpha, \beta, p)$. Hence, we can apply Proposition 12.3 to obtain the existence of the limit in (12.7). It remains to verify that the Γ-limit

$$F(u, U) = \Gamma(\mathrm{L}^p)\text{-}\lim_k F_{\varepsilon_{j_k}}(u, U)$$

satisfies the hypotheses of the Integral Representation Theorem 9.1. Conditions (i) and (iii)–(v) are trivially satisfied, while condition (ii) is given by Proposition 12.3. $\qquad\square$

Remark 12.6 The compactness result above can be obtained in an analogous way on classes of functions f satisfying a growth condition of Orlicz type

$$\Phi(|A|) \le f(x, A) \le c(1 + \Phi(|A|)),$$

provided that Φ is convex, superlinear, and $\Phi(2t) \le c(1 + \Phi(t))$. In fact, in these spaces it is possible to prove the necessary imbedding and approximation properties, as in Sobolev spaces (see Donaldson and Trudinger (1971)). In this case the integral representation is obtained in the corresponding Orlicz Sobolev space.

Remark 12.7 (Open problem) It is not known whether similar compactness theorems (and also the corresponding lower semicontinuity and relaxation results) hold in classes of functions f satisfying the conditions of Remark 12.4 for general g, under the only assumption that $u \mapsto \int_U g(x, Du)\, dx$ is lower semicontinuous.

12.4 Γ-limits of homogeneous functionals

In the case of functionals with integrands independent of the space variable Γ-convergence reduces to a pointwise convergence.

Proposition 12.8 *Let* $p > 1$, *and let* (f_ε) *be a family of continuous functions with* $f_\varepsilon : \mathbf{M}^{m \times n} \to [0, +\infty)$ *satisfying the estimate*

$$\alpha |A|^p \le f_\varepsilon(A) \le \beta(1 + |A|^p) \tag{12.8}$$

for all $A \in \mathbf{M}^{m \times n}$. *Let, for every bounded open set* U *of* $\mathbf{R}^n$,

$$F_\varepsilon(u, U) = \int_U f_\varepsilon(Du) \, dx \tag{12.9}$$

if $u \in \mathrm{W}^{1,p}(U; \mathbf{R}^m)$, and let (ε_j) be a sequence of positive real numbers converging to 0. We have that $F_{\varepsilon_j}(u, \Omega)$ $\Gamma(\mathrm{L}^p)$-converges to $F_0(u, \Omega)$ for all Ω bounded open sets of $\mathbf{R}^n$ and $u \in \mathrm{W}^{1,p}(\Omega; \mathbf{R}^m)$ if and only if $Qf_{\varepsilon_j} \to f$ pointwise and

$$F_0(u, \Omega) = \int_\Omega f(Du) \, dx \tag{12.10}$$

for all Ω bounded open sets of $\mathbf{R}^n$ and $u \in \mathrm{W}^{1,p}(\Omega; \mathbf{R}^m)$.

Proof Fix Ω as a bounded open set of $\mathbf{R}^n$. Suppose that $Qf_{\varepsilon_j} \to f$ pointwise. By Theorem 12.5 we can extract a subsequence (ε_{j_k}) such that

$$F(u, U) = \Gamma(\mathrm{L}^p)\text{-}\lim_k F_{\varepsilon_{j_k}}(u, U)$$

exists for all $u \in \mathrm{W}^{1,p}(\Omega; \mathbf{R}^m)$ and $U \in \mathcal{A}(\Omega)$. By Proposition 9.2 we can write

$$F(u, U) = \int_U \varphi(Du) \, dx$$

for some quasiconvex function φ, with

$$\alpha|A|^p \le \varphi(A) \le \beta(1 + |A|^p)$$

for all $A \in \mathbf{M}^{m \times n}$. We can suppose that $|\partial\Omega| = 0$ by the inner regularity of $F(u, \cdot)$. Let $A \in \mathbf{M}^{m \times n}$; by Definition 5.14, Remark 5.15, Proposition 11.7, applied with $\phi = Ax$, and Theorem 7.2, we have

$$
\begin{aligned}
|\Omega|\varphi(A) &= \min\left\{ \int_\Omega \varphi(Du) \, dx : \ u - Ax \in \mathrm{W}_0^{1,p}(\Omega; \mathbf{R}^m) \right\} \\
&= \min\left\{ F(u, \Omega) : \ u - Ax \in \mathrm{W}_0^{1,p}(\Omega; \mathbf{R}^m) \right\} \\
&= \lim_k \inf\left\{ F_{\varepsilon_{j_k}}(u, \Omega) : \ u - Ax \in \mathrm{W}_0^{1,p}(\Omega; \mathbf{R}^m) \right\} \\
&= \lim_k \inf\left\{ \int_\Omega f_{\varepsilon_{j_k}}(Du) \, dx : \ u - Ax \in \mathrm{W}_0^{1,p}(\Omega; \mathbf{R}^m) \right\} \\
&= |\Omega| \lim_k Qf_{\varepsilon_{j_k}}(A) = |\Omega| \lim_j Qf_{\varepsilon_j}(A) = |\Omega| f(A).
\end{aligned}
$$

Hence the limit does not depend on the subsequence, and we obtain (12.10). Conversely, if (12.10) holds, proceeding as above we obtain

$$
\begin{aligned}
|\Omega| f(A) &= \min\left\{ \int_\Omega f(Du) \, dx : \ u - Ax \in \mathrm{W}_0^{1,p}(\Omega; \mathbf{R}^m) \right\} \\
&= \min\left\{ F_0(u, \Omega) : \ u - Ax \in \mathrm{W}_0^{1,p}(\Omega; \mathbf{R}^m) \right\}
\end{aligned}
$$

$$= \lim_j \inf \left\{ F_{\varepsilon_j}(u, \Omega) : \ u - Ax \in W_0^{1,p}(\Omega; \mathbf{R}^m) \right\}$$

$$= \lim_j \inf \left\{ \int_\Omega f_{\varepsilon_j}(Du)\, dx : \ u - Ax \in W_0^{1,p}(\Omega; \mathbf{R}^m) \right\}$$

$$= |\Omega| \lim_j Q f_{\varepsilon_j}(A),$$

and the thesis. $\qquad\square$

Corollary 12.9 *If f is a continuous function satisfying the growth condition*

$$\alpha |A|^p \leq f(A) \leq \beta(1 + |A|^p)$$

for all $A \in \mathbf{M}^{m \times n}$, then the lower semicontinuous envelope of the functional

$$F(u) = \int_\Omega f(Du)\, dx, \qquad u \in W^{1,p}(\Omega; \mathbf{R}^m)$$

in the L^p-topology is given by

$$\overline{F}(u) = \int_\Omega Q f(Du)\, dx$$

for all $u \in W^{1,p}(\Omega; \mathbf{R}^m)$.

Proof It suffices to choose $f_\varepsilon = f$ in Proposition 12.8. $\qquad\square$

12.5 Exercises

Exercise 12.1 Prove the statement of Remark 12.4.

Exercise 12.2 Let $f_\varepsilon : \mathbf{M}^{m \times n} \to [0, +\infty)$ be a family of polyconvex functions satisfying (12.8), such that the functionals F_ε in (12.9) Γ-converge to the functional F_0 in (12.10) with integrand f. Prove that f is polyconvex (use Proposition 12.8 and Remark 5.8).

Part III

Basic homogenization

13

A 1-DIMENSIONAL EXAMPLE

This chapter is completely devoted to a single example, which on one hand contains some features of the problems that will be faced afterwards, and on the other hand may highlight some of the differences between the scalar case and the vector-valued case. The solution 'by hand' in this simple case requires the relative complexity of some approximations and of some passages to the limit. In the next chapters many of these complications will be overcome by the use of the direct methods of Γ-convergence.

We will study the Γ-convergence of functionals of the form

$$F_\varepsilon(u) = \int_0^1 f\left(\frac{t}{\varepsilon}, u'(t)\right) dt \qquad u \in W^{1,p}(0,1),$$

where $f : \mathbf{R}^2 \to [0, +\infty)$ is a Carathéodory function which satisfies the following conditions:

$$f(t, \cdot) \text{ is convex for all } t \in \mathbf{R}, \tag{13.1}$$

$$f(\cdot, \xi) \text{ is 1-periodic for all } \xi \in \mathbf{R}, \tag{13.2}$$

$$|\xi|^p \le f(t, \xi) \le \beta(1 + |\xi|^p) \tag{13.3}$$

for all $(t, \xi) \in \mathbf{R}^2$, with $p > 1$.

We want to show that there exists a convex function $f_0 : \mathbf{R} \to [0, +\infty)$ such that for every sequence (ε_j) we have

$$\int_0^1 f_0(u') \, dt = \Gamma(\mathrm{L}^p)\text{-}\lim_j F_{\varepsilon_j}(u) \tag{13.4}$$

for all $u \in W^{1,p}(0,1)$. This will be done directly using the definition of Γ-convergence. In the next sections we will see how to prove similar statements via the direct methods of Γ-convergence.

13.1 The cell-problem homogenization formula

The first step to perform is to guess the form of f_0. We remark that, using Jensen's inequality, the value $f_0(\xi)$ can be expressed as a minimum problem:

$$f_0(\xi) = \min\left\{\int_0^1 f_0(u' + \xi) \, dt : \ u(0) = u(1) = 0\right\}.$$

We can apply (13.4) at this point with $\varepsilon_j = \frac{1}{j}$, and Theorem 7.2 to deduce that

$$
\begin{aligned}
f_0(\xi) &= \lim_j \min\left\{ \int_0^1 f(jt, u' + \xi)\, dt \;:\; u(0) = u(1) = 0 \right\} \\
&= \lim_j \min\left\{ \int_0^1 f(jt, u' + \xi)\, dt \;:\; u(0) = u(1) \right\} \\
&= \lim_j \min\left\{ \frac{1}{j} \int_0^j f(t, u' + \xi)\, dt \;:\; u(0) = u(j) \right\} \\
&= \lim_j \min\left\{ \frac{1}{j} \int_0^j f(t, u' + \xi)\, dt \;:\; u \text{ } j\text{-periodic} \right\}.
\end{aligned}
$$

We now show that indeed the minimum values in the last limit above do not depend on j, using the convexity of f. First, note that trivially

$$
\inf\left\{ \frac{1}{j} \int_0^j f(t, u' + \xi)\, dt \;:\; u \text{ } j\text{-periodic} \right\} \le \inf\left\{ \int_0^1 f(t, u' + \xi)\, dt \;:\; u \text{ } 1\text{-periodic} \right\}.
$$

On the other hand, if u is j-periodic, then we can define the convex combination of all possible different integer translations of u

$$
v(t) = \sum_{i=0}^{j-1} \frac{1}{j} u(t + i).
$$

The function v is 1-periodic. Using v as a test function we see by the convexity of f that

$$
\begin{aligned}
\inf\left\{ \int_0^1 f(t, w' + \xi)\, dt \;:\; w \text{ } 1\text{-periodic} \right\} &\le \int_0^1 f(t, v' + \xi)\, dt \\
&= \frac{1}{j} \int_0^j f(t, v' + \xi)\, dt \\
&= \frac{1}{j} \int_0^j f\left(t, \sum_{i=0}^{j-1} \frac{1}{j}(u'(t + i) + \xi)\right) dt \\
&\le \frac{1}{j} \sum_{i=0}^{j-1} \frac{1}{j} \int_0^j f(t, u'(t + i) + \xi)\, dt \\
&= \frac{1}{j} \sum_{i=0}^{j-1} \frac{1}{j} \int_0^j f(t, u' + \xi)\, dt \\
&= \frac{1}{j} \int_0^j f(t, u' + \xi)\, dt.
\end{aligned}
$$

Hence, taking the infimum on all j-periodic u we have

$$\inf\left\{\int_0^1 f(t, u' + \xi)\,dt \,:\, u \text{ 1-periodic}\right\} \le \inf\left\{\frac{1}{j}\int_0^j f(t, u' + \xi)\,dt \,:\, u \text{ } j\text{-periodic}\right\},$$

so that

$$\inf\left\{\int_0^1 f(t, u' + \xi)\,dt \,:\, u \text{ 1-periodic}\right\}$$
$$= \inf\left\{\frac{1}{j}\int_0^j f(t, u' + \xi)\,dt \,:\, u \text{ } j\text{-periodic}\right\}. \tag{13.5}$$

The guess for the 'homogenized function' f_0 is then given by the *cell-problem formula*

$$f_0(\xi) = \min\left\{\int_0^1 f(t, u' + \xi)\,dt \,:\, u \text{ 1-periodic}\right\}. \tag{13.6}$$

13.2 The asymptotic homogenization formula

We can give another formula for f_0. This formula will be easier to handle in the proof of the Γ-convergence, and, above all, can be easily extended to the non-convex vector case, for which a cell-problem formula does not hold.

We introduce the notation

$$g_\xi(T) = \min\left\{\frac{1}{T}\int_0^T f(t, u' + \xi)\,dt \,:\, u(0) = u(T) = 0\right\}. \tag{13.7}$$

We can use as test functions for $g_\xi(T)$ functions satisfying $u(0) = u([T]) = 0$ extended to 0 on $([T], T)$, and as test functions for $g_\xi([T]+1)$ functions satisfying $u(0) = u(T) = 0$ extended to 0 on $(T, [T] + 1)$, obtaining, thanks to (13.3),

$$\frac{[T]+1}{T}g_\xi([T] + 1) - \beta\frac{1}{T}(1 + |\xi|^p) \le g_\xi(T) \le \frac{[T]}{T}g_\xi([T]) + \beta\frac{1}{T}(1 + |\xi|^p).$$

From (13.5) and (13.6) we have $g_\xi(j) = f_0(\xi)$ for all $j \in \mathbf{N}$, so that we have the error estimate

$$|g_\xi(T) - f_0(\xi)| \le \beta\frac{1}{T}(1 + |\xi|^p),$$

and we obtain the *asymptotic homogenization formula*

$$f_0(\xi) = \lim_{T \to +\infty} \min\left\{\frac{1}{T}\int_0^T f(t, u' + \xi)\,dt \,:\, u(0) = u(T) = 0\right\}. \tag{13.8}$$

Note that if we define for all $x \in \mathbf{R}$

$$g_\xi^x(T) = \min\left\{\frac{1}{T}\int_x^{x+T} f(t, u' + \xi)\,dt \,:\, u(x) = u(x + T)\right\}, \tag{13.9}$$

then the same error estimate holds

$$|g_\xi^x(T) - f_0(\xi)| \le \beta \frac{1}{T}(1 + |\xi|^p), \tag{13.10}$$

uniformly in x.

13.3   Proof of the Γ-convergence

We can conjecture that if we set $F_0(u) = \int_0^1 f_0(u')\,dt$ then we have (13.4) for all $u \in W^{1,p}(0,1)$.

We remark that by the local Lipschitz continuity of F_0 on $W^{1,p}(0,1)$, it suffices to prove (13.4) for a set X in $W^{1,p}(0,1)$ which is dense with respect to the strong $W^{1,p}$ metric, say on piecewise affine functions. In fact if $u \in W^{1,p}(0,1)$, and $(u_j) \subset X$ converges to u strongly in $W^{1,p}(0,1)$, then by the lower semicontinuity of the Γ-lim sup we have

$$\Gamma(L^p)\text{-}\limsup_j F_{\varepsilon_j}(u) \le \liminf_k \Gamma(L^p)\text{-}\lim_j F_{\varepsilon_j}(u_k)$$

$$= \lim_k \int_0^1 f_0(u_k')\,dt = \int_0^1 f_0(u')\,dt.$$

Conversely, if $v_j \in W^{1,p}(0,1)$, $v_j \to u$ in $L^p(0,1)$, and

$$\Gamma(L^p)\text{-}\liminf_j F_{\varepsilon_j}(u) = \liminf_j F_{\varepsilon_j}(v_j)$$

then we have, recalling that $|f(x,\xi) - f(x,\eta)| \le c(1 + |\xi|^{p-1} + |\eta|^{p-1})|\xi - \eta|$ by the convexity and the growth conditions on f,

$$\Gamma(L^p)\text{-}\liminf_j F_{\varepsilon_j}(u)$$

$$\ge \liminf_k \left(\liminf_j F_{\varepsilon_j}(v_j - u + u_k) + \liminf_j (F_{\varepsilon_j}(v_j) - F_{\varepsilon_j}(v_j - u + u_k)) \right)$$

$$\ge \liminf_k \left(\int_0^1 f_0(u_k')\,dt \right.$$

$$- c \limsup_j \left(1 + \int_0^1 |u_k'|^p dt + \int_0^1 |u'|^p dt + \int_0^1 |v_j'|^p dt \right)^{(p-1)/p}$$

$$\left. \times \left(\int_0^1 |u' - u_k'|^p dt \right)^{\frac{1}{p}} \right)$$

$$= \lim_k \int_0^1 f_0(u_k')\,dt = \int_0^1 f_0(u')\,dt,$$

so that (13.4) is proven on all $W^{1,p}(0,1)$.

We now prove (13.4) for $u \in W^{1,p}(0,1)$ piecewise affine. Let $a_0 = 0 < a_1 < \cdots < a_N = 1$, and let $u(t) = m_k t + q_k$ on (a_{k-1}, a_k). Let v_k be a 1-periodic function such that $v_k(0) = 0$ and

$$\int_0^1 f(t, v_k' + m_k)\,dt = f_0(m_k),$$

and let $b_k^j = \varepsilon_j \left[\frac{a_k}{\varepsilon_j} \right]$. We define

$$u_j(t) = \begin{cases} \varepsilon_j v_k(\frac{t}{\varepsilon_j}) + m_k t + q_k & \text{on } (b_{k-1}^j + \varepsilon_j, b_k^j), \ k = 1, \dots, N, \\ u(t) & \text{if } t \in (0,1) \setminus \bigcup_{k=1}^N (b_{k-1}^j + \varepsilon_j, b_k^j). \end{cases}$$

We have $u_j \in W^{1,p}(0,1)$, $u_j \to u$ in $L^p(0,1)$. Using Example 2.7 we get

$$\limsup_j \int_0^1 f\left(\frac{t}{\varepsilon_j}, u_j'\right) dt = \limsup_j \sum_{k=1}^N \int_{b_{k-1}^j + \varepsilon_j}^{b_k^j} f\left(\frac{t}{\varepsilon_j}, v_k'\left(\frac{t}{\varepsilon_j}\right) + m_k\right) dt$$

$$\leq \sum_{k=1}^N \lim_j \int_{a_{k-1}}^{a_k} f\left(\frac{t}{\varepsilon_j}, v_k'\left(\frac{t}{\varepsilon_j}\right) + m_k\right) dt$$

$$= \sum_{k=1}^N (a_k - a_{k-1}) f_0(m_k) = \int_0^1 f_0(u') \, dt;$$

hence

$$\Gamma(L^p)\text{-}\limsup_j F_{\varepsilon_j}(u) \leq \int_0^1 f_0(u') \, dt.$$

It remains to prove that for u piecewise affine as above

$$\Gamma(L^p)\text{-}\liminf_j F_{\varepsilon_j}(u) \geq \int_0^1 f_0(u') \, dt.$$

Let $u_j \in W^{1,p}(0,1)$, $u_j \to u$ in $L^p(0,1)$. We now modify the values of u_j near a_k for all k, so that they match the values of u. Consider the function $v : (0,1) \to \mathbf{R}$ defined by

$$v(t) = (t - a_{k-1}) \wedge (a_k - t) \qquad \text{on } [a_{k-1}, a_k],$$

and define

$$v_j = \big((u_j - u) \wedge v\big) \vee \big((u_j - u) \vee (-v)\big).$$

Note that $v_j \to 0$ in $L^p(0,1)$, and that

$$\liminf_j F_{\varepsilon_j}(u + v_j) \leq \liminf_j F_{\varepsilon_j}(u_j)$$

(see Exercise 11.1). On the other hand

$$F_{\varepsilon_j}(u + v_j) = \sum_{k=1}^N \int_{a_{k-1}}^{a_k} f\left(\frac{t}{\varepsilon_j}, v_j' + u'\right) dt$$

$$= \sum_{k=1}^N \varepsilon_j \int_{a_{k-1}/\varepsilon_j}^{a_k/\varepsilon_j} f(t, v_j'(t\varepsilon_j) + m_k) \, dt.$$

If we set $T_j^k = \frac{1}{\varepsilon_j}(a_k - a_{k-1})$, $x_j^k = \frac{1}{\varepsilon_j}a_{k-1}$, in the notation of (13.9) we have

$$F_{\varepsilon_j}(u + v_j) \geq \sum_{k=1}^{N}(a_k - a_{k-1})g_{m_k}^{x_j^k}(T_j^k),$$

so that, by (13.10),

$$\liminf_j F_{\varepsilon_j}(u_j) \geq \liminf_j F_{\varepsilon_j}(u + v_j)$$
$$\geq \sum_{k=1}^{N}(a_k - a_{k-1})\, f_0(m_k) = \int_0^1 f_0(u')\, dt.$$

By the arbitrariness of the sequence (u_j) the desired inequality is proven.

Note that in the proof above two key arguments of the direct methods of Γ-convergence are present, namely the density of affine functions (which is crucial, for example, in the Integral Representation Theorem), and the problem of modifying sequences of functions to match proper boundary values without introducing a large error (which is the object of the fundamental estimate).

13.4 Exercises

Exercise 13.1 Prove that if $f(x, z) = a(x)|z|^p$ with a 1-periodic and strictly positive then $f_0(z) = \left(\int_0^1 a^{1/(1-p)}\, dt\right)^{1-p}|z|^p$. Note that if $p = 2$ then the value $f_0(1) = \left(\int_0^1 \frac{1}{a}\, dt\right)^{-1}$ is the *harmonic mean* of a.

Hint: f_0 is positively homogeneous of degree p; hence, it suffices to compute $f_0(1) = \min\{\int_0^1 a|u'|^p\, dt : u(0) = 0,\ u(1) = 1\}$. Use Euler's equation $(a|u'|^{p-2}u')' = 0$ to compute the unique solution of this minimum problem.

Exercise 13.2 Let $G : \mathbf{R} \to \mathbf{R}$ be defined by

$$G(x) = \begin{cases} \alpha & \text{if } n < x \leq n + \frac{1}{2},\ n \in \mathbf{N} \\[2mm] \beta & \text{if } n - \frac{1}{2} < x \leq n,\ n \in \mathbf{N} \end{cases}$$

and $\alpha, \beta > 0$. If $f(x, z) = G(x)z^2$ then $f_{\mathrm{hom}} = \varphi_G$, where

$$\varphi_G(z) = \frac{2\alpha\beta}{\alpha + \beta}z^2 = \left(\frac{1}{2}\left(\frac{1}{\alpha} + \frac{1}{\beta}\right)\right)^{-1}z^2.$$

(This is a special case of the previous exercise.)

Exercise 13.3 Let G be defined as in Exercise 13.2 and let $k \in \mathbf{N}$. Compute the homogenized integrand f_{hom}^k related to $f(x, z) = f^k(x, z) = \left(G(x) + G(kx)\right)z^2$. Compute the limit $\lim_k f_{\mathrm{hom}}^k$.

Hint: use the cell-problem formula and Exercise 13.1 to get

$$
f^k_{\mathrm{hom}}(z) =
\begin{cases}
\left(\dfrac{1}{2(\alpha+\beta)} + \dfrac{1}{8\alpha} + \dfrac{1}{8\beta}\right)^{-1} z^2 & \text{if } k = 2j \\[2ex]
\left(\dfrac{j}{2j+1}\left(\dfrac{1}{\alpha+\beta}\right) + \dfrac{j+1}{4(2j+1)}\left(\dfrac{1}{\alpha} + \dfrac{1}{\beta}\right)\right)^{-1} z^2 & \text{if } k = 2j+1.
\end{cases}
$$

Exercise 13.4 Let $\mathbf{W}$ be defined as in Exercise 2.3, and for every $u \in \mathbf{W}$ let $F_\varepsilon(u) = \int_0^1 f(\frac{t}{\varepsilon}, u')\,dt + \sum_{t \in S(u)} g(\frac{t}{\varepsilon})$, where f is as above with $p = 2$ and g is a continuous 1-periodic function. Prove that the Γ-limit of F_ε in the $L^1(0,1)$-topology as $\varepsilon \to 0$ is given on $\mathbf{W}$ by $\int_0^1 f_0(u')\,dt + \min g\,\#(S(u))$.

Hint: if $u_j \to u$ in $L^1(0,1)$ and $\lim_j F_{\varepsilon_j}(u_j) < +\infty$, let $v_j \in \mathrm{W}^{1,2}(0,1)$ be such that $v_j' = u_j'$. Note that we can suppose that (v_j) is bounded in $\mathrm{W}^{1,2}(0,1)$; hence, up to a subsequence, $v_j \rightharpoonup v$ and $v' = u'$ by Exercise 2.3. Use the Γ-convergence result in Section 13.3 and the fact that $\#(S(u)) \le \liminf_j \#(S(u_j))$ (again by Exercise 2.3) to obtain the $\liminf$ inequality.

To construct a recovery sequence for the Γ-$\limsup$, with fixed $u \in \mathbf{W}$ choose $v \in \mathrm{W}^{1,2}(0,1)$ such that $v' = u'$, and apply the Γ-convergence result in Section 13.3 to find a recovery sequence (v_ε) for v. Construct a sequence (u_ε) in $\mathbf{W}$ converging to u with $u_\varepsilon' = v_\varepsilon'$ and $g(\frac{t}{\varepsilon}) = \min g$ for all $t \in S(u_\varepsilon)$: if u has a jump in t_0 introduce a jump of the same size for u_ε at a minimum point of $t \mapsto g(\frac{t}{\varepsilon})$ whose distance to t_0 is not greater than ε.

14

PERIODIC HOMOGENIZATION

In this chapter we use the direct methods of Γ-convergence to obtain a homogenization theorem for functionals

$$F_\varepsilon(u) = \int_\Omega f\left(\frac{x}{\varepsilon}, Du(x)\right)\, dx,$$

where the function f satisfies a standard growth condition of order $p \geq 1$ and $u \in W^{1,p}(\Omega; \mathbf{R}^m)$. In order to describe our results better we introduce the following notation.

Definition 14.1 *Let $(F_\varepsilon)_{\varepsilon>0}$ be a family of functionals $F_\varepsilon : L^p(\Omega; \mathbf{R}^m) \to \mathbf{R}$; we say that*

$$F_0 = \Gamma(L^p)\text{-}\lim_{\varepsilon \to 0} F_\varepsilon$$

if we have

$$F_0 = \Gamma(L^p)\text{-}\lim_j F_{\varepsilon_j}$$

for every sequence (ε_j) of positive real numbers converging to 0.

Let $f : \mathbf{R}^n \times \mathbf{M}^{m \times n} \to [0, +\infty)$ be a Borel function satisfying the following conditions:

(i) (periodicity)

$$f(\cdot, A) \text{ is 1-periodic for all } A \in \mathbf{M}^{m \times n}, \tag{14.1}$$

i.e. $f(x + e_i, A) = f(x, A)$ for all $x \in \mathbf{R}^n$, $A \in \mathbf{M}^{m \times n}$ and $i = 1, \ldots, n$;

(ii) (standard growth condition of order p) there exist $0 < \alpha \leq \beta$ such that

$$\alpha|A|^p \leq f(x, A) \leq \beta(1 + |A|^p) \tag{14.2}$$

for all $x \in \mathbf{R}^n$ and $A \in \mathbf{M}^{m \times n}$.

We localize our functionals in the usual way, setting

$$F_\varepsilon(u, \Omega) = \int_\Omega f\left(\frac{x}{\varepsilon}, Du(x)\right)\, dx \tag{14.3}$$

if Ω is a bounded open subset of $\mathbf{R}^n$, and $u \in W^{1,p}(\Omega; \mathbf{R}^m)$. The functionals can be extended to $+\infty$ on $L^p(\Omega; \mathbf{R}^m) \setminus W^{1,p}(\Omega; \mathbf{R}^m)$. We will study their behaviour only on $W^{1,p}(\Omega; \mathbf{R}^m)$; note that in the case $p \geq 1$, by the growth estimate (14.2), this gives a complete picture since the Γ-limits will still be equal to $+\infty$ on $L^p(\Omega; \mathbf{R}^m) \setminus W^{1,p}(\Omega; \mathbf{R}^m)$. In the case $p = 1$ in general we will limit our analysis to $W^{1,1}(\Omega; \mathbf{R}^m)$ since a complete description would require the study of the Γ-limits in the space of functions of bounded variation on Ω.

By Theorem 12.5 we have the following compactness result.

Theorem 14.2 *For every sequence (ε_j) of positive real numbers converging to 0 there exist a subsequence (ε_{j_k}) and a Carathéodory function $\varphi : \mathbf{R}^n \times \mathbf{M}^{m \times n} \to [0, +\infty)$ satisfying*

$$\alpha|A|^p \leq \varphi(x, A) \leq \beta(1 + |A|^p) \tag{14.4}$$

for all $x \in \mathbf{R}^n$ and $A \in \mathbf{M}^{m \times n}$, such that for all bounded open sets Ω in $\mathbf{R}^n$

$$\int_\Omega \varphi(x, Du)\, dx = \Gamma(\mathrm{L}^p)\text{-}\lim_k F_{\varepsilon_{j_k}}(u, \Omega) \tag{14.5}$$

for all $u \in W^{1,p}(\Omega; \mathbf{R}^m)$.

Proof With fixed Ω we can apply Theorem 12.5, and obtain the equality (14.5) for all $u \in W^{1,p}(\Omega; \mathbf{R}^m)$. The locality property of the Γ-limit shows that φ is indeed independent of Ω. $\qquad\square$

14.1 The asymptotic homogenization formula

In order to show that the whole family (F_ε) Γ-converges, we will prove that φ is independent of the first variable, and can be expressed by an asymptotic formula which does not depend on (ε_{j_k}).

Proposition 14.3 *The function φ in Theorem 14.2 can be chosen independent of the first variable.*

Proof By Proposition 9.2 it is sufficient to prove that if $A \in \mathbf{M}^{m \times n}$, $y, z \in \mathbf{R}^n$ and $\rho > 0$ then

$$\Gamma(\mathrm{L}^p)\text{-}\lim_k F_{\varepsilon_{j_k}}(Ax, B_\rho(y)) = \Gamma(\mathrm{L}^p)\text{-}\lim_k F_{\varepsilon_{j_k}}(Ax, B_\rho(z)).$$

By Proposition 11.7 there exists a sequence $(u_k) \subset W_0^{1,p}(B_\rho(y); \mathbf{R}^m)$ such that $u_k \to 0$ in $L^p(B_\rho(y); \mathbf{R}^m)$ and

$$\lim_k F_{\varepsilon_{j_k}}(Ax + u_k, B_\rho(y)) = \Gamma(\mathrm{L}^p)\text{-}\lim_k F_{\varepsilon_{j_k}}(Ax, B_\rho(y)).$$

We extend u_k to $\mathbf{R}^n$ by 0 outside $B_\rho(y)$. Let $r > 1$, let $\tau_k \in \mathbf{R}^n$ be given by

$$(\tau_k)_i = \varepsilon_{j_k}\left\lfloor \frac{z_i - y_i}{\varepsilon_{j_k}} \right\rfloor$$

and let $v_k(x) = u_k(x - \tau_k)$. Note that $\tau_k \to z - y$ and τ_k is a period for $x \mapsto f(x/\varepsilon_{j_k}, A)$ for all A, so that

$$F_{\varepsilon_{j_k}}(Ax + v_k, \tau_k + B_\rho(y)) = F_{\varepsilon_{j_k}}(Ax + u_k, B_\rho(y)).$$

Moreover, $v_k = 0$ outside $\tau_k + B_\rho(y)$. We have $v_k \to 0$ in $L^p(B_{r\rho}(z); \mathbf{R}^m)$; hence,

$$\begin{aligned}
&\Gamma(L^p)\text{-}\lim_k F_{\varepsilon_{j_k}}(Ax, B_\rho(z))\\
&\leq \Gamma(L^p)\text{-}\lim_k F_{\varepsilon_{j_k}}(Ax, B_{r\rho}(z))\\
&\leq \liminf_k F_{\varepsilon_{j_k}}(Ax + v_k, B_{r\rho}(z))\\
&\leq \liminf_k F_{\varepsilon_{j_k}}(Ax + u_k, B_\rho(y)) + |B_{r\rho} \setminus B_\rho|\beta(1 + |A|^p)\\
&= \Gamma(L^p)\text{-}\lim_k F_{\varepsilon_{j_k}}(Ax, B_\rho(y)) + |B_{r\rho} \setminus B_\rho|\beta(1 + |A|^p).
\end{aligned}$$

Letting $r \to 1$ we obtain the inequality

$$\Gamma(L^p)\text{-}\lim_k F_{\varepsilon_{j_k}}(Ax, B_\rho(z)) \leq \Gamma(L^p)\text{-}\lim_k F_{\varepsilon_{j_k}}(Ax, B_\rho(y));$$

the opposite inequality is obtained by a symmetry argument. $\qquad\square$

By Proposition 4.3 the function φ is $W^{1,p}$-quasiconvex, which is equivalent to quasiconvexity by Remark 5.15. Hence, we have

$$\varphi(A) = \min\left\{\int_{(0,1)^n} \varphi(A + Du(y))\, dy : \ u \in W_0^{1,p}((0,1)^n; \mathbf{R}^m)\right\} \tag{14.6}$$

for all $A \in \mathbf{M}^{m\times n}$. This remark, together with the following proposition, will give the Homogenization Theorem.

Proposition 14.4 *Let $f : \mathbf{R}^n \times \mathbf{M}^{m\times n} \to [0, +\infty)$ be a Borel function satisfying the periodicity condition (14.1) and such that $\sup\{f(x, A) : \ x \in \mathbf{R}^n\}$ is finite for all $A \in \mathbf{M}^{m\times n}$; then the limit*

$$\lim_{t \to +\infty} \frac{1}{t^n} \inf\left\{\int_{(0,t)^n} f(x, A + Du(x))\, dx : \ u \in W_0^{1,p}((0,t)^n; \mathbf{R}^m)\right\} \tag{14.7}$$

exists for all $A \in \mathbf{M}^{m\times n}$.

Proof Fix $A \in \mathbf{M}^{m\times n}$ and define for $t > 0$

$$g_t = \frac{1}{t^n} \inf\left\{\int_{(0,t)^n} f(x, A + Du(x))\, dx : \ u \in W_0^{1,p}((0,t)^n; \mathbf{R}^m)\right\}; \tag{14.8}$$

moreover, let $u_t \in W_0^{1,p}((0,t)^n; \mathbf{R}^m)$ satisfy

$$\frac{1}{t^n} \int_{(0,t)^n} f(x, A + Du_t(x))\, dx \leq g_t + \frac{1}{t}. \tag{14.9}$$

Let $s > t$; we can construct $u_s \in W_0^{1,p}((0,s)^n; \mathbf{R}^m)$ as follows. Let I be the set of indices $\mathbf{i} = (i_1, \ldots, i_n) \in \mathbf{Z}^n$ with $0 \leq ([t] + 1)i_j < s$ for all $j = 1, \ldots, n$; we define

$$u_s(x) = \begin{cases} u_t(x - \mathbf{i}) & \text{if } x \in \mathbf{i} + (0,t)^n, \ \mathbf{i} \in I \\ 0 & \text{otherwise.} \end{cases} \tag{14.10}$$

We also define $Q_s = (0,s)^n \setminus \bigcup_{\mathbf{i} \in I}(\mathbf{i} + (0,t)^n)$; we have

$$|Q_s| \le s^n - (s - t - 1)^n \left(\frac{t}{t+1}\right)^n.$$

We can estimate g_s by using u_s as a test function:

$$
\begin{aligned}
g_s &\le \frac{1}{s^n} \int_{(0,s)^n} f(x, A + Du_s(x)) \, dx \\
&= \frac{1}{s^n} \left(\sum_{\mathbf{i} \in I} \int_{\mathbf{i} + (0,t)^n} f(x, A + Du_t(x - \mathbf{i})) \, dx + \int_{Q_s} f(x, A) \, dx \right) \\
&\le \frac{1}{s^n} \left(\sum_{\mathbf{i} \in I} \int_{(0,t)^n} f(y + \mathbf{i}, A + Du_t(y)) \, dy + |Q_s| c \right) \\
&\le \frac{1}{s^n} \left(\sum_{\mathbf{i} \in I} \int_{(0,t)^n} f(y, A + Du_t(y)) \, dy + |Q_s| c \right) \\
&\le \frac{t^n}{([t]+1)^n} \left(g_t + \frac{1}{t} \right) + \left(1 - \left(\frac{s - t - 1}{s} \right)^n \left(\frac{t}{t+1} \right)^n \right) c.
\end{aligned}
$$

Taking first the upper limit as $s \to +\infty$ and then the lower limit as $t \to +\infty$ we get

$$\limsup_{s \to +\infty} g_s \le \liminf_{t \to +\infty} g_t,$$

and we conclude the proof. $\qquad\square$

14.2 The Homogenization Theorem

By applying the previous results we prove the following theorem.

Theorem 14.5 *Let $f : \mathbf{R}^n \times \mathbf{M}^{m \times n} \to [0, +\infty)$ be a Borel function satisfying the periodicity assumption (14.1) and the standard growth condition (14.2) of order $p \ge 1$. If Ω is a bounded open set of $\mathbf{R}^n$ and we set for all $\varepsilon > 0$*

$$F_\varepsilon(u) = \int_\Omega f\left(\frac{x}{\varepsilon}, Du(x)\right) \, dx \tag{14.11}$$

for all $u \in W^{1,p}(\Omega; \mathbf{R}^m)$, then we have

$$\Gamma(L^p)\text{-}\lim_{\varepsilon \to 0} F_\varepsilon(u) = \int_\Omega f_{\mathrm{hom}}(Du(x)) \, dx,$$

for all $u \in W^{1,p}(\Omega; \mathbf{R}^m)$, where $f_{\mathrm{hom}} : \mathbf{M}^{m \times n} \to [0, +\infty)$ is a quasiconvex function satisfying the asymptotic homogenization formula

$$f_{\text{hom}}(A) = \lim_{t \to +\infty} \frac{1}{t^n} \inf\left\{ \int_{(0,t)^n} f(x, A + Du(x))\, dx : \ u \in W_0^{1,p}((0,t)^n; \mathbf{R}^m) \right\}$$

$$(14.12)$$

for all $A \in \mathbf{M}^{m \times n}$

Proof Let (ε_{j_k}) be given by Theorem 14.2, and let φ be as in Proposition 14.3. In the case $p > 1$ a straightforward proof is given by applying Theorem 7.2 to the restriction of these functionals to $Ax + W_0^{1,p}((0,1)^n; \mathbf{R}^m)$, recalling Proposition 11.7. That is, by (14.6),

$$\varphi(A) = \min\left\{ \int_{(0,1)^n} \varphi(A + Du(y))\, dy : \ u \in W_0^{1,p}((0,1)^n; \mathbf{R}^m) \right\}$$

$$= \lim_k \inf\left\{ \int_{(0,1)^n} f\left(\frac{y}{\varepsilon_{j_k}}, A + Du(y) \right) dy : \ u \in W_0^{1,p}((0,1)^n; \mathbf{R}^m) \right\}$$

$$= \lim_k \inf\left\{ \frac{1}{T_k^n} \int_{(0,T_k)^n} f(x, A + Du(x))\, dx : \ u \in W_0^{1,p}((0,T_k)^n; \mathbf{R}^m) \right\},$$

where $T_k = 1/\varepsilon_{j_k}$. By Proposition 14.4 the proof is concluded.

An alternative proof, valid also for $p = 1$, can be obtained by a slightly modified argument as follows. By Proposition 11.7 there exists a sequence (u_k) in $W_0^{1,p}((0,1)^n; \mathbf{R}^m)$ converging to 0 in $L^p((0,1)^n; \mathbf{R}^m)$ such that

$$\varphi(A) = \lim_k \int_{(0,1)^n} f\left(\frac{x}{\varepsilon_{j_k}}, A + Du_k \right) dx$$

$$\geq \lim_k \varepsilon_{j_k}^n \inf\left\{ \int_{(0,1/\varepsilon_{j_k})^n} f(x, A + Du(x))\, dx : \ u \in W_0^{1,p}\left(\left(0, \frac{1}{\varepsilon_{j_k}}\right)^n; \mathbf{R}^m \right) \right\}$$

$$= f_{\text{hom}}(A).$$

To prove the converse inequality, for all $l \in \mathbf{N}$ let $v_l \in W_0^{1,p}((0,l)^n; \mathbf{R}^m)$ satisfy

$$\int_{(0,l)^n} f(x, A + Dv_l(x))\, dx$$

$$\leq \inf\left\{ \int_{(0,l)^n} f(x, A + Du(x))\, dx : \ u \in W_0^{1,p}((0,l)^n; \mathbf{R}^m) \right\} + 1.$$

Extend v_l by periodicity to the whole $\mathbf{R}^n$, and define $u_k(x) = \varepsilon_{j_k} v_l(x/\varepsilon_{j_k})$. Note that $u_k \to 0$ in $L^p((0,1)^n; \mathbf{R}^m)$. We then get

$$\varphi(A) = \int_{(0,1)^n} \varphi(A)\, dx \leq \liminf_k \int_{(0,1)^n} f\left(\frac{x}{\varepsilon_{j_k}}, A + Du_k \right) dx$$

$$= \liminf_k \int_{(0,1)^n} f\left(\frac{x}{\varepsilon_{j_k}}, A + Dv_l\left(\frac{x}{\varepsilon_{j_k}} \right) \right) dx$$

$$= \frac{1}{l^n} \int_{(0,l)^n} f(x, A + Dv_l(x))\, dx,$$

the last equality being given by Example 2.7. We then have

$$\varphi(A) \leq \frac{1}{l^n} \inf\left\{ \int_{(0,l)^n} f(x, A + Du(x))\, dx : u \in W_0^{1,p}((0,l)^n; \mathbf{R}^m)\right\} + \frac{1}{l^n},$$

and the desired inequality follows by taking the limit as $l \to +\infty$. $\qquad\square$

Remark 14.6 (*An asymptotic formula on periodic functions*) The function f_{hom} in Theorem 14.5 also satisfies

$$f_{\text{hom}}(A) = \inf_{j \in \mathbf{N}} \inf\left\{ \frac{1}{j^n} \int_{(0,j)^n} f(x, A + Du(x))\, dx : u \in W_\#^{1,p}((0,j)^n; \mathbf{R}^m)\right\},$$
$$(14.13)$$

where

$$W_\#^{1,p}((0,j)^n; \mathbf{R}^m) = \{u \in W_{\text{loc}}^{1,p}(\mathbf{R}^n; \mathbf{R}^m) : u \ j\text{-periodic}\} \qquad (14.14)$$

is the closure of smooth j-periodic functions in $W_{\text{loc}}^{1,p}(\mathbf{R}^n; \mathbf{R}^m)$.

Note that (14.13) is equivalent to

$$f_{\text{hom}}(A) = \lim_j \inf\left\{ \int_{(0,1)^n} f(jx, A + Du(x))\, dx : u \in W_\#^{1,p}((0,1)^n; \mathbf{R}^m)\right\}.$$
$$(14.15)$$

Indeed, it can be immediately seen that the limit in (14.15) is an infimum since

$$\inf\left\{ \int_{(0,1)^n} f(kjx, A + Du(x))\, dx : u \in W_\#^{1,p}((0,1)^n; \mathbf{R}^m)\right\}$$
$$\leq \inf\left\{ \int_{(0,1)^n} f(jx, A + Du(x))\, dx : u \in W_\#^{1,p}((0,1)^n; \mathbf{R}^m)\right\}$$

for all $k = 1, 2, \ldots$. This formula follows by a change of variables.

Clearly (14.15) is proved if we show that

$$\varphi(A) \leq \lim_j \inf\left\{ \int_{(0,1)^n} f(jx, A + Du(x))\, dx : u \in W_\#^{1,p}((0,1)^n; \mathbf{R}^m)\right\}$$
$$= \lim_j \inf\left\{ \frac{1}{j^n} \int_{(0,j)^n} f(x, A + Du(x))\, dx : u \in W_\#^{1,p}((0,j)^n; \mathbf{R}^m)\right\},$$

since the opposite inequality follows by the inclusion of $W_0^{1,p}((0,1)^n; \mathbf{R}^m)$ in $W_\#^{1,p}((0,1)^n; \mathbf{R}^m)$. This follows as in the last part of the proof of Theorem 14.5, choosing for every $l \in \mathbf{N}$ $v_l \in W_\#^{1,p}((0,l)^n; \mathbf{R}^m)$ satisfying

$$\int_{(0,l)^n} f(x, A + Dv_l(x))\, dx$$

$$\leq \inf\left\{ \int_{(0,l)^n} f(x, A + Du(x))\, dx : \; u \in W^{1,p}_{\#}((0,l)^n; \mathbf{R}^m) \right\} + 1.$$

Note that in the case $p > 1$ the proof becomes easier. Remarking that, by quasiconvexity and Remark 6.11,

$$f_{\mathrm{hom}}(A) = \inf\left\{ \int_{(0,1)^n} f_{\mathrm{hom}}(A + Du(x))\, dx : \; u \in W^{1,p}_{\#}((0,1)^n; \mathbf{R}^m) \right\},$$

we immediately obtain (14.15) by Theorem 7.2. In fact, by Proposition 11.7 we deduce that the functionals

$$G_j(u) = \begin{cases} \displaystyle\int_{(0,1)^n} f(jx, A + Du)\, dx & \text{if } u \in W^{1,p}_{\#}((0,1)^n; \mathbf{R}^m) \\[2mm] +\infty & \text{otherwise} \end{cases}$$

Γ-converge to the functional

$$G_{\mathrm{hom}}(u) = \begin{cases} \displaystyle\int_{(0,1)^n} f_{\mathrm{hom}}(A + Du)\, dx & \text{if } u \in W^{1,p}_{\#}((0,1)^n; \mathbf{R}^m) \\[2mm] +\infty & \text{otherwise.} \end{cases}$$

14.3 Convex homogenization

14.3.1 *The cell-problem formula*

In the convex case we can give an alternative formula for the 'homogenized' function f_{hom}, which consists of a single periodic minimization problem.

Theorem 14.7 *Let $f : \mathbf{R}^n \times \mathbf{M}^{m\times n} \to [0, +\infty)$ be a Borel function satisfying the periodicity assumption (14.1), the standard growth condition (14.2) of order $p \geq 1$, and in addition let $f(x, \cdot)$ be convex for all $x \in \mathbf{R}^n$. Then the conclusions of Theorem 14.5 hold with $f_{\mathrm{hom}} : \mathbf{M}^{m\times n} \to [0, +\infty)$ given by the cell-problem formula*

$$f_{\mathrm{hom}}(A) = \inf\left\{ \int_{(0,1)^n} f(y, A + Du(y))\, dy : \; u \in W^{1,p}_{\#}((0,1)^n; \mathbf{R}^m) \right\} \quad (14.16)$$

for all $A \in \mathbf{M}^{m\times n}$.

Proof Note as in Remark 14.6 that

$$f_{\mathrm{hom}}(A) = \lim_j \inf\left\{ \frac{1}{j^n} \int_{(0,j)^n} f(y, A + Du(y))\, dy : \; u \in W^{1,p}_{\#}((0,j)^n; \mathbf{R}^m) \right\}.$$

Trivially

$$\inf\left\{ \frac{1}{j^n} \int_{(0,j)^n} f(y, A + Du(y))\, dy : \; u \in W^{1,p}_{\#}((0,j)^n; \mathbf{R}^m) \right\}$$

$$\leq \inf\left\{ \int_{(0,1)^n} f(y, A + Du(y))\, dy : \ u \in \mathrm{W}^{1,p}_{\#}((0,1)^n; \mathbf{R}^m) \right\}.$$

On the other hand, if $v \in \mathrm{W}^{1,p}_{\#}((0,j)^n; \mathbf{R}^m)$ then we can define the convex combination

$$u(x) = \sum_{\mathbf{i} \in I} \frac{1}{j^n} v(x + \mathbf{i}),$$

where $I = \{0, 1, \ldots, j-1\}^n$. The function u belongs to $\mathrm{W}^{1,p}_{\#}((0,1)^n; \mathbf{R}^m)$. Moreover, by the convexity and the periodicity of f,

$$\begin{aligned}
\int_{(0,1)^n} f(x, A + Du)\, dx &= \frac{1}{j^n} \int_{(0,j)^n} f(x, A + Du)\, dx \\
&= \frac{1}{j^n} \int_{(0,j)^n} f\left(x, \sum_{\mathbf{i} \in I} \frac{1}{j^n}\big(4 + Dv(x + \mathbf{i})\big)\right) dx \\
&\leq \frac{1}{j^n} \sum_{\mathbf{i} \in I} \frac{1}{j^n} \int_{(0,j)^n} f(x, A + Dv(x + \mathbf{i}))\, dx \\
&= \frac{1}{j^n} \sum_{\mathbf{i} \in I} \frac{1}{j^n} \int_{(0,j)^n} f(x, A + Dv(x))\, dx \frac{1}{j^n} \int_{(0,j)^n} f(x, A + Dv(x))\, dx,
\end{aligned}$$

so that

$$\begin{aligned}
\inf\ & \left\{ \int_{(0,1)^n} f(y, A + Du(y))\, dy : \ u \in \mathrm{W}^{1,p}_{\#}((0,1)^n; \mathbf{R}^m) \right\} \\
&\leq \inf\left\{ \frac{1}{j^n} \int_{(0,j)^n} f(y, A + Dv(y))\, dy : \ v \in \mathrm{W}^{1,p}_{\#}((0,j)^n; \mathbf{R}^m) \right\}.
\end{aligned}$$

This inequality completes the proof. $\qquad\square$

14.3.2 *Non-coercive convex homogenization*

We can generalize Theorem 14.7 slightly to the case of non-coercive functionals. Note, however, that in this case, in general we do not describe the limit functional on the whole $\mathrm{L}^p(\Omega; \mathbf{R}^m)$. Note, moreover, that even in the case when $\int_\Omega f_{\mathrm{hom}}(Du)\, dx$ is coercive, in general Theorem 7.2 cannot be applied and we do not have a convergence result for the solutions of minimum problems involving F_ε on Ω. In some cases, though, it will be possible to choose proper precompact minimizing sequences in $\mathrm{W}^{1,p}(\Omega; \mathbf{R}^m)$ (see Chapters 19 and 20).

Theorem 14.8 (Homogenization Theorem for Non-coercive Functionals) *Let Ω be a bounded open subset of $\mathbf{R}^n$ with Lipschitz boundary. Let f be a convex function satisfying the hypotheses of Theorem 14.7, with condition (11.5) in place of (14.2). Then we have*

$$\Gamma(\mathrm{L}^p)\text{-}\lim_{\varepsilon \to 0} F_\varepsilon(u) = \int_\Omega f_{\mathrm{hom}}(Du(x))\, dx,$$

for all $u \in \mathrm{W}^{1,1}(\Omega; \mathbf{R}^m)$, where $f_{\mathrm{hom}} : \mathbf{M}^{m \times n} \to [0, +\infty)$ is the convex function given by the cell-problem formula (14.16). Moreover, if f_{hom} satisfies

$$\lim_{|A| \to +\infty} \frac{f_{\mathrm{hom}}(A)}{|A|} = +\infty, \tag{14.17}$$

then the Γ-limit exists on the whole $\mathrm{L}^p(\Omega; \mathbf{R}^m)$, and it takes the value $+\infty$ on $\mathrm{L}^p(\Omega; \mathbf{R}^m) \setminus \mathrm{W}^{1,1}(\Omega; \mathbf{R}^m)$.

Proof By Example 11.4 the conclusions of Theorem 14.7 still hold. It remains to extend the representation of F_{hom} outside $\mathrm{W}^{1,p}(\Omega; \mathbf{R}^m)$.

In the course of the proof (ρ_j) denotes a sequence of mollifiers with $\mathrm{spt}\, \rho_j \subset B(0, 1/j)$, and $\rho_j * v$ is the convolution between ρ_j and v. Note that since Ω has Lipschitz boundary, by the standard reflection technique near $\partial\Omega$ (see for instance Adams (1975), Theorems 4.26, 4.28 and Section 4.29 for details) all functions can be extended to some $\Omega' \supset\supset \Omega$, so that we can suppose that each $\rho_j * v$ is defined on the whole Ω'. Such an extension will not influence the validity of our arguments.

For every $u \in \mathrm{L}^p(\Omega; \mathbf{R}^m)$, let $F_\varepsilon(u, U)$ denote the localization of the functionals F_ε to the set $U \in \mathcal{A}(\Omega)$. With fixed (ε_j) a sequence of positive numbers converging to 0, we use the notation F' and F'' for the lower and upper Γ-limit of F_{ε_j}, respectively.

Step 1: $F'(u, U) \geq \int_U f_{\mathrm{hom}}(Du)\, dx$ for all $U \in \mathcal{A}(\Omega)$ and $u \in \mathrm{W}^{1,1}(\Omega; \mathbf{R}^m)$.

Note that $F'(\cdot, U)$ is convex for all U, and $F'(u, \cdot)$ is an increasing set function for all u. From the definition of Γ-$\liminf$, it can be immediately checked by a translation argument that for all $U, U' \in \mathcal{A}(\Omega)$, $v \in \mathrm{L}^p(\Omega; \mathbf{R}^m)$, and $y \in \mathbf{R}^n$, if $U' \subset\subset y + U$ then $F'(v^y, U') \leq F'(v, U)$, where $v^y(x) = v(x - y)$.

Let $U' \subset\subset U$. Using Jensen's inequality (see Exercise 5.2) and the properties of F' recalled above, we get, for j large enough as to have $U' \subset\subset y + U$ for all $y \in B(0, 1/j)$,

$$F'(\rho_j * u, U') \leq \int_{B(0, 1/j)} \rho_j(y)\, F'(u^y, U')\, dy$$

$$\leq \int_{B(0, 1/j)} \rho_j(y)\, F'(u, U)\, dy = F'(u, U).$$

On the other hand, by the representation of F' on $\mathrm{W}^{1,p}(\Omega; \mathbf{R}^m)$, we have

$$F'(\rho_j * u, U') = \int_{U'} f_{\mathrm{hom}}(D(\rho_j * u))\, dx.$$

Since the functional $v \mapsto \int_{U'} f_{\mathrm{hom}}(Dv)\, dx$ is lower semicontinuous with respect to the L^p-convergence, and $\rho_j * u \to u$ in $\mathrm{L}^p(U', \mathbf{R}^m)$, we get by the previous formulae

$$\int_{U'} f_{\mathrm{hom}}(Du)\, dx \leq \liminf_j \int_{U'} f_{\mathrm{hom}}(D(\rho_j * u))\, dx \leq F'(u, U).$$

By the arbitrariness of $U' \subset\subset U$ the step is concluded.

Step 2: if f_{hom} satisfies (14.17) then $F'(u, \Omega) = +\infty$ for all $u \in \mathrm{L}^p(\Omega; \mathbf{R}^m) \setminus \mathrm{W}^{1,1}(\Omega; \mathbf{R}^m)$.

We proceed exactly as in the previous step, noting that (14.17) implies that

$$\liminf_j \int_{U'} f_{\mathrm{hom}}(D(\rho_j * u)) \, dx = +\infty.$$

Step 3: $F''(u, U) \leq \int_U f_{\mathrm{hom}}(Du) \, dx$ for all $U \in \mathcal{A}(\Omega)$ and $u \in \mathrm{W}^{1,1}(\Omega; \mathbf{R}^m)$.
We have, using the lower semicontinuity of F'' and Jensen's inequality,

$$
\begin{aligned}
F''(u, U) &\leq \liminf_j F''(\rho_j * u, U) = \liminf_j F(\rho_j * u, U) \\
&= \liminf_j \int_U f_{\mathrm{hom}}(D(\rho_j * u)) \, dx \\
&\leq \liminf_j \int_U \int_{B(0,1/j)} \rho_j(y) f_{\mathrm{hom}}(D(u(x - y))) \, dy \, dx \\
&= \liminf_j \int_{B(0,1/j)} \rho_j(y) \int_{U+y} f_{\mathrm{hom}}(Du) \, dx \, dy \\
&\leq \liminf_j \int_{B(0,1/j)} \rho_j(y) \int_{U'} f_{\mathrm{hom}}(Du) \, dx \, dy = \int_{U'} f_{\mathrm{hom}}(Du) \, dx,
\end{aligned}
$$

for all $U' \supset\supset U$. By the arbitrariness of U' the proof is achieved. $\quad\square$

Example 14.9 (Fiber materials) Let $n \geq 3$. For all $i = 1, \ldots, n$ choose $\Omega_i \subset (0, 1)^{n-1} \subset \mathbf{R}^{n-1}$ and set

$$E_i = \{x \in \mathbf{R}^n : \widehat{x}_i \in \Omega_i + \mathbf{Z}^{n-1}\},$$

where $\widehat{x}_i = (x_1, \ldots, x_{i-1}, x_{i+1}, \ldots, x_n)$. The set E_i represents the 'fibers' in the direction e_i. Let f be a convex function satisfying the hypotheses of Theorem 14.8, and, in addition,

$$f(x, A) \geq \Big(\sum_{j=1}^m |A_{ji}|^2\Big)^{p/2} \qquad \text{if } x \in E_i,$$

so that

$$\int_{\Omega \cap E_i} f(x, Du) \, dx \geq \int_{\Omega \cap E_i} |D_i u|^p \, dx.$$

Note that we may have $E_i \cap E_j = \emptyset$ if $i \neq j$, so that f may be degenerate at every x. It is easy to see that $f_{\mathrm{hom}}(A) \geq c|A|^p$ for all $A \in \mathbf{M}^{m \times n}$. In fact, if u is a test function for (14.16) then we have, denoting by A_i the vector in $\mathbf{R}^m$ given by the i-th column of A (so that $D_i(Ax) = A_i$),

$$\int_{(0,1)^n} f(x, Du + A) \, dx \geq c \sum_{i=1}^n \int_{E_i \cap (0,1)^n} |D_i u + A_i|^p \, dx$$

$$= c \sum_{i=1}^{n} \int_{\Omega_i} \int_0^1 \left| \frac{du(y_1, \ldots, y_i, t, y_{i+1}, \ldots, y_{n-1})}{dt} + A_i \right|^p dt \, dy$$

$$\geq c \sum_{i=1}^{n} |\Omega_i| |A_i|^p \geq c|A|^p,$$

where we have used Jensen's inequality applied to $\xi \mapsto |\xi|^p$, and we have taken into account that for each $y \in (0,1)^{n-1}$,

$$u(y_1, \ldots, y_i, 0, y_{i+1}, \ldots, y_{n-1}) = u(y_1, \ldots, y_i, 1, y_{i+1}, \ldots, y_{n-1}).$$

Hence, we can apply Theorem 14.8 and obtain that the homogenized functional exists on the whole $L^p(\Omega; \mathbf{R}^m)$, and is finite and coercive on $W^{1,p}(\Omega; \mathbf{R}^m)$.

Example 14.10 (Porous media) As a particular case, we can consider in Theorem 14.8

$$f(x, A) = |A|^p \chi_E(x), \tag{14.18}$$

where E is a 1-periodic set. If E contains a connected open set, then, using the argument of the previous example, it is possible to prove that $f_{\mathrm{hom}}(A) \geq c|A|^p$, so that the homogenized functional is completely described. To check this, note that for all $i = 1, \ldots, n$ it is possible to find a 1-periodic smooth path γ_i contained in E together with an η-tubular neighbourhood $T_\eta(\gamma_i)$ (we can suppose that $\eta > 0$ is independent of i). Moreover, we can choose η small enough so that we have a smooth bijection $\Phi_i : T_\eta(\gamma_i) \to T_i$, where $T_i = \{x \in \mathbf{R}^n : |\widehat{x}_i| < \eta\}$. We suppose that Φ_i can be extended by periodicity to a smooth bijection, for which we use the same notation, $\Phi_i : T_\eta(\gamma_i) + \mathbf{Z}^n \to E_i$, where

$$E_i = \{x \in \mathbf{R}^n : \widehat{x}_i \in B(z, \eta) + \mathbf{Z}^{n-1}\}$$

and $z \in (0,1)^{n-1}$ is chosen in such a way that $B(z, \eta) \subset (0,1)^{n-1}$. Note that the notation for E_i corresponds to that of the previous example. If $u \in W^{1,p}_{\#}((0,1)^n; \mathbf{R}^m)$ we have

$$\int_{(0,1)^n} f(x, Du + A) \, dx \geq \frac{1}{n} \sum_{i=1}^{n} \int_{(T_\eta(\gamma_i) + \mathbf{Z}^n) \cap (0,1)^n} |Du + A|^p \, dx$$

$$\geq c \sum_{i=1}^{n} \int_{E_i \cap (0,1)^n} |D_i(u \circ \Phi_i^{-1}) + A_i|^p \, dx \geq c|A|^p,$$

by the same argument as in Example 14.9.

Example 14.11 (Fissured media) From Theorem 14.8 we can derive homogenization results in different frameworks. As an example, we can consider the case of media with a periodic array of small cracks. We outline here a simple model for the convex case.

Let $n > 1$, let $K \subset (0,1)^n$ be a closed subset of a smooth hypersurface, and let $\mathcal{K} = K + \mathbf{Z}^n$. Let f be a Borel function as in Theorem 14.7, and let Ω be a bounded open subset of $\mathbf{R}^n$ with Lipschitz boundary. For every $\varepsilon > 0$ we set

$$F_\varepsilon(u) = \int_{\Omega \backslash \varepsilon \mathcal{K}} f\left(\frac{x}{\varepsilon}, Du\right) dx \qquad u \in W^{1,p}(\Omega \backslash \varepsilon \mathcal{K}; \mathbf{R}^m) \tag{14.19}$$

(extended to $+\infty$ on $L^p(\Omega; \mathbf{R}^m) \backslash W^{1,p}(\Omega \backslash \varepsilon \mathcal{K}; \mathbf{R}^m)$). Then we have

$$\Gamma\text{-}\lim_{\varepsilon \to 0} F_\varepsilon(u) = F_{\text{hom}}(u), \tag{14.20}$$

where $F_{\text{hom}} : L^p(\Omega; \mathbf{R}^m) \to [0, +\infty]$ is defined as

$$F_{\text{hom}}(u) = \begin{cases} \int_\Omega f_{\text{hom}}(Du)\, dx & \text{if } u \in W^{1,p}(\Omega; \mathbf{R}^m) \\ +\infty & \text{otherwise,} \end{cases} \tag{14.21}$$

and

$$f_{\text{hom}}(A) = \inf\left\{ \int_{(0,1)^n \backslash K} f(y, Du + A)\, dy : u \in W^{1,p}_\#((0,1)^n \backslash K; \mathbf{R}^m) \right\}.$$

To check this convergence result we can use a comparison argument with a sequence of functionals, each one satisfying the hypotheses of Theorem 14.8. For every $j \in \mathbf{N}$, we introduce the sets

$$K^j = \left\{ x \in (0,1)^n : \text{dist}\,(x, K) < \frac{1}{j} \right\}, \qquad \mathcal{K}^j = K^j + \mathbf{Z}^n.$$

Correspondingly, we can define the functions

$$f^j(x, A) = \begin{cases} f(x, A) & \text{if } x \notin \mathcal{K}^j \\ 0 & \text{otherwise,} \end{cases}$$

and the functionals

$$F^j_\varepsilon(u) = \int_\Omega f^j\left(\frac{x}{\varepsilon}, Du\right) dx \qquad u \in W^{1,p}(\Omega; \mathbf{R}^m)$$

(extended to $+\infty$ on $L^p(\Omega; \mathbf{R}^m) \backslash W^{1,p}(\Omega; \mathbf{R}^m)$). The functions f^j satisfy the hypotheses of Theorem 14.8. We define their homogenized functions, given by the cell-problem formula,

$$f^j_{\text{hom}}(A) = \inf\left\{ \int_{(0,1)^n \backslash K^j} f(y, Du + A)\, dy : u \in W^{1,p}_\#((0,1)^n; \mathbf{R}^m) \right\}.$$

Using Example 14.9 it follows that $f^j_{\text{hom}}(A) \geq c|A|^p$, with c independent of j, for j large enough. It can be immediately verified that $\overline{F}^j_\varepsilon \leq F_\varepsilon$, where $\overline{F}^j_\varepsilon$

denotes the lower semicontinuous envelope of F_ε^j with respect to the $\mathrm{L}^p(\Omega;\mathbf{R}^m)$ convergence. Then, using Proposition 7.13, we have that for any sequence (ε_k) of positive numbers converging to 0

$$\Gamma\text{-}\liminf_k F_{\varepsilon_k} \geq \Gamma\text{-}\liminf_k \overline{F}_{\varepsilon_k}^j = \Gamma\text{-}\liminf_k F_{\varepsilon_k}^j = \Gamma\text{-}\lim_{\varepsilon\to 0} F_\varepsilon^j.$$

Since f^j converges increasingly to f on $\mathbf{R}^n \setminus \mathcal{K}$, it is possible to prove that $f_{\mathrm{hom}}(A) = \sup_j f_{\mathrm{hom}}^j(A) = \lim_j f_{\mathrm{hom}}^j(A)$, using the fact that (up to translations by a constant) minimizing sequences for the minimum problems defining $f_{\mathrm{hom}}^j(A)$ are compact in $\mathrm{W}_{\mathrm{loc}}^{1,p}((0,1)^n \setminus K;\mathbf{R}^m)$ (recall Remark 7.4(ii) and Theorem 7.2). Hence, $\sup_j \Gamma\text{-}\lim_{\varepsilon\to 0} F_\varepsilon^j = F_{\mathrm{hom}}$ by Remark 3.4(b). We then have $\Gamma\text{-}\liminf_k F_{\varepsilon_k} \geq F_{\mathrm{hom}}$.

The opposite inequality for $F(u) = \Gamma\text{-}\limsup_k F_{\varepsilon_k}(u)$ needs to be checked only for $u \in \mathrm{W}^{1,p}(\Omega;\mathbf{R}^m)$. We can apply the direct methods of Γ-convergence to the localized functionals

$$F_\varepsilon(u,U) = \int_{U\setminus\varepsilon\mathcal{K}} f\left(\frac{x}{\varepsilon}, Du\right) dx.$$

The functionals F_ε satisfy the L^p-fundamental estimate, with the same proof as for functionals with standard growth conditions, so that we can use the same method as in Chapter 12, and, up to passing to a subsequence we can suppose that $F(\cdot, U)$ is indeed a Γ-limit and

$$F(u,U) = \int_U \varphi(x, Du)\, dx$$

for $u \in \mathrm{W}^{1,p}(\Omega;\mathbf{R}^m)$ (the details are left to the reader). Setting $u_k = \varepsilon_k u(x/\varepsilon_k) + Ax$, with $u \in \mathrm{W}_{\#}^{1,p}((0,1)^n \setminus K;\mathbf{R}^m)$ we have $u_k \to Ax$, and

$$F(Ax,U) \leq \liminf_k F_{\varepsilon_k}(u_k,U) = |U| \int_{(0,1)^n\setminus K} f(y, Du + A)\, dy$$

by Exercise 2.7, so that $F(Ax,U) \leq f_{\mathrm{hom}}(A)$. From this inequality it can be immediately deduced that $\varphi(x,A) \leq f_{\mathrm{hom}}(A)$ a.e., and $F \leq F_{\mathrm{hom}}$ as desired.

14.4   A counterexample to the cell-problem formula

In the general vector-valued non-convex case, the following example by S. Müller (1987) shows that formula (14.16) does not hold, even if $f(x,\cdot)$ is polyconvex for all x. This means that in the description of the homogenized integrand we cannot restrict our analysis to a single-cell problem.

Example 14.12 Let $f_0 : \mathbf{M}^{2\times 2} \to \mathbf{R}$ be the polyconvex function defined by

$$f_0(A) = |A|^4 + h(\det A),$$

where $h : \mathbf{R} \to \mathbf{R}$ is given by

$$
h(r) = \begin{cases} \dfrac{8(1+a)^2}{r+a} - 8(1+a) - 4 & \text{if } r > 0 \\[2ex] \dfrac{8(1+a)^2}{a} - 8(1+a) - 4 - \dfrac{8(1+a)^2}{a^2}r & \text{if } r \leq 0. \end{cases}
$$

Here $0 < a < 1/2$ is an arbitrary real number.

Note that $f_0 \geq 0$, and $f_0(A) = 0$ if and only if $A^t A = I$ and $\det A > 0$ (A^t is the *transpose* matrix of A). In fact, it can be immediately checked that $h(r) > 0$ for $r \leq 0$, while for $\det A > 0$ we can rewrite f_0 in terms of the eigenvalues $\nu_i = \nu_i(A)$ of the symmetric matrix (with positive eigenvalues) $(A^t A)^{1/2}$

$$
f_0(A) = (\nu_1^2 + \nu_2^2)^2 + h(\nu_1 \nu_2), \tag{14.22}
$$

and it can also be verified that the minimum value is 0 and is attained only at $\nu_1 = \nu_2 = 1$.

We remark that from (14.22) we have $f_0(A) = \psi(A^t A)$ (when $\det A > 0$) with ψ of class C^2 which attains its minimum value 0 on the identity matrix. This implies that

$$
f_0(A) = O(|A^t A - I|^2) \qquad \text{as } |A^t A - I| \to 0 \tag{14.23}
$$

provided that $\det A > 0$.

With fixed $\alpha > 0$, we define $f : [0,1)^2 \times \mathbf{M}^{2\times 2} \to [0, +\infty)$, extended by periodicity to $\mathbf{R}^2 \times \mathbf{M}^{2\times 2}$, by

$$
f(x, A) = \begin{cases} f_0(A) & \text{if } x \in [0, 1/2) \times [0, 1) \\ \alpha\, f_0(A) & \text{if } x \in [1/2, 1) \times [0, 1). \end{cases}
$$

Note that, up to a translation by a constant, f satisfies the hypotheses of Theorem 14.5, with $p = 4$. Let f_{hom} be given by (14.12), and let

$$
\widehat{f}(A) = \inf\left\{ \int_{(0,1)^n} f(y, A + Du(y))\, dy : \; u \in W^{1,p}_{\#}((0,1)^n; \mathbf{R}^2) \right\}.
$$

We want to check that

$$
f_{\mathrm{hom}}(\overline{A}) < \widehat{f}(\overline{A}) \tag{14.24}
$$

for $\overline{A} = \operatorname{diag}(1, d)$, if $\pi/4 < d < 1$ and α sufficiently small.

We first prove that a constant $K > 0$ independent of α exists such that

$$
f_{\mathrm{hom}}(\overline{A}) \leq K\alpha. \tag{14.25}
$$

Let R be the unique positive solution of $2R\sin(1/(2R)) = d$, which is well defined since d is large enough, and set

$$C(t) = \cos\left(\frac{1}{R}\left(t - \frac{1}{2}\right)\right), \qquad S(t) = \sin\left(\frac{1}{R}\left(t - \frac{1}{2}\right)\right).$$

Let (u_j) be the sequence in $\mathrm{W}^{1,\infty}((0,1)^2; \mathbf{R}^2)$ defined by

$$u_j^1(x) = x_1 + \left(R - \left|x_1 - \frac{1}{j}\left[jx_1 + \frac{1}{2}\right]\right|\right)(1 - C(x_2))$$

$$u_j^2(x) = \frac{d}{2} + \left(R - \left|x_1 - \frac{1}{j}\left[jx_1 + \frac{1}{2}\right]\right|\right)S(x_2).$$

We have $u_j \to u$ in $\mathrm{L}^\infty((0,1)^2; \mathbf{R}^2)$, where $u = (u^1, u^2)$ is defined by

$$u^1(x) = x_1 + R\left(1 - C(x_2)\right)$$

$$u^2(x) = \frac{d}{2} + R\,S(x_2).$$

Note that $u \in \mathrm{W}^{1,\infty}((0,1)^2; \mathbf{R}^2)$. Moreover, from the definition of R we have

$$
\begin{aligned}
u(t,1) - u(t,0) &= (0,d) && \text{for all } t \in (0,1), \\
u(1,t) - u(0,t) &= (1,0) && \text{for all } t \in (0,1);
\end{aligned}
$$

hence we can write

$$u(x_1, x_2) = (x_1, dx_2) + \overline{u}(x_1, x_2) = \overline{A}x + \overline{u}(x_1, x_2), \qquad (14.26)$$

where $\overline{u} \in \mathrm{W}^{1,\infty}((0,1)^2; \mathbf{R}^2)$ can be extended to a function in $\mathrm{W}_{\#}^{1,p}((0,1)^2; \mathbf{R}^2)$.
We now check that

$$\liminf_j \int_{(0,1)^2} f(jy, Du_j(y))\, dy \le K\alpha, \qquad (14.27)$$

for some $K > 0$ independent of α. If $0 \le x_1 - \frac{1}{j}\left[jx_1 + \frac{1}{2}\right] \le \frac{1}{2j}$ (i.e. if $f(jx, A) = f_0(A)$) we have

$$Du_j(x) = \begin{pmatrix} C(x_2) & S(x_2) \\ -S(x_2) & C(x_2) \end{pmatrix} - \frac{1}{R}\left(x_1 - \frac{1}{j}\left[jx_1 + \frac{1}{2}\right]\right)\begin{pmatrix} 0 & S(x_2) \\ 0 & C(x_2) \end{pmatrix}.$$

Hence $\|(Du_j)^t(Du_j) - I\|_\infty \le \frac{c}{j}$. By (14.23) this ensures that $f_0(Du_j) \le c/j^2$. If $\frac{1}{2j} \le x_1 - \frac{1}{j}\left[jx_1 + \frac{1}{2}\right] \le \frac{1}{j}$ (i.e., if $f(jx, A) = \alpha f_0(A)$) by the equiboundedness of $\|Du_j\|_\infty$ we have $f_0(Du_j) \le c$. Therefore

$$\int_{(0,1)^2} f(jy, Du_j(y))\, dy \le c\left(\frac{1}{j^2} + \alpha\right), \qquad (14.28)$$

and (14.27) is proved. From the Homogenization Theorem 14.5 and (7.3) we get

$$\int_{(0,1)^2} f_{\text{hom}}(\overline{A} + D\overline{u})\, dy \le K\alpha$$

so that by quasiconvexity and Remark 6.11 we obtain (14.25).

We now prove that

$$\widehat{f}(\overline{A}) \ge \delta, \tag{14.29}$$

with $\delta > 0$ independent of α. We denote by V the closure of the set

$$\{v \in C^\infty((0,1/2) \times (0,1); \mathbf{R}^2) :\ v(t,0) = v(t,1) \text{ for all } t \in (0,1/2)\}$$

in the $\mathrm{W}^{1,p}((0,1/2) \times (0,1); \mathbf{R}^2)$-norm, and define

$$m = \inf\Big\{ \int_{(0,1/2)\times(0,1)} f_0(\overline{A} + Dv)\, dy :\ v \in V \Big\}. \tag{14.30}$$

As the functional in (14.30) is mildly coercive on V and is weakly lower semi-continuous by Theorem 5.10, the infimum in (14.30) is achieved by some $\overline{v} \in V$. By the properties of f_0 we have $m \ge 0$, and m equals 0 if and only if

$$(\overline{A} + D\overline{v})^t(\overline{A} + D\overline{v}) = I \qquad \text{and} \qquad \det(\overline{A} + D\overline{v}) > 0 \tag{14.31}$$

a.e. in $(0,1/2) \times (0,1)$. Suppose that $\overline{v}$ satisfies (14.31). This implies that $\overline{A} + D\overline{v}$ is a rotation a.e., and therefore $\psi = \overline{A}x + \overline{v}$ is a rigid motion (see Appendix C). In particular ψ is an isometry. Since $\psi(t,1) - \psi(t,0) = (0,d)$ we thus reach a contradiction, as $d < 1$. Hence $m > 0$. Since $v \in \mathrm{W}^{1,p}_{\#}((0,1)^2; \mathbf{R}^2)$ implies $v_{|(0,1/2)\times(0,1)} \in V$, we obtain the inequality $\widehat{f}(\overline{A}) \ge m > 0$, i.e. (14.29), as desired, with $\delta = m$.

Taking $\alpha < \delta/K$, (14.24) follows from (14.25) and (14.29), and the example is concluded.

14.5 An application: homogenization of elliptic equations in divergence form

In Chapter 8 we noted that the Γ-convergence of quadratic functionals can be deduced from the convergence of solutions of elliptic problems. In this section we prove the converse in the case of homogenization, and we give a formula for the 'homogenized coefficients' of the limit elliptic operator.

We consider a symmetric matrix (a_{ij}) of $\mathrm{L}^\infty(\mathbf{R}^n)$ 1-periodic functions satisfying the ellipticity condition

$$\alpha|\xi|^2 \le \sum_{i,j=1}^n a_{ij}(x)\xi_i\xi_j \le \beta|\xi|^2 \tag{14.32}$$

for all $\xi \in \mathbf{R}^n$ and $x \in \mathbf{R}^n$. The corresponding quadratic functionals

$$F_\varepsilon(u) = \int_\Omega \sum_{i,j=1}^n a_{ij}\left(\frac{x}{\varepsilon}\right) D_i u D_j u \, dx, \tag{14.33}$$

defined on $W^{1,2}(\Omega)$, satisfy the hypotheses of Theorem 14.7. Hence the Γ-limit $F_{\mathrm{hom}} = \Gamma\text{-}\lim_{\varepsilon \to 0} F_\varepsilon$ exists on $W^{1,2}(\Omega)$, and

$$F_{\mathrm{hom}}(u) = \int_\Omega f_{\mathrm{hom}}(Du) \, dx. \tag{14.34}$$

By Remark 7.12(ii) there exists a symmetric constant matrix (q_{ij}) such that

$$f_{\mathrm{hom}}(\xi) = \sum_{i,j=1}^n q_{ij}\xi_i\xi_j \tag{14.35}$$

for all $\xi \in \mathbf{R}^n$. Computing the corresponding Euler equations, we see, by Theorem 7.2, that Γ-convergence implies the convergence of the solutions of problems (8.5) to the solution of problem (8.6).

On the other hand, $f_{\mathrm{hom}}(\xi)$ satisfies the homogenization formula (14.16), which takes the form

$$f_{\mathrm{hom}}(\xi) = \inf\Big\{ \int_{(0,1)^n} \sum_{i,j=1}^n a_{ij}(y)(\xi_i + D_i u(y))(\xi_j + D_j u(y)) \, dy :$$

$$u \in W^{1,2}_\#((0,1)^n)\Big\}. \tag{14.36}$$

In particular, we can easily compute

$$q_{ll} = f_{\mathrm{hom}}(e_l)$$

$$= \inf\Big\{ \int_{(0,1)^n} \sum_{i,j=1}^n a_{ij}(y)D_i(y_l + u(y))D_j(y_l + u(y)) \, dy : \ u \in W^{1,2}_\#((0,1)^n)\Big\}.$$

$$\tag{14.37}$$

By the strict convexity of $\xi \mapsto \sum_{i,j=1}^n a_{ij}(x)\xi_i\xi_j$ the unique minimum point, up to an additive constant, in (14.37) can be expressed via Euler equations as the solution P^l of the 'cell problem'

$$\begin{cases} -\sum_{i,j=1}^n D_j(a_{ij}D_i P^l) = \sum_{j=1}^n D_j a_{lj} \\ P^l \in W^{1,2}_\#((0,1)^n), \end{cases} \tag{14.38}$$

and we have

$$q_{ll} = \int_{(0,1)^n} \sum_{i,j=1}^n a_{ij}(y)D_i(y_l + P^l(y))D_j(y_l + P^l(y)) \, dy. \tag{14.39}$$

More generally, we notice that from the equality $f_{\hom}(e_l + e_k) = q_{ll} + q_{kk} + 2q_{kl}$ it is easy to obtain

$$q_{kl} = \int_{(0,1)^n} \sum_{i,j=1}^{n} a_{ij}(y) D_i(y_k + P^k(y)) D_j(y_l + P^l(y)) \, dy, \qquad (14.40)$$

for $k, l = 1, \ldots, n$. Hence, we have obtained with (14.40) and (14.38) a description of the homogenized operator in differential terms.

Remark 14.13 It is useful to note that homogenized integrands often possess symmetries due to the averaging process. In the case of quadratic forms, these symmetries often simplify calculations. For example, if ψ is a quadratic form on $\mathbf{R}^n$, and it is symmetric with respect to the coordinate axes, then it is diagonal, i.e. $\psi(z) = \sum_{i=1}^{n} a_i z_i^2$.

14.6 Exercises

Exercise 14.1 Let f satisfy the hypotheses of Theorem 14.5 with $p = m = n$. If $\widetilde{f}(x, A) = f(x, A) + \det A$ defines a strictly positive function then prove that the related homogenized function $\widetilde{f}_{\hom}$ is of the form $\widetilde{f}_{\hom}(A) = f_{\hom}(A) + \det A$ (use (6.1) and (14.12)).

Exercise 14.2 Let $a : \mathbf{R} \to [\alpha, \beta]$ be 1-periodic. Compute the homogenized functional given by Theorem 14.7 with $m = 1$ and $f(x, z) = a(x_1)|z|^2$.

Hint: using Remark 14.13 note that $f_{\hom}(z) = \sum_{i=1}^{n} a_i z_i^2$. Since $f(\cdot, z)$ depends only on x_1 show that the solution to the minimum problem defining $f_{\hom}(e_1)$ is of the form $u(x) = v(x_1)$. Deduce from Exercise 13.1 that $a_1 = \left(\int_0^1 a^{-1} \, ds\right)^{-1}$. If $i > 1$, note that the problem defining $f_{\hom}(e_i)$ is minimized by $u = 0$, and hence $a_i = \int_0^1 a(s) \, ds$.

Exercise 14.3 (Homogenization of layered materials) Let $0 < \theta < 1$ and let
$$G^\theta(s) = \begin{cases} \alpha & \text{if } 0 < s \leq \theta \\ \beta & \text{if } \theta < s \leq 1 \end{cases}$$
(extended to $\mathbf{R}$ by 1-periodicity). Compute the homogenized functional as in Theorem 14.7 with $m = 1$ and $f(x, z) = G^\theta(x_1)|z|^2$.

Hint: use the previous exercise to show that $f_{\hom}(z) = \sum_{i=1}^{n} a_i z_i^2$, with

$$a_i = \begin{cases} \left(\dfrac{\theta}{\alpha} + \dfrac{(1-\theta)}{\beta}\right)^{-1} & \text{if } i = 1 \\[2ex] \theta\alpha + (1-\theta)\beta & \text{if } i > 1. \end{cases}$$

Exercise 14.4 Let $b_i : \mathbf{R} \to [\alpha, \beta]$, $i = 1, \ldots, n$, be 1-periodic. Compute the homogenized functional given by Theorem 14.7 when $m = 1$ and $f(x, z) = \prod_{i=1}^{n} b_i(x_i)|z|^2$.

Hint: (this is a generalization of Exercise 14.2) show that the solution to the problem defining $f_{\hom}(e_i)$ depends only on x_i. Deduce that $f_{\hom}(z) = \sum_{i=1}^{n} a_i z_i^2$, with

$$a_i = \left(\int_0^1 b_i^{-1}\, ds \right)^{-1} \prod_{j \neq i} \left(\int_0^1 b_j(s)\, ds \right).$$

Compare with the result of Exercise 14.2.

Exercise 14.5 Prove that if $f : \mathbf{R}^n \times \mathbb{M}^{n \times n} \to [0, +\infty)$ is a strictly convex elliptic quadratic form $f(x, A) = \sum_{i,j,k,l=1}^n a_{ij}^{kl}(x) A_{ki} A_{lj}$ with coefficients in $\mathrm{L}^\infty(\Omega)$ and such that $a_{ij}^{kl} = a_{ji}^{lk}$, then the related homogenized function f_{hom} is of the form $f_{\mathrm{hom}}(A) = \sum_{i,j,k,l=1}^n q_{ij}^{kl} A_{ki} A_{lj}$, with

$$q_{zw}^{rs} = \int_{(0,1)^n} \sum_{i,j,k,l=1}^n a_{ij}^{kl}(y) D_i \big((e_r \otimes e_z y)_k + (e_s \otimes e_w y)_k + P_{rz}^k(y) \big)$$

$$\times D_j \big((e_r \otimes e_z y)_l + (e_s \otimes e_w y)_l + P_{sw}^l(y) \big)\, dy,$$

where $P_{rz} = (P_{rz}^1, \ldots, P_{rz}^n)$ is the solution of

$$\begin{cases} -\displaystyle\sum_{i,j,l=1}^n D_j(a_{ji}^{hl} D_i P_{rz}^l) = \sum_{j=1}^n D_j a_{zj}^{rh} & h = 1, \ldots, n \\[2mm] P_{rz} \in \mathrm{W}_{\#}^{1,2}((0,1)^n; \mathbf{R}^n). \end{cases}$$

(Repeat the reasoning of Section 14.5.)

Exercise 14.6 (Parameterized homogenization) Consider a function $f : \mathbf{R}^n \times \mathbf{R}^n \times \mathbf{M}^{m \times n} \to [0, +\infty)$ such that:

    (i) $f(x, \cdot, A)$ is a measurable function for all $(x, A) \in \mathbf{R}^n \times \mathbf{M}^{m \times n}$;

    (ii) $f(x, y, \cdot)$ is a convex function for all $(x, y) \in \mathbf{R}^n \times \mathbf{R}^n$;

    (iii) $|f(x, y, A) - f(x', y, A)| \leq \omega(|x - x'|)(a(y) + f(x, y, A))$ for all x, x', $y \in \mathbf{R}^n, A \in \mathbf{M}^{m \times n}$, where ω is a continuous positive real function with $\omega(0) = 0$;

    (iv) the functions $f(\cdot, y, A)$ and $f(x, \cdot, A)$ are 1-periodic;

    (v) $|A|^p \leq f(x, y, A) \leq C(1 + |A|^p)$ for all $x, y \in \mathbf{R}^n, A \in \mathbf{M}^{m \times n}$.

For every Ω bounded open subset of $\mathbf{R}^n$ and $\varepsilon > 0$ define

$$F_\varepsilon(u, \Omega) = \begin{cases} \displaystyle\int_\Omega f\left(x, \frac{x}{\varepsilon}, Du(x)\right) dx & \text{if } u \in \mathrm{W}^{1,p}(\Omega; \mathbf{R}^m) \\[3mm] +\infty & \text{otherwise} \end{cases}$$

on $\mathrm{L}^p(\Omega; \mathbf{R}^m)$. Prove that there exists the Γ-limit

$$\Gamma(\mathrm{L}^p)\text{-}\lim_{\varepsilon \to 0} F_\varepsilon(u, \Omega) = \begin{cases} \displaystyle\int_\Omega \varphi(x, Du(x)) dx & \text{if } u \in \mathrm{W}^{1,p}(\Omega; \mathbf{R}^m) \\[3mm] +\infty & \text{otherwise,} \end{cases}$$

where the function $\varphi : \mathbf{R}^n \times \mathbf{M}^{m \times n} \to [0, +\infty)$ is convex and satisfies

$$\varphi(x, A) = \min\left\{ \int_{(0,1)^n} f(x, y, Du(y) + A)\, dy : u \in W_{\#}^{1,p}((0,1)^n; \mathbf{R}^m) \right\}.$$

Hint: with fixed Ω and $\eta > 0$ let $(A_i^\eta)_{i \in I^\eta}$ be a finite family of disjoint open sets with diameter less than η and such that $|\Omega \setminus \bigcup_{i \in I^\eta} A_i^\eta| = 0$. For every $i \in I^\eta$ let $x_i^\eta \in A_i^\eta$ and define $f_\eta(x, y, A) = \sum_{i \in I^\eta} \chi_{A_i^\eta} f(x_i^\eta, y, A)$. By Theorem 14.7 there exists the limit

$$\Gamma(L^p)\text{-}\lim_{\varepsilon \to 0} \int_\Omega f_\eta \left(x, \frac{x}{\varepsilon}, Du(x) \right) dx = \sum_{i \in I^\eta} \int_{A_i^\eta} \varphi(x_i^\eta, Du(x)) dx.$$

Thanks to (iii) we have

$$\left| \int_\Omega f_\eta \left(x, \frac{x}{\varepsilon}, Du(x) \right) dx - F_\varepsilon(u, \Omega) \right| \le \omega(\eta) \left(\int_\Omega a(x) dx + \int_\Omega f_\eta \left(x, \frac{x}{\varepsilon}, Du(x) \right) dx \right).$$

Eventually, pass to the Γ-limit as $\varepsilon \to 0$ and then let $\eta \to 0$.

Exercise 14.7 Let f satisfy the hypotheses of Theorem 14.7 with $m = 1$, and for all $x \in \mathbf{R}^n$ let the function $\xi \mapsto f(x, \xi)$ be strictly convex and of class C^1 on $\mathbf{R}^n$. Prove that the solution u to the minimum problem (14.16) is characterized as being the unique $u \in W_{\#}^{1,p}((0,1)^n)$ such that

$$\int_{(0,1)^n} \sum_{i=1}^n \frac{\partial f(x, A + Du)}{\partial \xi_i} D_i v\, dx = 0$$

for every $v \in W_{\#}^{1,p}((0,1)^n)$.

Hint: use the strict convexity of f to get the uniqueness of u, and write the Euler equations for the minimum problem (14.16).

Exercise 14.8 Let f and u be as in Exercise 14.7, and let f_ξ denote the vector whose i-th component is the derivative of $\xi \mapsto f(x, \xi)$ with respect to ξ_i. Suppose that we have $|f_\xi(x, \xi)| \le c(1 + |\xi|^{p-1})$. Prove that $u \in W_{\text{loc}}^{1,p}(\mathbf{R}^n)$, Du is 1-periodic, and $\operatorname{div}(f_\xi(x, \xi + Du)) = 0$ on $\mathbf{R}^n$ in the sense of distributions.

Hint: the only thing to show is the last one. To prove this show that

$$\int_{z+(0,1)^n} \langle f_\xi(y, \xi + Du(y)), D\varphi \rangle\, dy = 0$$

for all $z \in \mathbf{R}^n$ and $\varphi \in C_0^\infty(z + (0,1)^n)$ (note that such a φ can be extended to a function in $W_{\#}^{1,p}((0,1)^n)$, and use the previous exercise). Conclude by remarking that all functions in $C_0^\infty(\mathbf{R}^n)$ can be written as a finite sum of C^∞ functions, each one with support in a cube of the form $z + (0,1)^n$.

15

ALMOST-PERIODIC HOMOGENIZATION

In this chapter we generalize the Homogenization Theorem to uniformly almost-periodic functions, and to integrands of the form $f(x, u, A)$. Note that in this case, even if f is 1-periodic in the first two variables, the function $x \mapsto f(x, Ax, A)$ is in general non-periodic; this remark suggests that almost-periodic methods are particularly suited also for some apparently periodic situations.

We begin by recalling the definition of a uniformly almost-periodic function (see also Appendix A).

Definition 15.1 *Let $(X, \| \ \|)$ be a complex Banach space. We say that a measurable function $v : \mathbf{R}^N \to X$ is uniformly almost periodic (u.a.p. for short), and we write $v \in UAP(\mathbf{R}^N; X)$, if it is the uniform limit of a sequence of trigonometric polynomials on X, i.e. $\lim_k \|P_k - v\|_\infty = 0$ for some functions of the form*

$$P_k(y) = \sum_{j=1}^{r_k} \mathbf{x}_j^k \, e^{i\langle \lambda_j^k, y \rangle},$$

with $\mathbf{x}_j^k \in X$, $\lambda_j^k \in \mathbf{R}^N$, and $r_k \in \mathbf{N}$. The definition easily extends to real Banach spaces. If $X = \mathbf{R}$, this definition is the usual definition of uniformly almost-periodic functions in the sense of Bohr.

Remark 15.2 If $u : \mathbf{R}^N \to \mathbf{R}$ is a $\mathrm{L}^1_{\mathrm{loc}}$ function, we define the *mean value* of u (over $\mathbf{R}^N$) as

$$\fint u \, dx = \limsup_{T \to +\infty} \frac{1}{(2T)^N} \int_{[-T,T]^N} u(x) \, dx.$$

The mean value of u.a.p. functions is finite (Proposition A.3).

15.1 Homogenization of uniformly almost-periodic functionals

In this section we prove a homogenization theorem with a hypothesis of uniform almost periodicity on the function $f = f(x, s, A)$, besides a growth condition of the form

$$\alpha |A|^p \le f(x, s, A) \le \beta(1 + |A|^p) \tag{15.1}$$

for all $(x, s, A) \in \mathbf{R}^n \times \mathbf{R}^m \times \mathbf{M}^{m \times n}$, with $p > 1$. Our hypotheses will directly cover some interesting cases: for example, $f(s, A) = H(s) + |A|^p$, with H periodic and bounded, or

$$f(u, Du) = \sum_{i,j=1}^{n} \sum_{k,l=1}^{m} a_{ij}^{kl}(u) \frac{\partial u^k}{\partial x_i} \frac{\partial u^l}{\partial x_j},$$

with a_{ij}^{kl} u.a.p. and such that (15.1) is satisfied with $p = 2$. Note also that the case when a_{ij}^{kl} are periodic is not covered by the results of the previous section.

In view of the characterization of u.a.p. functions in Theorem A.6 of Appendix A, and the growth hypothesis (15.1), we will make the following assumption on the Borel function f: for every $A \in \mathbf{M}^{m \times n}$ and $\eta > 0$ the sets

$$T_\eta^A = \{\tau \in \mathbf{R}^n \ : \ |f(x + \tau, s + A\tau, Y) - f(x, s, Y)| < \eta(1 + |Y|^p)$$
$$\text{for every } (x, s, Y) \in \mathbf{R}^n \times \mathbf{R}^m \times \mathbf{M}^{m \times n}\}$$

$$(15.2)$$

$$T_\eta^0 = \{\tau \in \mathbf{R}^m \ : \ |f(x, s + \tau, Y) - f(x, s, Y)| < \eta(1 + |Y|^p)$$
$$\text{for every } (x, s, Y) \in \mathbf{R}^n \times \mathbf{R}^m \times \mathbf{M}^{m \times n}\}$$

are relatively dense in $\mathbf{R}^n$ and $\mathbf{R}^m$, respectively. We recall that a set $T \subset \mathbf{R}^N$ is *relatively dense* in $\mathbf{R}^N$ if there exists an *inclusion length* $L > 0$ such that $T + [0, L)^N = \mathbf{R}^N$, i.e. for every $z \in \mathbf{R}^N$ there exists $\tau \in T \cap (z + [0, L)^N)$. Note that we do not assume f to be continuous.

We can now state the Homogenization Theorem for a function f satisfying (15.1) and such that the sets in (15.2) are relatively dense.

Theorem 15.3 *Let $p > 1$, let $f : \mathbf{R}^n \times \mathbf{R}^m \times \mathbf{M}^{m \times n} \to \mathbf{R}$ satisfy (15.1) and let the sets in (15.2) be relatively dense. Then there exists a quasiconvex function $f_{\mathrm{hom}} : \mathbf{M}^{m \times n} \to \mathbf{R}$ such that for every bounded open subset Ω of $\mathbf{R}^n$ and every $u \in \mathrm{W}^{1,p}(\Omega; \mathbf{R}^m)$ the limit*

$$\Gamma(\mathrm{L}^p)\text{-}\lim_{\varepsilon \to 0} \int_\Omega f\left(\frac{x}{\varepsilon}, \frac{u(x)}{\varepsilon}, Du(x)\right) dx = \int_\Omega f_{\mathrm{hom}}(Du(x)) \, dx \qquad (15.3)$$

exists, and the function f_{hom} satisfies the asymptotic homogenization formula

$$f_{\mathrm{hom}}(A) = \lim_{t \to +\infty} \inf\left\{\frac{1}{t^n} \int_{(0,t)^n} f(x, u(x) + Ax, Du(x) + A) \, dx \ : \right.$$
$$\left. u \in \mathrm{W}_0^{1,p}((0,t)^n; \mathbf{R}^m)\right\} \qquad (15.4)$$

for all $A \in \mathbf{M}^{m \times n}$.

The first step is to prove a compactness result.

Proposition 15.4 *For every sequence (ε_j) of positive real numbers converging to 0, there exist a subsequence (ε_{j_k}) and a quasiconvex function $\varphi : \mathbf{M}^{m \times n} \to \mathbf{R}$ such that for every bounded open set Ω in $\mathbf{R}^n$ the Γ-limit*

$$\Gamma(\mathrm{L}^p)\text{-}\lim_k \int_\Omega f\left(\frac{x}{\varepsilon_{j_k}}, \frac{u(x)}{\varepsilon_{j_k}}, Du(x)\right) dx = \int_\Omega \varphi(Du(x)) \, dx \qquad (15.5)$$

exists for every $u \in \mathrm{W}^{1,p}(\Omega; \mathbf{R}^m)$.

Proof Let Ω be fixed. By applying Proposition 12.3 to the family of functionals $F_\varepsilon : \mathrm{W}^{1,p}(\Omega; \mathbf{R}^m) \times \mathcal{A}(\Omega) \to [0, +\infty)$ defined by

$$F_\varepsilon(u, U) = \int_U f\left(\frac{x}{\varepsilon}, \frac{u(x)}{\varepsilon}, Du(x)\right) dx,$$

we obtain the existence of a subsequence (ε_{j_k}) such that the limit

$$F(u, U) = \Gamma(\mathrm{L}^p)\text{-}\lim_k \int_U f\left(\frac{x}{\varepsilon_{j_k}}, \frac{u(x)}{\varepsilon_{j_k}}, Du(x)\right) dx$$

exists for every $u \in \mathrm{W}^{1,p}(\Omega; \mathbf{R}^m)$ and $U \in \mathcal{A}(\Omega)$, and it satisfies the following properties:

(i) for every $U \in \mathcal{A}(\Omega)$ and for every $u \in \mathrm{W}^{1,p}(\Omega; \mathbf{R}^m)$

$$\alpha \int_U |Du|^p \, dx \leq F(u, U) \leq \beta \int_U (1 + |Du|^p) \, dx;$$

(ii) F is local, i.e. $F(u, U) = F(v, U)$, whenever $U \in \mathcal{A}(\Omega)$ and $u = v$ a.e. on U;

(iii) for every fixed $u \in \mathrm{W}^{1,p}(\Omega; \mathbf{R}^m)$ the set function $F(u, \cdot)$ is the restriction of a Borel measure to $\mathcal{A}(\Omega)$.

Moreover:

(iv) for every $U \in \mathcal{A}(\Omega)$, $u \in \mathrm{W}^{1,p}(\Omega; \mathbf{R}^m)$ and $a \in \mathbf{R}^m$,

$$F(u + a, U) = F(u, U).$$

In fact, let (u_k) be a sequence in $\mathrm{W}^{1,p}(\Omega; \mathbf{R}^m)$ such that $u_k \to u$ in $\mathrm{L}^p(U; \mathbf{R}^m)$ and

$$F(u, U) = \lim_k \int_U f\left(\frac{x}{\varepsilon_{j_k}}, \frac{u_k(x)}{\varepsilon_{j_k}}, Du_k(x)\right) dx.$$

Fix $\eta > 0$, let $a_k \in \mathbf{R}^m$ be such that $a_k \to a$ and $\tau_k = a_k/\varepsilon_{j_k} \in T_\eta^0$, i.e.

$$|f(x, s + \tau_k, A) - f(x, s, A)| \leq \eta(1 + |A|^p)$$

for every $x \in \mathbf{R}^n, s \in \mathbf{R}^m, A \in \mathbf{M}^{m \times n}$. Then we have

$$
\begin{aligned}
F(u + a, U) &\leq \liminf_k \int_U f\left(\frac{x}{\varepsilon_{j_k}}, \frac{u_k(x) + a_k}{\varepsilon_{j_k}}, Du_k(x)\right) dx \\
&= \liminf_k \int_U f\left(\frac{x}{\varepsilon_{j_k}}, \frac{u_k(x)}{\varepsilon_{j_k}} + \tau_k, Du_k(x)\right) dx \\
&= \lim_k \int_U f\left(\frac{x}{\varepsilon_{j_k}}, \frac{u_k(x)}{\varepsilon_{j_k}}, Du_k(x)\right) dx \\
&\qquad + \eta \liminf_k (|U| + \|Du_k\|_{\mathrm{L}^p(U; \mathbf{M}^{m \times n})}^p)
\end{aligned}
$$

$$\leq F(u, U) + \eta(|U| + \sup_k \|Du_k\|^p_{L^p(U;\mathbf{M}^{m \times n})}).$$

By the arbitrariness of η, we have $F(u + a, U) \leq F(u, U)$. In the same way we prove the opposite inequality.

By (i)–(iv) and the lower semicontinuity of the Γ-limit, we can apply Theorem 9.1 to obtain the existence of a quasiconvex Carathéodory function φ : $\mathbf{R}^n \times \mathbf{M}^{m \times n} \to \mathbf{R}$ such that

$$F(u, \Omega) = \int_\Omega \varphi(x, Du(x))\, dx$$

for every $u \in W^{1,p}(\Omega; \mathbf{R}^m)$. In order to complete the proof, by Proposition 9.2 it suffices to prove that for every $y, z \in \mathbf{R}^n$, $\rho > 0$ and $A \in \mathbf{M}^{m \times n}$, we have $F(Ax, B(y, \rho)) = F(Ax, B(z, \rho))$. This can be proven similarly to Proposition 14.3. In fact, let $(u_k) \subset W_0^{1,p}(B(y, \rho); \mathbf{R}^m)$ be a sequence such that $u_k \to 0$ in $L^p(\Omega; \mathbf{R}^m)$ and

$$\lim_k F_{\varepsilon_{j_k}}(Ax + u_k, B(y, \rho)) = F(Ax, B(y, \rho)).$$

We extend u_k to $\mathbf{R}^n$ by 0 outside $B(y, \rho)$. Let $r > 1$, let $\eta > 0$ be fixed and let (τ_k) be a sequence converging to $z - y$ such that $\tau_k/\varepsilon_{j_k} \in T_\eta^A$. Let $v_k(x) = u_k(x - \tau_k)$. Note that $v_k = 0$ outside $\tau_k + B(y, \rho)$, and that by (15.2) and (15.1)

$$F_{\varepsilon_{j_k}}(Ax + v_k, \tau_k + B(y, \rho))$$

$$\leq F_{\varepsilon_{j_k}}(Ax + u_k, B(y, \rho)) + \eta \int_{B(y,\rho)} (1 + |Du_k + A|^p)\, dx$$

$$\leq F_{\varepsilon_{j_k}}(Ax + u_k, B(y, \rho)) + \eta\Big(|B(y, \rho)| + \frac{1}{\alpha} F_{\varepsilon_{j_k}}(Ax + u_k, B(y, \rho))\Big).$$

We have $v_k \to 0$ in $W^{1,p}(B(z, r\rho); \mathbf{R}^m)$; hence,

$$F(Ax, B(z, \rho)) \leq F(Ax, B(z, r\rho))$$

$$\leq \liminf_k F_{\varepsilon_{j_k}}(Ax + v_k, B(z, r\rho))$$

$$\leq \Big(1 + \frac{\eta}{\alpha}\Big) \liminf_k F_{\varepsilon_{j_k}}(Ax + u_k, B(y, \rho))$$
$$+ \eta|B(y, \rho)| + |B(z, r\rho) \setminus B(z, \rho)|\beta(1 + |A|^p)$$

$$= \Big(1 + \frac{\eta}{\alpha}\Big) F(Ax, B(y, \rho)) + \eta|B(y, \rho)| + |B(z, r\rho) \setminus B(z, \rho)|\beta(1 + |A|^p).$$

Letting $r \to 1$, and then $\eta \to 0$, we obtain the inequality

$$F(Ax, B(z, \rho)) \leq F(Ax, B(y, \rho));$$

the opposite inequality is obtained by a symmetry argument. $\qquad\square$

Taking $\Omega = (0,1)^n$ we can apply Proposition 11.7 and Theorem 7.2 to the functionals $F_{\varepsilon_{j_k}}(\cdot,\Omega)$ on $Ax + W_0^{1,p}(\Omega;\mathbf{R}^m)$ so that

$$\min\left\{\int_{(0,1)^n} \varphi(A+Du)dx \;:\; u \in W_0^{1,p}((0,1)^n;\mathbf{R}^m)\right\}$$

$$= \lim_k \inf\left\{\int_{(0,1)^n} f\left(\frac{x}{\varepsilon_{j_k}}, \frac{Ax+u}{\varepsilon_{j_k}}, A+Du\right)dx \;:\; u \in W_0^{1,p}((0,1)^n;\mathbf{R}^m)\right\}.$$

By the quasiconvexity of φ and Remark 5.15, the left-hand side is equal to $\varphi(A)$, while, making a change of variables in the right-hand side, we obtain

$$\varphi(A) = \lim_k \inf\left\{\varepsilon_{j_k}^n \int_{(0,\varepsilon_{j_k}^{-1})^n} f(x, Ax+u, A+Du)dx : u \in W_0^{1,p}((0,\varepsilon_{j_k}^{-1})^n;\mathbf{R}^m)\right\}.$$

The proof of Theorem 15.3 will be completed by the following proposition, which shows that the function φ is independent of the subsequence (ε_{j_k}).

Proposition 15.5 *The limit*

$$f_{\mathrm{hom}}(A) = \lim_{t\to+\infty} \inf\left\{\frac{1}{t^n}\int_{(0,t)^n} f(x, Ax+u, A+Du)\,dx : u \in W_0^{1,p}((0,t)^n;\mathbf{R}^m)\right\}$$

$$\tag{15.6}$$

exists for every $A \in \mathbf{M}^{m\times n}$.

Proof The matrix $A \in \mathbf{M}^{m\times n}$ will remain fixed throughout the proof. Define, for every $t > 0$, the quantity

$$g_t = \inf\left\{\frac{1}{t^n}\int_{(0,t)^n} f(x, u(x) + Ax, Du(x) + A)\,dx : u \in W_0^{1,p}((0,t)^n;\mathbf{R}^m)\right\}.$$

For all $t > 0$, let $u_t \in W_0^{1,p}((0,t)^n;\mathbf{R}^m)$ be such that

$$\frac{1}{t^n}\int_{(0,t)^n} f(x, u_t(x) + Ax, Du_t(x) + A)\,dx \le g_t + \frac{1}{t}.$$

We want to estimate g_s, for $s \gg t$, in terms of g_t. For this purpose, we construct $u_s \in W_0^{1,p}((0,s)^n;\mathbf{R}^m)$, by a patchwork procedure, exploiting the 'uniform almost periodicity' of the function f.

Fix $\eta > 0$, let T_η^A be as in (15.2), and let L_η be the inclusion length related to T_η^A. Let $t > 0$ be fixed, and let $s > t + L_\eta$. Define I_s as the set of all $\mathbf{z} = (z_1,\ldots,z_n) \in \mathbf{Z}^n$ such that

$$0 \le z_j \le \left[\frac{s}{t+L_\eta}\right] - 1 \qquad j = 1,\ldots,n,$$

and, for every $\mathbf{z} \in I_s$, choose

$$\tau_{\mathbf{z}} \in \big((t + L_\eta)\mathbf{z} + [0, L_\eta)^n\big) \cap T_\eta^A.$$

We then define

$$u_s(x) = \begin{cases} u_t(x - \tau_{\mathbf{z}}) & \text{if } x \in \tau_{\mathbf{z}} + (0, t)^n \\ 0 & \text{otherwise,} \end{cases}$$

and

$$Q_s' = (0, s)^n \setminus \bigcup_{\mathbf{z} \in I_s} (\tau_{\mathbf{z}} + (0, t)^n).$$

We have

$$|Q_s'| = s^n - \left(\left[\frac{s}{t + L_\eta}\right] - 1\right)^n t^n \leq s^n \left(1 - \left(\frac{t}{t + L_\eta} - \frac{t}{s}\right)^n\right).$$

By again using the relative density of the sets in (15.2) we can now estimate g_s $(s > t + L_\eta)$:

$$\begin{aligned}
g_s &\leq \frac{1}{s^n} \int_{(0,s)^n} f(x, u_s(x) + Ax, Du_s(x) + A)\, dx \\
&= \frac{1}{s^n} \left(\sum_{\mathbf{z} \in I_s} \int_{\tau_{\mathbf{z}} + (0,t)^n} f(x, u_t(x - \tau_{\mathbf{z}}) + Ax, Du_t(x - \tau_{\mathbf{z}}) + A)\, dx \right. \\
&\qquad\qquad \left. + \int_{Q_s'} f(x, Ax, A)\, dx \right) \\
&= \frac{1}{s^n} \left(\sum_{\mathbf{z} \in I_s} \int_{(0,t)^n} f(x + \tau_{\mathbf{z}}, u_t(x) + Ax + A\tau_{\mathbf{z}}, Du_t(x) + A)\, dx \right. \\
&\qquad\qquad \left. + \int_{Q_s'} f(x, Ax, A)\, dx \right) \\
&= \frac{1}{s^n} \left(\sum_{\mathbf{z} \in I_s} \int_{(0,t)^n} \big(f(x + \tau_{\mathbf{z}}, u_t(x) + Ax + A\tau_{\mathbf{z}}, Du_t(x) + A) \right. \\
&\qquad\qquad - f(x, u_t(x) + Ax, Du_t(x) + A)\big)\, dx \\
&\qquad\qquad + \left[\frac{s}{t + L_\eta}\right]^n \int_{(0,t)^n} f(x, u_t(x) + Ax, Du_t(x) + A)\, dx \\
&\qquad\qquad \left. + \int_{Q_s'} f(x, Ax, A)\, dx \right)
\end{aligned}$$

$$\leq \frac{1}{s^n}\left(\sum_{z\in I_s}\eta\int_{(0,t)^n}(1+|Du_t+A|^p)\,dx + \left[\frac{s}{t+L_\eta}\right]^n t^n\left(g_t+\frac{1}{t}\right)\right.$$

$$\left. +\beta|Q'_s|(1+|A|^p)\right)$$

$$\leq \eta\left(\frac{t}{t+L_\eta}\right)^n\left(1+\frac{1}{\alpha}\left(g_t+\frac{1}{t}\right)\right) + \left(\frac{t}{t+L_\eta}\right)^n\left(g_t+\frac{1}{t}\right)$$

$$+\beta\left(1-\left(\frac{t}{t+L_\eta}-\frac{t}{s}\right)^n\right)(1+|A|^p).$$

Thus we have

$$g_s \leq g_t\left(\frac{t}{t+L_\eta}\right)^n\left(1+\frac{\eta}{\alpha}\right)$$

$$+\left(\frac{t}{t+L_\eta}\right)^n\left(\eta\left(1+\frac{1}{\alpha t}\right)+\frac{1}{t}\right) + \beta\left(1-\left(\frac{t}{t+L_\eta}-\frac{t}{s}\right)^n\right)(1+|A|^p).$$

Taking the limit, first as $s\to+\infty$, and then as $t\to+\infty$, we obtain

$$\limsup_{s\to+\infty} g_s \leq \left(1+\frac{\eta}{\alpha}\right)\liminf_{t\to+\infty} g_t + \eta.$$

By the arbitrariness of η we have the desired result. $\qquad\square$

Remark 15.6 As in Remark 14.6, we can give an alternative formula for f_{hom} considering periodic boundary conditions in the place of homogeneous Dirichlet boundary conditions, e.g.

$$f_{\mathrm{hom}}(A) = \lim_{t\to+\infty}\inf\left\{\frac{1}{t^n}\int_{(0,t)^n} f(x,u(x)+Ax,Du(x)+A)\,dx\right.$$

$$\left. : u\in W^{1,p}_{\#}((0,t)^n;\mathbf{R}^m)\right\} \tag{15.7}$$

for all $A\in\mathbf{M}^{m\times n}$.

Example 15.7 The functionals

$$F_\varepsilon(u) = \int_0^1\left|u'-f\left(\frac{t}{\varepsilon},\frac{u}{\varepsilon}\right)\right|^2 dt$$

are related to the systems of ordinary equations

$$u' = f\left(\frac{t}{\varepsilon},\frac{u}{\varepsilon}\right).$$

If f is periodic and continuous, Theorem 15.3 can be applied, obtaining a limit functional of the form (15.3), with the convex function f_{hom} given by (15.4) as integrand. In the case $m = 1$, the connection between f_{hom} and the rotation number of the Poincaré map related to the equation $u' = f(t, u)$ has been studied by Migorski *et al.* (1992).

15.2 An example: loss of smoothness by homogenization

We consider the special case of functionals of the form

$$F_\varepsilon(u) = \int_\Omega f\left(\frac{u}{\varepsilon}, |Du|\right) dx \qquad (15.8)$$

depending on scalar-valued u, with f satisfying condition (15.2). In particular we will consider the case when $f(\cdot, z)$ is 1-periodic for all $z \geq 0$. At the end of the section we will present an example of a smooth f such that the corresponding f_{hom} is only Lipschitz continuous.

We can apply Theorem 15.3 with $f(x, s, A) = f(s, |A|)$ and $m = 1$. In this case the Γ-limit is clearly isotropic, and we can write

$$f_{\mathrm{hom}}(\xi) = \varphi(|\xi|)$$

for all $\xi \in \mathbf{R}^n$. We can give a first characterization of φ as in the following remark.

Remark 15.8 If $f(s, \cdot)$ is increasing for all $s \in \mathbf{R}$, then, with fixed $z \geq 0$, we have, choosing $A = ze_1$ in (15.7),

$$\varphi(z)$$
$$= \lim_{t \to +\infty} \inf\left\{ \frac{1}{t^n} \int_{(0,t)^n} f(u(x) + zx_1, |Du(x) + ze_1|) \, dx : u \in \mathrm{W}^{1,p}_\#((0,t)^n) \right\}$$
$$\geq \limsup_{t \to +\infty} \inf\left\{ \frac{1}{t^n} \int_{(0,t)^n} f(u(x) + zx_1, |D_1u(x) + ze_1|) \, dx : u \in \mathrm{W}^{1,p}_\#((0,t)^n) \right\}$$
$$\geq \limsup_{t \to +\infty} \inf\left\{ \frac{1}{t} \int_0^t f(v(\tau) + z\tau, |v'(\tau) + z|) d\tau : v \in \mathrm{W}^{1,p}_\#(0,t) \right\}.$$

On the other hand, for all $v \in \mathrm{W}^{1,p}_\#(0,t)$, if we set $u(x) = v(x_1)$, then $u \in \mathrm{W}^{1,p}_\#((0,t)^n)$, and

$$\frac{1}{t} \int_0^t f(v(\tau) + z\tau, |v'(\tau) + z|) \, d\tau = \frac{1}{t^n} \int_{(0,t)^n} f(u(x) + zx_1, |Du(x) + ze_1|) \, dx.$$

Hence,

$$\liminf_{t \to +\infty} \inf\left\{ \frac{1}{t} \int_0^t f(v(\tau) + z\tau, |v'(\tau) + z|) \, d\tau : v \in \mathrm{W}^{1,p}_\#(0,t) \right\}$$

$$\geq \lim_{t \to +\infty} \inf\left\{ \frac{1}{t^n} \int_{(0,t)^n} f(u(x) + zx_1, |Du(x) + ze_1|)\, dx : u \in W^{1,p}_{\#}((0,t)^n) \right\}$$
$$= \varphi(z),$$

and we obtain

$$\varphi(z) = \lim_{t \to +\infty} \inf\left\{ \frac{1}{t} \int_0^t f(v(\tau) + z\tau, |v'(\tau) + z|)\, d\tau : v \in W^{1,p}_{\#}(0,t) \right\} \tag{15.9}$$

for all $z \geq 0$.

Note that if $z = 0$ then, choosing all constants as test functions, we get $\varphi(0) \leq \inf_s f(s,0)$. Since

$$\inf_s f(s,0) = \inf\{f(s,z) : \ s \in \mathbf{R},\ z \geq 0\} \leq \varphi(0),$$

we have indeed

$$\varphi(0) = \inf_s f(s,0). \tag{15.10}$$

Since periodic boundary conditions can be replaced by homogeneous Dirichlet boundary conditions φ is also given by

$$\varphi(z) = \lim_{t \to +\infty} \inf\left\{ \frac{1}{t} \int_0^t f(v(\tau) + z\tau, |v'(\tau) + z|)\, d\tau : v \in W^{1,p}_0(0,t) \right\} \tag{15.11}$$

for all $z \geq 0$.

We show now that in the periodic case it is possible to describe the homogenized integrand φ by means of a single minimum problem. Note that this does not follow directly as in the cell-problem formula, since the functionals in (15.8) are not convex.

Proposition 15.9 *Let $f : \mathbf{R} \times [0, +\infty) \to [0, +\infty)$ be a continuous function, 1-periodic in the first variable, convex and non-decreasing in the second one, satisfying a standard growth condition of order $p > 1$. Then the Γ-limit*

$$\Gamma(L^p)\text{-}\lim_{\varepsilon \to 0} \int_\Omega f\left(\frac{u(x)}{\varepsilon}, |Du(x)| \right) dx = \int_\Omega \varphi(|Du(x)|)\, dx \tag{15.12}$$

exists for every bounded open subset Ω of $\mathbf{R}^n$ and every $u \in W^{1,p}(\Omega)$, and the function φ satisfies

$$\varphi(z) = \inf\left\{ z \int_0^{1/z} f(v(\tau) + z\tau, |v'(\tau) + z|)\, d\tau : \ v \in W^{1,p}_0(0, 1/z) \right\} \tag{15.13}$$

for all $z > 0$, and is given by (15.10) if $z = 0$.

Proof By Theorem 15.3 and Remark 15.8 the only thing to show is that formula (15.11) coincides with (15.13). This will be done in the propositions below. $\qquad\square$

Let $\Phi : [0, +\infty) \to [0, +\infty)$ be a convex increasing function such that

$$\liminf_{z \to +\infty} \frac{\Phi(z)}{z^p} > 0. \tag{15.14}$$

We define the set $\mathcal{G}_\Phi$ as the family of all functions $g \in C^2(\mathbf{R}^2)$ satisfying the following properties:

(a) for all $z \in \mathbf{R}$ the function $g(\cdot, z)$ is 1-periodic on $\mathbf{R}$;

(b) $\partial^2 g(s, z)/\partial z^2 > 0$ for all $s, z \in \mathbf{R}$;

(c) $\gamma > 0$ exists such that $\Phi(|z|) \le g(s, z) \le \gamma(1 + |z|^p)$ for all $s, z \in \mathbf{R}$.

Proposition 15.10 *If $g \in \mathcal{G}_\Phi$ then, with fixed $a > 0$ and $|b| > 1$, the minimum problem*

$$\min\left\{ \int_0^a g(v, v')\, dt : \ v \in W^{1,p}(0, a), \ v(0) = 0, v(a) = b \right\} \tag{15.15}$$

admits a solution $u \in C^2([0, a])$ which satisfies the following:

(i) there exists a constant k independent of $g \in \mathcal{G}_\Phi$ such that $0 < |u'| \le k$ on $[0, a]$;

(ii) if we denote by τ the unique point in $(0, a)$ such that $|u(\tau)| = 1$, then $u'(x + \tau) = u'(x)$ for all $x \in [0, a - \tau]$.

Proof The existence of a C^2 solution follows from classical results, noting that u satisfies the Euler equation

$$\frac{d}{dt} \frac{\partial g(u, u')}{\partial z} = \frac{\partial g(u, u')}{\partial s}, \tag{15.16}$$

which can be written in normal form thanks to (b). From (15.16) we obtain that there exists a constant C

$$u' \frac{\partial g(u, u')}{\partial z} - g(u, u') = C. \tag{15.17}$$

Trivially, by the Mean Value Theorem a point $\bar{t} \in (0, a)$ exists such that

$$u'(\bar{t}) = \frac{b}{a}, \tag{15.18}$$

which by (15.17) implies that

$$|C| \le \sup\left\{ \left| \frac{b}{a} \frac{\partial g\left(s, \frac{b}{a}\right)}{\partial z} - g\left(s, \frac{b}{a}\right) \right| : \ s \in \mathbf{R} \right\} \le c\left(1 + \left|\frac{b}{a}\right|^p\right). \tag{15.19}$$

Note that for all $s \in \mathbf{R}$ the function

$$z \mapsto z\frac{\partial g(s, z)}{\partial z} - g(s, z) \tag{15.20}$$

is strictly increasing on $[0, +\infty)$ and strictly decreasing on $(-\infty, 0]$ by (b). Hence, for all $t \in [0, a]$ we have

$$C = u'(t)\frac{\partial g(u(t), u'(t))}{\partial z} - g(u(t), u'(t)) \geq -g(u(t), 0).$$

As $|b| > 1$ the range of u covers the whole interval of periodicity $[0, 1]$ and we have

$$C \geq -\inf\{g(s, 0) : s \in [0, 1]\} = -\inf\{g(s, 0) : s \in \mathbf{R}\}.$$

If a point $t_0 \in [0, a]$ exists such that $u'(t_0) = 0$, then $C = -g(u(t_0), 0)$ so that $g(u(t_0), 0) = \min\{g(s, 0) : s \in \mathbf{R}\}$ and

$$\frac{\partial g(u(t_0), 0)}{\partial s} = 0.$$

This implies that the constant function $v = u(t_0)$ solves (15.16) with Cauchy conditions $v(t_0) = u(t_0)$ and $v'(t_0) = u'(t_0)$. By the uniqueness of the solutions for the Cauchy problem for (15.16) we then have $u = u(t_0)$. In particular, $u'(\bar{t}) = 0$, which contradicts (15.18). Hence $|u'| > 0$. By the hypotheses on g we have

$$\lim_{|z| \to +\infty} \inf\left\{z\frac{\partial g(s, z)}{\partial z} - g(s, z) : s \in \mathbf{R}, \ g \in \mathcal{G}_\Phi\right\} = +\infty;$$

hence by (15.19) we obtain (i).

In order to prove (ii) it is not a restriction to consider the case $b > 1$. In this case $u(\tau) = 1$, and we have

$$u'(0)\frac{\partial g(0, u'(0))}{\partial z} - g(0, u'(0)) = u'(\tau)\frac{\partial g(1, u'(\tau))}{\partial z} - g(1, u'(\tau))$$

$$= u'(\tau)\frac{\partial g(0, u'(\tau))}{\partial z} - g(0, u'(\tau))$$

as $g(\cdot, z)$ is 1-periodic. Now, since $u' > 0$ on $[0, a]$ and the function in (15.20) is one-to-one on $[0, +\infty)$, we have $u'(0) = u'(\tau)$. The function $w(t) = u(t + \tau) - 1$, defined in $[0, a - \tau]$, is a solution to the Cauchy problem related to (15.16) with initial data $w(0) = u(0)$ and $w'(0) = u'(\tau) = u'(0)$. By uniqueness we have $w = u$ on $[0, a - \tau]$, i.e. (ii). $\square$

Proposition 15.11 *Let $a > 0$ and $|b| > 1$. Then there exists a solution u of the problem*

$$\min\left\{\int_0^a f(v, |v'|)\, dt : v \in \mathrm{W}^{1,p}(0, a), \ v(0) = 0, \ v(a) = b\right\} \tag{15.21}$$

which is monotone and such that $\tau \in [0, a]$ exists such that $|u(x + \tau)| = |u(x)| + 1$ on $[0, a - \tau]$.

Proof The function $(s, z) \mapsto f(s, |z|)$ can be approximated uniformly on compact subsets of $\mathbf{R}^2$ by a sequence (g_j) in $\mathcal{G}_\Phi$ for some Φ satisfying (15.14). For each j let u_j be the solution to problem (15.15) with g_j in place of g. We have that u_j is monotone and $\tau_j \in [0, a]$ exist such that $|u_j(x + \tau_j)| = |u_j(x)| + 1$ on $[0, a - \tau_j]$. By Proposition 15.10(i) the sequence (u_j) is equicontinuous so that we can find a subsequence (not relabelled) uniformly converging to $u \in W^{1,p}(0, a)$ and at the same time such that τ_j converges to some $\tau \in [0, a]$. The function u satisfies the required properties. $\square$

Remark 15.12 Let $a > 0$; then we have

$$\min\left\{ \int_0^{|k|a} f(v, |v'|)\, dt : \; v \in W^{1,p}(0, |k|a), \; v(0) = 0, \; v(|k|a) = k \right\}$$

$$= |k| \min\left\{ \int_0^a f(v, |v'|)\, dt : \; v \in W^{1,p}(0, a), \; v(0) = 0, \; v(a) = 1 \right\}$$

for all $k \in \mathbf{Z} \setminus \{0\}$. In fact, from Proposition 15.11 the solution u to the first problem is such that $u(t) - (t/\tau)$ is τ-periodic, with $u(\tau) = 1$, and from the condition $u(|k|a) = k$ we get $\tau = a$, so that u is also a solution to the second minimum problem.

Proposition 15.13 *Equation* (15.13) *holds for all $z > 0$.*

Proof For $a > 0$ and $z > 0$ set

$$\psi_z(a) = \min\left\{ \int_0^a f(v, |v'|)\, dt : \; v \in W^{1,p}(0, a), \; v(0) = 0, \; v(a) = za \right\}.$$

By (15.11) we have

$$\varphi(z) = \lim_{a \to +\infty} \frac{1}{a} \psi_z(a) = \lim_k \frac{z}{k} \psi_z\left(\frac{k}{z}\right).$$

On the other hand, by Remark 15.12 for all $k \in \mathbf{N} \setminus \{0\}$

$$\frac{z}{k} \psi_z\left(\frac{k}{z}\right) = z\, \psi_z\left(\frac{1}{z}\right),$$

and the proof is concluded. $\square$

We finally remark that the homogenized function φ may lose smoothness, as in the following example.

Example 15.14 Let $n = 1$ and $f(s, z) = z^2 + g(s)$ with g continuous and 1-periodic. Then we have

$$\lim_{z \to 0+} \frac{\varphi(z) - \varphi(0)}{z} = 2 \int_0^1 \sqrt{g(s) - \min g}\, ds. \tag{15.22}$$

In particular, φ is not differentiable at 0.

To prove (15.22), up to a translation argument, it suffices to consider the case

$$\min g = 0 = g(0) = g(1),$$

so that by (15.13) and (15.10), (15.22) reduces to

$$\lim_{z \to 0+} \inf \left\{ \int_0^{1/z} (|u'|^2 + g(u)) \, d\tau : \; u(0) = 0, u\left(\frac{1}{z}\right) = 1 \right\} = 2 \int_0^1 \sqrt{g(s)} \, ds.$$

To prove one inequality for this limit, it suffices to remark that for each test function we have

$$\int_0^{1/z} (|u'|^2 + g(u)) \, d\tau \geq \int_0^{1/z} 2\sqrt{g(u)}|u'| \, d\tau$$

$$\geq \left| \int_0^{1/z} 2\sqrt{g(u)}u' \, d\tau \right| = 2 \int_0^1 \sqrt{g(s)} \, ds.$$

To obtain the converse inequality, for every $\eta > 0$ let $v_\eta : [0, +\infty) \to [0, +\infty)$ be the solution to

$$\begin{cases} v'_\eta = \sqrt{g(v_\eta) + \eta} \\ v_\eta(0) = 0, \end{cases}$$

and set $u_\eta = v_\eta \wedge 1$ and $\delta_\eta = \min\{t : u_\eta(t) = 1\}$. If $z \leq 1/\delta_\eta$, then u_η is a test function for the minimum problem above, and we have

$$\int_0^{1/z} (|u'_\eta|^2 + g(u_\eta)) \, d\tau = \int_0^{\delta_\eta} (|v'_\eta|^2 + g(v_\eta)) \, d\tau$$

$$\leq \int_0^{\delta_\eta} (|v'_\eta|^2 + g(v_\eta) + \eta) \, d\tau$$

$$= 2 \int_0^{\delta_\eta} \sqrt{g(v_\eta) + \eta} \, v'_\eta \, d\tau = 2 \int_0^1 \sqrt{g(s) + \eta} \, ds.$$

By the arbitrariness of η we easily get the desired inequality.

15.3   Exercises

Exercise 15.1 Let $a : \mathbf{R} \to [\alpha, \beta]$ be u.a.p., and let $f(x, z) = a(x)z^2$. Use the asymptotic homogenization formula to prove that $f_{\text{hom}}(z) = (\fint a^{-1} \, dt)^{-1}$.

Exercise 15.2 Let G be defined as in Exercise 13.2 and let $r \in \mathbf{R} \setminus \mathbf{Q}$. Prove that the homogenized integrand related to $f(x, z) = f_r(x, z) = (G(x) + G(rx))z^2$ is

$$f_{\text{hom}}(z) = \left(\frac{1}{2(\alpha + \beta)} + \frac{1}{8\alpha} + \frac{1}{8\beta} \right)^{-1} z^2,$$

independent of r. Compare the result with Exercise 13.3.

Hint: use the asymptotic formula, the fact that

$$\oint \frac{1}{G(t) + G(rt)}\, dt = \int_0^1 \int_0^1 \frac{1}{G(x) + G(y)}\, dx\, dy$$

(use Theorem A.13 and an approximation argument of G by continuous functions), and Exercise 13.2.

16

TWO APPLICATIONS

16.1 Homogenization of Riemannian metrics

We consider as an application of Theorem 15.3 the limits of functionals related to Riemannian metrics and show that in the limit we may obtain a Finsler metric.

Proposition 16.1 *Let $N \in \mathbf{N}$, (a_{ij}) be a 1-periodic $N \times N$ matrix of bounded measurable functions such that*

$$\alpha|\xi|^2 \leq \sum_{i,j=1}^{N} a_{ij}(s)\xi_i\xi_j \leq \beta|\xi|^2 \tag{16.1}$$

for all $\xi \in \mathbf{R}^N$ and $s \in \mathbf{R}^N$. Then there exists a convex function $\varphi : \mathbf{R}^N \to [0, +\infty)$ such that for every bounded open subset I of $\mathbf{R}$ and $u = (u_1, \ldots, u_N) \in \mathrm{W}^{1,p}(I; \mathbf{R}^N)$ the limit

$$\Gamma(\mathrm{L}^p)\text{-}\lim_{\varepsilon \to 0} \int_I \sum_{i,j=1}^{N} a_{ij}\left(\frac{u(t)}{\varepsilon}\right) u_i' u_j' \, dt = \int_I \varphi(u'(t)) \, dt \tag{16.2}$$

exists, and the function φ satisfies

$$\varphi(\xi) = \lim_{t \to +\infty} \inf\left\{\frac{1}{t} \int_0^t \sum_{i,j=1}^{N} a_{ij}(u + \xi\tau)(u_i' + \xi_i)(u_j' + \xi_j) \, d\tau\right.$$

$$\left. : u \in \mathrm{W}_0^{1,p}((0,t); \mathbf{R}^N)\right\} \tag{16.3}$$

for all $\xi \in \mathbf{R}^N$

Proof It suffices to apply Theorem 15.3 with $n = 1$, $m = N$, and $f(x, s, \xi) = \sum_{i,j=1}^{N} a_{ij}(s)\xi_i\xi_j$. $\qquad\square$

We show by an explicit calculation that the function φ may not be a quadratic function. In this case the limit energy can be interpreted as related to a Finsler metric.

Example 16.2 We take $N = 2$ in the previous proposition, and $a_{ij}(s) = a(s)\delta_{ij}$, where $a : \mathbf{R}^2 \to \{\alpha, \beta\}$ is the 1-periodic function defined on $[0,1)^2$ by

$$a(s) = \begin{cases} \beta & \text{if } s \in (0, \tfrac{1}{2}) \times (\tfrac{1}{2}, 1) \text{ or } s \in (\tfrac{1}{2}, 1) \times (0, \tfrac{1}{2}) \\ \alpha & \text{otherwise,} \end{cases} \tag{16.4}$$

where

$$4\alpha < \beta. \tag{16.5}$$

The matrix (a_{ij}) is related to a chessboard-type structure on $\mathbf{R}^2$. The corresponding functionals take the form

$$\int_I a\left(\frac{u(t)}{\varepsilon}\right) |u'(t)|^2 \, dt, \tag{16.6}$$

and their Γ-limit is given by (16.2) and (16.3).

Fix $\xi \in \mathbf{Q}^2$. Let $j \in \mathbf{N}$ such that $j\xi \in \mathbf{Z}^2$. By (16.3) we have

$$\varphi(\xi) = \lim_k \min\left\{ \int_0^1 a(kju)|u'|^2 \, d\tau \, : \, u - \xi\tau \in W_0^{1,2}((0,1);\mathbf{R}^2) \right\} \tag{16.7}$$

for all $\xi \in \mathbf{R}^2$. For all $k \in \mathbf{N}$, let $u_k \in W_0^{1,2}((0,1);\mathbf{R}^2)$ be a solution of the minimum problem in (16.7).

Step 1: for every $k \in \mathbf{N}$ the function u_k is piecewise affine on $(0,1)$.

Let $\mathcal{Q}$ be the family of squares of the form either $[0,\frac{1}{2j}]^2 + \frac{1}{j}\mathbf{Z}^2$ or $[-\frac{1}{2j},0]^2 + \frac{1}{j}\mathbf{Z}^2$. Suppose that $Q \in \mathcal{Q}$ and that, if we set $T_Q = \{\tau \in [0,1] : u_k(\tau) \in Q\}$, and $s = \min T_Q$, $t = \max T_Q$ (we consider the continuous representative of u_k), we have $s < t$. Define the function

$$v(\tau) = \begin{cases} u_k(\tau) & \text{if } \tau \le s \text{ or } \tau \ge t \\[2mm] u_k(s) + \dfrac{u_k(t) - u_k(s)}{t - s}(\tau - s) & \text{otherwise.} \end{cases}$$

We have

$$\int_s^t \alpha|v'|^2 d\tau \le \int_s^t a(kju_k)|u_k'|^2 \, d\tau,$$

and equality holds only if $u_k = v$. This argument shows that u_k is affine on each T_Q. Note that since $\|u_k\|_\infty$ is finite we have $\#\{Q \in \mathcal{Q} : T_Q \ne \emptyset\} < +\infty$.

If Q is an open square of the form $(0,\frac{1}{2j}) \times (\frac{1}{2j},\frac{1}{j}) + \frac{1}{j}\mathbf{Z}^2$ or $(\frac{1}{2j},\frac{1}{j}) \times (0,\frac{1}{2j}) + \frac{1}{j}\mathbf{Z}^2$, then $T_Q = u_k^{-1}(Q)$ is an open subset of $(0,1)$. The previous argument shows that we can write T_Q as the union of at most three open intervals. If (s,t) is one of these open intervals and we define v as above, then we have by convexity that

$$\int_s^t a(jkv)|v'|^2 d\tau \le \int_s^t \beta|v'|^2 d\tau \le \int_s^t \beta|u_k'|^2 \, d\tau = \int_s^t a(kju_k)|u_k'|^2 \, d\tau,$$

and again equality holds only if $v = u_k$. This completes the proof of the step.

Step 2: the set $I = \{\tau : u_k(\tau) \notin Q, \ Q \in \mathcal{Q}\}$ is empty.

Choose a maximal open subinterval (s,t) of I on which u_k is affine. Since 0 and 1 belong to $[0,1] \setminus I$, we have that $s, t \notin I$; hence we can find a minimum point v for the problem

$$\min\left\{\int_s^t |v'|^2 \, d\tau : v \in W^{1,2}((s,t); \mathbf{R}^2), \ v(s) = u_k(s), \ v(t) = u_k(t), \ a(jkv) = \alpha\right\}.$$

The function v is piecewise affine, and

$$\int_s^t |v'|^2 \, d\tau \leq 4 \int_s^t |u_k'|^2 \, d\tau.$$

By (16.5) this implies that

$$\int_s^t a(jkv)|v'|^2 \, d\tau < \int_s^t a(jku_k)|u_k'|^2 \, d\tau,$$

contradicting the minimality of u_k.

Step 3: for every $k \in \mathbf{N}$ we have that $|u_k'|$ is constant on $(0,1)$.

Let $[s_1, t_1]$ and $[s_2, t_2]$ be two disjoint intervals of $(0,1)$ on each of whom u_k is affine. We assume $s_1 < t_1 \leq s_2 < t_2$. Set

$$c = \frac{|u_k(t_1) - u_k(s_1)| + |u_k(t_2) - u_k(s_2)|}{|t_1 - s_1| + |t_2 - s_2|}$$

$$t_1' = s_1 + \frac{|u_k(t_1) - u_k(s_1)|}{c}, \qquad s_2' = t_1' + (s_2 - t_1),$$

and define

$$v(\tau) = \begin{cases} u_k(\tau) & \text{if } \tau \leq s_1 \\[2ex] u_k(s_1) + \dfrac{u_k(t_1) - u_k(s_1)}{t_1' - s_1}(\tau - s_1) & \text{if } s_1 < \tau \leq t_1' \\[2ex] u_k(\tau - t_1' + t_1) & \text{if } t_1' < \tau \leq s_2' \\[2ex] u_k(s_2) + \dfrac{u_k(t_2) - u_k(s_2)}{t_2 - s_2'}(\tau - s_2') & \text{if } s_2' < \tau \leq t_2 \\[2ex] u_k(\tau) & \text{if } t_2 < \tau \,. \end{cases}$$

We have

$$\int_0^1 a(kjv)|v'|^2 d\tau \leq \int_0^1 a(kju_k)|u_k'|^2 d\tau,$$

and equality holds only if $u_k = v$; that is, $|u_k'|$ has the same value on both intervals. This shows that $|u_k'|$ is constant on $(0,1)$.

Step 4: computation of $\varphi(\xi)$.
Since $a(jku_k) = \alpha$ and $|u'_k|$ is constant on $(0,1)$, we have

$$\int_0^1 a(jku_k)|u'_k|^2 d\tau = \alpha\Big(\min\Big\{\int_0^1 |v'|d\tau : \ v(0) = 0, \ v(1) = \xi, \ a(jkv) = \alpha\Big\}\Big)^2,$$

i.e. we have to minimize the length among all curves joining 0 and ξ which lie in the set $\{\zeta : \ a(jk\zeta) = \alpha\}$. Elementary geometric reasonings show that (for $\xi_1 > \xi_2 \geq 0$) a minimal (piecewise affine) curve is given by the segment joining $(0,0)$ and (ξ_2, ξ_2) and the segment joining (ξ_2, ξ_2) and (ξ_1, ξ_2). In the general case symmetry arguments show that

$$\varphi(\xi) = \alpha\Big((\sqrt{2} - 1)\min\{|\xi_1|, |\xi_2|\} + \max\{|\xi_1|, |\xi_2|\}\Big)^2.$$

The same formula for φ holds on the whole $\mathbf{R}^2$ since φ is continuous.

Step 5: conclusions.
Since $\varphi(1,0) = \varphi(0,1) = \varphi(\frac{1}{\sqrt{2}}, \frac{1}{\sqrt{2}}) = \alpha$, if φ were a quadratic form we would have $\varphi(\xi) = \alpha|\xi|^2$, which is not the case. Hence the example is complete.

16.2 Homogenization of Hamilton Jacobi equations

We can derive from Theorem 15.3 a homogenization result for Hamilton Jacobi equations, via the Γ-convergence of functionals related to the Legendre transforms of the Hamiltonians. More precisely, let $H : \mathbf{R} \times \mathbf{R}^n \times \mathbf{R}^n \to \mathbf{R}$ be a continuous function satisfying:

(i) $H(t, x, \cdot)$ is convex for every (t, x);

(ii) there exists $1 < p < +\infty$ and $C > 0$ such that

$$|\xi|^p \leq H(t, x, \xi) \leq C(1 + |\xi|^p)$$

for every (t, x, ξ);

(iii) the function $t \mapsto H(t, \cdot, \xi)$ is uniformly almost periodic with values in $UAP(\mathbf{R}^n; \mathbf{R})$, in the sense of Definition 15.1.

In particular we can consider the case when $(t, x) \mapsto H(t, x, \xi)$ is 1-periodic.

We will study the limiting behaviour of the viscosity solutions of the Cauchy problem

$$\begin{cases} \dfrac{\partial u_\varepsilon}{\partial t} + H\left(\dfrac{t}{\varepsilon}, \dfrac{x}{\varepsilon}, Du_\varepsilon\right) = 0 & \text{in } \mathbf{R}^n \times [0, +\infty) \\[2mm] u_\varepsilon(x, 0) = \varphi(x) & \text{in } \mathbf{R}^n, \end{cases} \tag{16.8}$$

where φ is a given bounded and uniformly continuous function in $\mathbf{R}^n$. We recall the following definitions.

Definition 16.3 *The* Legendre transform *of H is defined as*

$$L(t,x,\xi) = \sup_{\xi' \in \mathbf{R}^n} \{\langle \xi, \xi' \rangle - H(t,x,\xi')\},$$

for every (t,x,ξ).

Definition 16.4 *Let H be a continuous function satisfying* (i) *and* (ii). *Then, the (unique)* viscosity solution *to the* Hamilton Jacobi equation

$$\begin{cases} \dfrac{\partial v}{\partial t} + H(t,x,Dv) = 0 & \text{in } \mathbf{R}^n \times [0,+\infty) \\[2mm] v(x,0) = \varphi(x) & \text{in } \mathbf{R}^n, \end{cases} \tag{16.9}$$

where φ is a given bounded and uniformly continuous function in $\mathbf{R}^n$, is the function constructed as follows. Let L be the Legendre transform of H. We define for $x,y \in \mathbf{R}^n$ and $0 \le s < t$

$$\begin{aligned} S(x,t;y,s) = \inf\Big\{ &\int_s^t L(\tau,u(\tau),u'(\tau))\,d\tau : \\ & u(s) = y,\ u(t) = x,\ u \in W^{1,\infty}((s,t);\mathbf{R}^n) \Big\} \\ = \inf\Big\{ &\int_s^t L(\tau,u(\tau),u'(\tau))\,d\tau : \\ & u(\tau) - \left(\frac{y-x}{s-t}(\tau-s)+y\right) \in W_0^{1,p'}((s,t);\mathbf{R}^n) \Big\}. \end{aligned}$$

Then v is given by

$$v(x,t) = \inf\{\varphi(y) + S(x,t;y,s) \ : \ y \in \mathbf{R}^n, 0 \le s < t\}.$$

This last equality is usually referred to as the Lax formula.

We have the following proposition.

Proposition 16.5 *The limit*

$$\overline{L}(\xi) = \lim_{T \to +\infty} \frac{1}{T} \inf\Big\{ \int_0^T L(\tau,u(\tau)+\xi\tau,u'(\tau)+\xi)\,d\tau : \ u \in W_0^{1,p'}((0,T);\mathbf{R}^n) \Big\} \tag{16.10}$$

exists for every $\xi \in \mathbf{R}^n$.

Proof By the growth hypothesis (ii), if we set

$$C_1 = \frac{p-1}{p^{p'}}\frac{1}{C^{1/(p-1)}} \qquad C_2 = \frac{p-1}{p^{p'}},$$

we have that for every (t,x,ξ)

$$C_1|\xi|^{p'} - C \le L(t, x, \xi) \le C_2|\xi|^{p'}.$$

The existence of the limit in (16.10) follows by Theorem 15.3 with $m = n$, taking $f(t, x, \xi) = L(t, x, \xi) + C$. $\qquad\square$

We can now define the *homogenized Hamiltonian* $\overline{H}$ as the Legendre transform of $\overline{L}$, i.e.

$$\overline{H}(\xi) = \sup_{\xi' \in \mathbf{R}^n} \{\langle \xi, \xi' \rangle - \overline{L}(\xi')\},$$

and state the convergence result as follows.

Theorem 16.6 *Let φ be a given bounded and uniformly continuous function in $\mathbf{R}^n$, and let u_ε be the unique viscosity solution of (16.8). Then as $\varepsilon \to 0$, the family (u_ε) converges uniformly on compact sets to the unique viscosity solution of the Cauchy problem*

$$\begin{cases} \dfrac{\partial u}{\partial t} + \overline{H}(Du) = 0 & \text{in } \mathbf{R}^n \times [0, +\infty) \\[2mm] u(x, 0) = \varphi(x) & \text{in } \mathbf{R}^n. \end{cases} \tag{16.11}$$

Proof Following Definition 16.4, we set for $x, y \in \mathbf{R}^n$ and $0 \le s < t$

$$\begin{aligned} S_\varepsilon(x, t; y, s) &= \inf\Big\{ \int_s^t L\left(\frac{\tau}{\varepsilon}, \frac{u(\tau)}{\varepsilon}, u'(\tau)\right) d\tau \ : \\ &\qquad\qquad u(s) = y, \ u(t) = x, \ u \in W^{1,\infty}((s, t); \mathbf{R}^n) \Big\} \\ &= \inf\Big\{ \int_s^t L\left(\frac{\tau}{\varepsilon}, \frac{u(\tau)}{\varepsilon}, u'(\tau)\right) d\tau \ : \\ &\qquad\qquad u(\tau) - \left(\frac{y - x}{s - t}(\tau - s) + y\right) \in W_0^{1,p'}((s, t); \mathbf{R}^n) \Big\}. \end{aligned}$$

Then the unique viscosity solution to problem (16.8) is

$$u_\varepsilon(x, t) = \inf\{\varphi(y) + S_\varepsilon(x, t; y, s) \ : \ y \in \mathbf{R}^n, 0 \le s < t\}.$$

By Theorem 15.3 and Proposition 11.7, we have that for every $x, y \in \mathbf{R}^n$ and $0 \le s < t$

$$\begin{aligned} \lim_{\varepsilon \to 0} S_\varepsilon(x, t; y, s) &= \min\Big\{ \int_s^t \overline{L}(u'(\tau)) \, d\tau \ : \\ &\qquad u(\tau) - \left(\frac{y - x}{s - t}(\tau - s) + y\right) \in W_0^{1,p'}((s, t); \mathbf{R}^n) \Big\} \\ &= (t - s)\overline{L}\left(\frac{y - x}{s - t}\right), \end{aligned}$$

the last equality following from the convexity of $\overline{L}$ and Jensen's inequality. By the growth hypothesis on L we obtain that the functions $S_\varepsilon(x,t;\cdot,\cdot)$ are equicontinuous in $\{y \in \mathbf{R}^n, 0 \le s \le t - \eta\}$, and then

$$u_\varepsilon(x,t) \to u(x,t)$$

pointwise, where

$$u(x,t) = \inf\left\{\varphi(y) + (t-s)\,\overline{L}\left(\frac{y-x}{s-t}\right) \;:\; y \in \mathbf{R}^n, 0 \le s < t\right\}.$$

Since the functions u_ε are equicontinuous on compact sets, the convergence is uniform on bounded sets. Again by the Lax formula, and by the definition of $\overline{H}$, u is the unique viscosity solution of (16.11). $\qquad\square$

Example 16.7 Let $H = H(x,\xi)$ be 1-periodic in the first variable, and uniformly continuous on $\mathbf{R}^n \times B(0,R)$ for all $R > 0$. We can give an alternative definition of $\overline{H}$: for every $\xi \in \mathbf{R}^n$, $\overline{H}(\xi)$ is the unique constant λ such that the stationary problem

$$H(x, \xi + Du(x)) = \lambda \tag{16.12}$$

has a C^1 periodic solution.

This fact can be proven using some results of the theory of viscosity solutions. We outline here a proof, referring to Lions (1982) for details. To prove the existence of u and λ satisfying (16.12), consider the approximate equations

$$H(x, \xi + Du_\eta(x)) + \eta u_\eta = 0 \tag{16.13}$$

for $\eta > 0$. By the hypotheses on H there exists a unique viscosity solution $u_\eta \in \mathrm{W}^{1,\infty}(\mathbf{R}^n)$ of (16.13). By the uniqueness we also get that u_η is 1-periodic. Furthermore we have

$$-\sup\{H(y,\xi) : y \in \mathbf{R}^n\} \le \eta u_\eta \le -\inf\{H(y,\xi) : y \in \mathbf{R}^n\}. \tag{16.14}$$

The growth conditions on H, (16.13) and (16.14) imply that $\|Du_\eta\|_\infty \le c$, with c independent of η. Set $\hat{u}_\eta = u_\eta - \min u_\eta$. Up to a subsequence, we may assume that $(\hat{u}_\eta, -\eta u_\eta)$ converge uniformly to some pair $(u,\lambda) \in \mathrm{W}^{1,\infty}(\mathbf{R}^n) \times \mathbf{R}$ as $\eta \to 0$. By the properties of viscosity solutions u solves (16.12). The uniqueness of λ can be proved by a comparison principle.

To prove that the constant λ defined above equals $\overline{H}(\xi)$, it suffices to check the convergence of viscosity solutions in a particular case. If we choose $u_\varepsilon(x,t) = \langle \xi, x \rangle - \lambda t + \varepsilon u(\frac{x}{\varepsilon})$ (with u defined above), then u_ε is a viscosity solution of

$$\begin{cases} \dfrac{\partial u_\varepsilon}{\partial t} + H\left(\dfrac{x}{\varepsilon}, Du_\varepsilon\right) = 0 & \text{in } \mathbf{R}^n \times [0, +\infty) \\[2mm] u_\varepsilon(x,0) = u^\varepsilon(x) & \text{in } \mathbf{R}^n, \end{cases} \tag{16.15}$$

where $u^\varepsilon(x) = \langle \xi, x \rangle + \varepsilon u(\frac{x}{\varepsilon})$. As u^ε converge to $u^0(x) = \langle \xi, x \rangle$, and $u_\varepsilon(x,t)$ converge to $u_0(x,t) = \langle \xi, x \rangle - \lambda t$, it is possible to deduce that u_0 is a viscosity solution of

$$\begin{cases} \dfrac{\partial u_0}{\partial t} + \overline{H}(Du_0) = 0 & \text{in } \mathbf{R}^n \times [0, +\infty) \\[2mm] u_0(x,0) = u^0(x) & \text{in } \mathbf{R}^n, \end{cases} \qquad (16.16)$$

i.e. $\lambda = \overline{H}(\xi)$.

We can apply this remark to obtain an explicit formula for $\overline{H}$, when $n = 1$,

$$H(x,\xi) = |\xi|^2 - V(x),$$

V is 1-periodic, continuous, and $\min V = 0$. When $H(\xi) > 0$, from equation (16.12) we have

$$|u'(x) + \xi|^2 = V(x) + \overline{H}(\xi) > 0;$$

hence, by the requirement that u' be continuous,

$$u'(x) = -\xi + \sqrt{V(x) + \overline{H}(\xi)} \ \text{ or } \ u'(x) = -\xi - \sqrt{V(x) + \overline{H}(\xi)}.$$

The function u is then periodic if and only if the mean value of u' is zero, i.e.

$$|\xi| = \int_0^1 \sqrt{V(y) + \overline{H}(\xi)}\, dy.$$

Since $\overline{H}$ is positive and convex, we obtain the formula

$$\overline{H}(\xi) = \begin{cases} 0 & \text{if } |\xi| \leq \int_0^1 \sqrt{V(y)}\, dy \\[3mm] \alpha & \text{if } |\xi| = \int_0^1 \sqrt{V(y) + \alpha}\, dy. \end{cases}$$

The flat piece in the graph of $\overline{H}$ corresponds to the lack of differentiability of $\overline{L}$ at 0, as already observed in Section 15.2.

17

A CLOSURE THEOREM FOR THE HOMOGENIZATION

In this chapter we show that homogenization is stable under some kinds of convergence of the integrands, and that in particular it is not affected by perturbations with compact support. As an application we will prove a homogenization theorem under very mild assumptions of almost periodicity.

17.1 A closure theorem

The closure theorem that we are going to prove relies on the following partial regularity result.

Theorem 17.1 (Meyers Regularity Theorem) *Let $p > 1$, and let $f : \mathbf{R}^n \times \mathbf{M}^{m \times n} \to \mathbf{R}$ be a Borel function, quasiconvex in the second variable, satisfying*

$$\alpha |A|^p \le f(x, A) \le \beta(1 + |A|^p). \tag{17.1}$$

Let B be an open ball in $\mathbf{R}^n$ and let $\overline{u} \in C^\infty(\overline{B}; \mathbf{R}^m)$. Then there exists $\eta > 0$ depending only on α, β, B and $\overline{u}$ such that for all $\lambda \ge 0$ the minimum points u_λ of

$$G_\lambda(u) = \int_B f(x, Du + D\overline{u}) \, dx + \lambda \int_B |u|^p \, dx \tag{17.2}$$

on $\mathrm{W}_0^{1,p}(B; \mathbf{R}^m)$ belong to $\mathrm{W}^{1,p+\eta}(B; \mathbf{R}^m)$, and

$$\|u_\lambda\|_{\mathrm{W}^{1,p+\eta}(B; \mathbf{R}^m)} \le C, \tag{17.3}$$

with the constant C depending only on $\lambda, \alpha, \beta, B$ and $\overline{u}$.

Definition 17.2 *Let (f_j) and (φ_j) be two sequences of real Borel functions defined on $\mathbf{R}^n \times \mathbf{M}^{m \times n}$. We say that the two sequences above are* equivalent sequences *if we have*

$$\lim_j \fint \sup_{|A| \le y} |f_j(x, A) - \varphi_j(x, A)| \, dx = 0 \tag{17.4}$$

for all $y \ge 0$.

Remark 17.3 (i) If (f_j) and (φ_j) are equivalent sequences and we define

$$g_j(x, y) = \sup_{|A| \le y} |f_j(x, A) - \varphi_j(x, A)|, \tag{17.5}$$

then for every bounded open set Ω we have

$$\lim_j \limsup_{\varepsilon \to 0+} \int_\Omega g_j\left(\frac{x}{\varepsilon}, y\right) dx = 0 \tag{17.6}$$

for all $y \geq 0$.

(ii) If all functions f_j and φ_j satisfy the growth condition (17.1) uniformly, then

$$0 \leq g_j(x, y) \leq 2\beta(1 + y^p) \tag{17.7}$$

for all $y \geq 0$.

Proposition 17.4 *Let (f_j) and (φ_j) be two equivalent sequences of functions uniformly satisfying the growth condition (17.1) with $p > 1$ and quasiconvex in the second variable. Let B be an open ball of $\mathbf{R}^n$ and let $\overline{u} \in C^\infty(\overline{B}; \mathbf{R}^m)$. We define the functionals*

$$F_\varepsilon^j(u, B) = \int_B f_j\left(\frac{x}{\varepsilon}, Du\right) dx, \qquad \Phi_\varepsilon^j(u, B) = \int_B \varphi_j\left(\frac{x}{\varepsilon}, Du\right) dx \tag{17.8}$$

for $u \in \overline{u} + W_0^{1,p}(B; \mathbf{R}^m)$, extended to $+\infty$ outside $\overline{u} + W_0^{1,p}(B; \mathbf{R}^m)$. Then we have

$$\lim_{\lambda \to +\infty} \liminf_j \limsup_{\varepsilon \to 0+} T_\lambda^p F_\varepsilon^j(\overline{u}, B) = \lim_{\lambda \to +\infty} \liminf_j \limsup_{\varepsilon \to 0+} T_\lambda^p \Phi_\varepsilon^j(\overline{u}, B), \tag{17.9}$$

$$\lim_{\lambda \to +\infty} \limsup_j \liminf_{\varepsilon \to 0+} T_\lambda^p F_\varepsilon^j(\overline{u}, B) = \lim_{\lambda \to +\infty} \limsup_j \liminf_{\varepsilon \to 0+} T_\lambda^p \Phi_\varepsilon^j(\overline{u}, B) \tag{17.10}$$

(T_λ^p denotes the Yosida transform with respect to the L^p metric).

Proof Fix $j \in \mathbf{N}$ and $\varepsilon > 0$. By the definition of the Yosida transform we have

$$T_\lambda^p \Phi_\varepsilon^j(\overline{u}, B) = \min\left\{\int_B \varphi_j\left(\frac{x}{\varepsilon}, Du + D\overline{u}\right) dx + \lambda \int_B |u|^p dx : u \in W_0^{1,p}(B; \mathbf{R}^m)\right\}; \tag{17.11}$$

hence, by Theorem 17.1 there exist η and $C(\lambda)$ such that every minimum point w_λ of (17.11) satisfies

$$\|w_\lambda + \overline{u}\|_{W^{1,p+\eta}(B; \mathbf{R}^m)} \leq C(\lambda). \tag{17.12}$$

We can estimate

$$T_\lambda^p F_\varepsilon^j(\overline{u}, B) \leq \int_B f_j\left(\frac{x}{\varepsilon}, Dw_\lambda + D\overline{u}\right) dx + \lambda \int_B |w_\lambda|^p dx$$

$$\leq T_\lambda^p \Phi_\varepsilon^j(\overline{u}, B) + \int_B \left|f_j\left(\frac{x}{\varepsilon}, Dw_\lambda + D\overline{u}\right) - \varphi_j\left(\frac{x}{\varepsilon}, Dw_\lambda + D\overline{u}\right)\right| dx$$

$$\leq T_\lambda^p \Phi_\varepsilon^j(\overline{u}, B) + \int_B g_j\left(\frac{x}{\varepsilon}, |Dw_\lambda + D\overline{u}|\right) dx. \tag{17.13}$$

For every $K > 0$ let $B_K = \{x \in B : |Dw_\lambda(x) + D\overline{u}(x)| > K\}$; clearly we have

$$\begin{aligned}
|B_K| K^p &= \int_{B_K} K^p \, dx \leq \int_{B_K} |Dw_\lambda + D\overline{u}|^p \, dx \\
&\leq \int_B |Dw_\lambda + D\overline{u}|^p \, dx \\
&\leq \frac{1}{\alpha} \int_B \varphi_j\left(\frac{x}{\varepsilon}, Dw_\lambda + D\overline{u}\right) dx \\
&\leq \frac{1}{\alpha} T_\lambda^p \Phi_\varepsilon^j(\overline{u}, B) \leq \frac{1}{\alpha} \int_B \varphi_j\left(\frac{x}{\varepsilon}, D\overline{u}\right) dx \\
&\leq \frac{\beta}{\alpha} \int_B (1 + |D\overline{u}|^p) \, dx,
\end{aligned}$$

so that

$$|B_K| \leq C_0 K^{-p}, \tag{17.14}$$

with C_0 depending only on α, β, B and $\overline{u}$. By Hölder's inequality and (17.12)

$$\begin{aligned}
\int_{B_K} |Dw_\lambda + D\overline{u}|^p \, dx &\leq |B_K|^{\eta/\eta+p}\left(\int_B |Dw_\lambda + D\overline{u}|^{p+\eta} \, dx\right)^{p/p+\eta} \\
&\leq C_1(\lambda) K^{-p\eta/p+\eta}, \tag{17.15}
\end{aligned}$$

where $C_1(\lambda) = C_0^{\eta/\eta+p} C(\lambda)^p$. From (17.13)–(17.15) and (17.7) we obtain

$$\begin{aligned}
T_\lambda^p F_\varepsilon^j(\overline{u}, B) &\leq T_\lambda^p \Phi_\varepsilon^j(\overline{u}, B) + \int_B g_j\left(\frac{x}{\varepsilon}, |Dw_\lambda + D\overline{u}|\right) dx \\
&\leq T_\lambda^p \Phi_\varepsilon^j(\overline{u}, B) + \int_{B \setminus B_K} g_j\left(\frac{x}{\varepsilon}, |Dw_\lambda + D\overline{u}|\right) dx \\
&\quad + 2\beta \int_{B_K} (1 + |Dw_\lambda + D\overline{u}|^p) \, dx \\
&\leq T_\lambda^p \Phi_\varepsilon^j(\overline{u}, B) + \int_{B \setminus B_K} g_j\left(\frac{x}{\varepsilon}, K\right) dx \\
&\quad + 2\beta(C_0 K^{-p} + C_1(\lambda) K^{-p\eta/p+\eta}).
\end{aligned}$$

We can pass to the limit as $\varepsilon \to 0+$, and obtain

$$\begin{aligned}
\limsup_{\varepsilon \to 0+} T_\lambda^p F_\varepsilon^j(\overline{u}, B) &\leq \limsup_{\varepsilon \to 0+} T_\lambda^p \Phi_\varepsilon^j(\overline{u}, B) \\
&\quad + \limsup_{\varepsilon \to 0+} \int_B g_j\left(\frac{x}{\varepsilon}, K\right) dx + 2\beta(C_0 K^{-p} + C_1(\lambda) K^{-p\eta/p+\eta});
\end{aligned}$$

then, we take the limit in j, recalling (17.6), to get

$$\liminf_j \limsup_{\varepsilon \to 0+} T_\lambda^p F_\varepsilon^j(\overline{u}, B)$$

$$\leq \liminf_j \limsup_{\varepsilon \to 0+} T_\lambda^p \Phi_\varepsilon^j(\overline{u}, B) + 2\beta(C_0 K^{-p} + C_1(\lambda)K^{-p\eta/p+\eta}).$$

Eventually, we take the limit first as $K \to +\infty$, and then as $\lambda \to +\infty$, and we obtain

$$\lim_{\lambda \to +\infty} \liminf_j \limsup_{\varepsilon \to 0+} T_\lambda^p F_\varepsilon^j(\overline{u}, B) \leq \lim_{\lambda \to +\infty} \liminf_j \limsup_{\varepsilon \to 0+} T_\lambda^p \Phi_\varepsilon^j(\overline{u}, B). \quad (17.16)$$

These limits exist since the functions are increasing with λ.

As F_ε^j and Φ_ε^j play symmetric roles, inequality (17.16) can be reversed to obtain (17.9). In the same way we can prove (17.10). $\qquad\square$

Theorem 17.5 (Homogenization Closure Theorem) *Let $p > 1$, let $f : \mathbf{R}^n \times \mathbf{M}^{m \times n} \to [0, +\infty)$ be a Borel function and let (φ_j) be a sequence of Borel functions equivalent to the constant sequence f; let all functions f and φ_j be quasiconvex in the second variable and satisfy the growth condition (17.1). If for all $j \in \mathbf{N}$ the limit*

$$\Gamma(\mathrm{L}^p)\text{-}\lim_{\varepsilon \to 0+} \int_\Omega \varphi_j\left(\frac{x}{\varepsilon}, Du\right) dx = \int_\Omega \psi_j(Du)\, dx \quad (17.17)$$

exists for all Ω bounded open subsets of $\mathbf{R}^n$ and $u \in \mathrm{W}^{1,p}(\Omega; \mathbf{R}^m)$, then also the limit

$$\Gamma(\mathrm{L}^p)\text{-}\lim_{\varepsilon \to 0+} \int_\Omega f\left(\frac{x}{\varepsilon}, Du\right) dx = \int_\Omega f_{\mathrm{hom}}(Du)\, dx \quad (17.18)$$

exists for all Ω bounded open subsets of $\mathbf{R}^n$ and $u \in \mathrm{W}^{1,p}(\Omega; \mathbf{R}^m)$, with f_{hom} as in Theorem 14.5. Moreover,

$$f_{\mathrm{hom}}(A) = \lim_j \psi_j(A) \quad (17.19)$$

for all $A \in \mathbf{M}^{m \times n}$.

Proof We denote $F_\varepsilon(u, \Omega) = \int_\Omega f(\frac{x}{\varepsilon}, Du)\, dx$. By Theorem 12.5, from every sequence (ε_k) we can extract a further subsequence, which we still denote by (ε_k), such that the Γ-limit

$$G(u, \Omega) = \Gamma(\mathrm{L}^p)\text{-}\lim_k F_{\varepsilon_k}(u, \Omega) = \int_\Omega \psi(x, Du)\, dx$$

exists for all Ω bounded open subsets of $\mathbf{R}^n$ and $u \in \mathrm{W}^{1,p}(\Omega; \mathbf{R}^m)$. Moreover, by Proposition 11.7 in order to evaluate $G(u, \Omega)$ it suffices to restrict all functionals to the space $u + \mathrm{W}_0^{1,p}(\Omega; \mathbf{R}^m)$. Hence, with fixed u we can consider the Yosida transforms with respect to the L^p metric related to the restriction of F_ε to $u + W_0^{1,p}(\Omega; \mathbf{R}^m)$, which we denote by

$$\widetilde{T}_\lambda^p F_\varepsilon(u, \Omega) = \min\left\{ \int_\Omega f\left(\frac{x}{\varepsilon}, Dv + Du\right) dx + \lambda \int_\Omega |v|^p \, dx : \ v \in W_0^{1,p}(\Omega; \mathbf{R}^m)\right\}.$$

By Proposition 8.1 we have

$$\begin{aligned} G(u, \Omega) &= \lim_{\lambda \to +\infty} \liminf_k \widetilde{T}_\lambda^p F_{\varepsilon_k}(u, \Omega) \\ &= \lim_{\lambda \to +\infty} \limsup_k \widetilde{T}_\lambda^p F_{\varepsilon_k}(u, \Omega). \end{aligned} \tag{17.20}$$

Consider $A \in \mathbf{M}^{m \times n}$, $z \in \mathbf{R}^n$ and $\rho > 0$, and fix the notation $\overline{u} = Ax$ and $B = B_\rho(z)$. Let Φ_ε^j be defined by (17.8), and define

$$\Psi^j(v, B) = \begin{cases} \displaystyle\int_B \psi_j(Dv) \, dx & \text{if } v - \overline{u} \in W_0^{1,p}(\Omega; \mathbf{R}^m) \\[2ex] +\infty & \text{otherwise.} \end{cases}$$

Again, by Proposition 11.7 $\Psi^j(\overline{u}, B) = \Gamma(L^p)\text{-}\lim_{\varepsilon \to 0+} \Phi_\varepsilon^j(\overline{u}, B)$, and by Theorem 7.2 and Remark 7.3(v) we have

$$T_\lambda^p \Psi^j(Ax, B_\rho(z)) = \lim_{\varepsilon \to 0+} T_\lambda^p \Phi_\varepsilon^j(Ax, B_\rho(z)). \tag{17.21}$$

By Proposition 17.4 applied with $f_j = f$ and by (17.21) we obtain the following estimate:

$$\begin{aligned} &\lim_{\lambda \to +\infty} \liminf_j T_\lambda^p \Psi^j(Ax, B_\rho(z)) \\ =\ &\lim_{\lambda \to +\infty} \liminf_j \lim_{\varepsilon \to 0+} T_\lambda^p \Phi_\varepsilon^j(Ax, B_\rho(z)) \\ =\ &\lim_{\lambda \to +\infty} \liminf_j \limsup_{\varepsilon \to 0+} \widetilde{T}_\lambda^p F_\varepsilon(Ax, B_\rho(z)) \\ =\ &\lim_{\lambda \to +\infty} \limsup_{\varepsilon \to 0+} \widetilde{T}_\lambda^p F_\varepsilon(Ax, B_\rho(z)) \\ \geq\ &\lim_{\lambda \to +\infty} \limsup_k \widetilde{T}_\lambda^p F_{\varepsilon_k}(Ax, B_\rho(z)) \\ \geq\ &\lim_{\lambda \to +\infty} \liminf_k \widetilde{T}_\lambda^p F_{\varepsilon_k}(Ax, B_\rho(z)) \\ \geq\ &\lim_{\lambda \to +\infty} \liminf_{\varepsilon \to 0+} \widetilde{T}_\lambda^p F_\varepsilon(Ax, B_\rho(z)) \\ =\ &\lim_{\lambda \to +\infty} \limsup_j \liminf_{\varepsilon \to 0+} \widetilde{T}_\lambda^p F_\varepsilon(Ax, B_\rho(z)) \\ =\ &\lim_{\lambda \to +\infty} \limsup_j \liminf_{\varepsilon \to 0+} T_\lambda^p \Phi_\varepsilon^j(Ax, B_\rho(z)) \\ =\ &\lim_{\lambda \to +\infty} \limsup_j T_\lambda^p \Psi^j(Ax, B_\rho(z)) \end{aligned}$$

$$\geq \lim_{\lambda \to +\infty} \liminf_j T_\lambda^p \Psi^j (Ax, B_\rho(z)). \tag{17.22}$$

Since the first and the last term coincide, all inequalities in (17.22) are indeed equalities, and in particular by (17.20), Proposition 8.1 and Proposition 12.8

$$G(Ax, B_\rho(z)) = \Gamma(L^p)\text{-}\lim_j \Psi^j (Ax, B_\rho(z)) = |B_\rho(z)| \lim_j \psi_j(A),$$

which proves that $\psi(x, A) = \lim_j \psi_j(A)$. In particular ψ does not depend on x, and G does not depend on (ε_k). The equality $\psi = f_{\text{hom}}$ follows as in the proof of Theorem 14.5. $\qquad\square$

Corollary 17.6 (Homogenization stability) *Let $f, \varphi : \mathbf{R}^n \times \mathbf{M}^{m \times n} \to [0, +\infty)$ be Borel functions, quasiconvex in the second variable and satisfying (17.1). If*

$$\fint \sup_{|A| \leq y} |f(x, A) - \varphi(x, A)| \, dx = 0 \tag{17.23}$$

for all $y \geq 0$, and the limit

$$\Gamma(L^p)\text{-}\lim_{\varepsilon \to 0+} \int_\Omega \varphi \left(\frac{x}{\varepsilon}, Du \right) dx = \int_\Omega \varphi_{\text{hom}}(Du) \, dx \tag{17.24}$$

exists for all Ω bounded open subsets of $\mathbf{R}^n$ and $u \in \mathrm{W}^{1,p}(\Omega; \mathbf{R}^m)$, then so does the limit in (17.18), with f_{hom} as in Theorem 14.5. Moreover, $f_{\text{hom}} = \varphi_{\text{hom}}$.

Remark 17.7 Condition (17.23) is satisfied if the difference $\varphi(x, A) - f(x, A)$ has compact support in x; in this sense 'homogenization is stable under compact support perturbations'.

Example 17.8 The growth conditions are essential in the closure and stability properties shown above. A trivial example is obtained by taking $f, \varphi : \mathbf{R} \times \mathbf{R} \to [0, +\infty)$

$$f(x, z) = \begin{cases} 0 & \text{if } |x| \in \bigcup_{j \in \mathbf{N}} [j, j + 2^{-j}) \\ z^2 & \text{otherwise,} \end{cases}$$

and $\varphi(x, z) = z^2$. It is easy to see that (17.23) is satisfied and that the limit in (17.18) exists and is equal to 0, so that $f_{\text{hom}} \neq \varphi_{\text{hom}}$, contrary to the conclusions of Corollary 17.6.

We can push this example further by taking $g : \mathbf{R} \times \mathbf{R} \to [0, +\infty)$ defined by

$$g(x, z) = \begin{cases} f(x, z) & \text{if } |x| \in \bigcup_{k \in \mathbf{N}} [2^{2k+1}, 2^{2k+2}) \\ z^2 & \text{otherwise.} \end{cases}$$

Clearly, (17.23) is still satisfied, with g in place of f. In this case, though, we have, taking $\varepsilon_j = \frac{1}{2} 4^{-j}$,

$$\Gamma(\mathrm{L}^2)\text{-}\lim_j \int_{(1,2)} g\left(\frac{t}{\varepsilon_j}, u'\right) dt = \Gamma(\mathrm{L}^2)\text{-}\lim_j \int_{(1,2)} f\left(\frac{t}{\varepsilon_j}, u'\right) dt = 0,$$

while, taking $\varepsilon_j = 4^{-j}$,

$$\Gamma(\mathrm{L}^2)\text{-}\lim_j \int_{(1,2)} g\left(\frac{t}{\varepsilon_j}, u'\right) dt = \Gamma(\mathrm{L}^2)\text{-}\lim_j \int_{(1,2)} |u'|^2 \, dt = \int_{(1,2)} |u'|^2 \, dt.$$

Hence, in this case not even the Γ-limit

$$\Gamma(\mathrm{L}^2)\text{-}\lim_{\varepsilon \to 0} \int_{(1,2)} g\left(\frac{t}{\varepsilon}, u'\right) dt$$

exists.

17.2 An application: homogenization of Besicovitch almost-periodic functionals

We now give a definition of an almost-periodic function that is much weaker than Definition 15.1. Note that with the following definition all L^1 functions are almost periodic, and indeed 'equivalent' to the constant zero.

Definition 17.9 *Let $(X, \| \ \|)$ be a complex Banach space. We say that a measurable function $v : \mathbf{R}^N \to X$ is (Besicovitch) almost periodic, and we write $v \in AP(\mathbf{R}^N; X)$, if it is the limit in the mean of a sequence of trigonometric polynomials on X, i.e.*

$$\lim_k \int \| P_k(x) - v(x) \| \, dx = 0$$

for some

$$P_k(y) = \sum_{j=1}^{r_k} \mathbf{x}_j^k \, e^{i\langle \lambda_j^k, y \rangle},$$

with $\mathbf{x}_j^k \in X$, $\lambda_j^k \in \mathbf{R}^N$ and $r_k \in \mathbf{N}$. The definition easily extends to real Banach spaces. If $X = \mathbf{R}$, this definition is the usual one of almost-periodic functions in the sense of Besicovitch, and we write AP_N for $AP(\mathbf{R}^N; \mathbf{R})$.

We will prove the following homogenization result.

Theorem 17.10 *Let $f : \mathbf{R}^n \times \mathbf{M}^{m \times n} \to [0, +\infty)$ be a Carathéodory function, quasiconvex in the second variable and satisfying (17.1) with $p > 1$, such that*

$$f(\cdot, A) \in AP_n \text{ for all } A \in \mathbf{M}^{m \times n}. \tag{17.25}$$

Then the limit in (17.18) exists for all Ω bounded open subsets of $\mathbf{R}^n$ and $u \in W^{1,p}(\Omega; \mathbf{R}^m)$, with f_{hom} as in Theorem 14.5.

Proof The proof is obtained by an approximation procedure of f by means of a sequence (φ_j) which satisfies the hypotheses of Theorems 15.3 and 17.5. It will be achieved in several steps.

Step 1: definition of the approximating functions.
Let (A_j) be a dense sequence in $\mathbf{M}^{m \times n}$. For every $k \in \mathbf{N}$ and $j \in \{1, \ldots, k\}$ let P_k^j be a trigonometric polynomial such that

$$\fint |P_k^j(x) - f(x, A_j)| \, dx \le \frac{1}{k^2}.$$

We set

$$f_k(x, A) = \begin{cases} f(x, A) & \text{if } A \in \{A_1, \ldots, A_k\} \\ \beta(1 + |A|^p) & \text{otherwise,} \end{cases} \tag{17.26}$$

$$\psi_k(x, A) = \begin{cases} \left(P_k^j(x) \vee \alpha |A|^p\right) \wedge \left(\beta(1 + |A|^p)\right) & \text{if } A \in \{A_1, \ldots, A_k\} \\ \beta(1 + |A|^p) & \text{otherwise.} \end{cases} \tag{17.27}$$

Moreover, we define

$$\varphi_k = Q(\psi_k) \tag{17.28}$$

(Q stands for quasiconvexification as in Definition 6.3).

Step 2: the function φ_k is uniformly almost periodic (in the sense of (15.2)).
Since φ_k does not depend on 'u' we have to show that for every $A \in \mathbf{M}^{m \times n}$ and $\eta > 0$ the set

$$\{\tau \in \mathbf{R}^n : \ |\varphi_k(x + \tau, A) - \varphi_k(x, A)| < \eta(1 + |A|^p) \text{ for every } x \in \mathbf{R}^n\} \tag{17.29}$$

is relatively dense in $\mathbf{R}^n$. Since the set of uniformly almost-periodic functions is closed under addition and lattice operations, the functions

$$R_k^j(x) = \left(P_k^j(x) \vee \alpha |A_j|^p\right) \wedge \left(\beta(1 + |A_j|^p)\right) = \psi_k(x, A_j)$$

are uniformly almost-periodic, and there exists $L_\eta > 0$ such that for all $a \in \mathbf{R}^n$ there exists $\tau \in a + [0, L_\eta)^n$ such that

$$|R_k^j(x + \tau) - R_k^j(x)| \le \eta(1 + |A_j|^p) \tag{17.30}$$

for all $x \in \mathbf{R}^n$, $j = 1, \ldots, k$; this condition trivially extends to ψ_k. For such a τ we have (B any open ball)

$$\varphi_k(x + \tau, A) \le \frac{1}{|B|} \int_B \psi_k(x + \tau, A + Du(y)) dy$$

$$\leq \frac{1}{|B|} \int_B \psi_k(x, A + Du(y))dy + \eta \frac{1}{|B|} \int_B (1 + |Du + A|^p)dy$$

for all $u \in W_0^{1,p}(B; \mathbf{R}^m)$. If we take a minimizing sequence for $Q\psi_k(x, A)$, then the last integral is bounded by $(1 + \frac{\beta}{\alpha})(1 + |A|^p)$; hence we obtain

$$\varphi_k(x + \tau, A) - \varphi_k(x, A) \leq \eta \left(1 + \frac{\beta}{\alpha}\right)(1 + |A|^p).$$

By a symmetry argument we also obtain

$$|\varphi_k(x + \tau, A) - \varphi_k(x, A)| \leq \eta \left(1 + \frac{\beta}{\alpha}\right)(1 + |A|^p).$$

This proves (15.2) for φ_k.

Step 3. Since we have $f \leq Qf_k \leq f_k$ we obtain

$$Qf_k(x, A_j) = f(x, A_j) \text{ for all } j = 1, \ldots, k. \tag{17.31}$$

Step 4: *the sequence* (Qf_k) *and the constant sequence* f *are equivalent* (in the sense of Definition 17.2).

In fact, the functions $f(x, \cdot)$ and $Qf_k(x, \cdot)$ are Lipschitz functions on $\{|A| \leq y\}$ with Lipschitz constant $L(y) = c(1 + 2y^{p-1})$, where c does not depend on k. Hence

$$\sup_{|A| \leq y} |f(x, A) - Qf_k(x, A)| \leq 2L(y) \sup_{|A| \leq y} \inf\{|A - A_j| : |A_j| \leq y, \ j = 1, \ldots, k\},$$

and (17.4) follows by the density of (A_j).

Step 5: *the sequences* (Qf_k) *and* (φ_k) *are equivalent.*

In fact, let (u_l) be a minimizing sequence for

$$\varphi_k(A) = \inf\left\{\frac{1}{|B|} \int_B \psi_k(x, A + Du(y)) \, dy : \ u \in W_0^{1,p}(B; \mathbf{R}^m)\right\}.$$

We have

$$Qf_k(x, A) \leq \frac{1}{|B|} \int_B f_k(x, A + Du_l(y)) \, dy$$

$$\leq \frac{1}{|B|} \int_B |f_k(x, A + Du_l(y)) - \psi_k(x, A + Du_l(y))| \, dy$$

$$+ \frac{1}{|B|} \int_B \psi_k(x, A + Du_l(y)) \, dy$$

$$\leq \|f_k(x, \cdot) - \psi_k(x, \cdot)\|_\infty + \frac{1}{|B|} \int_B \psi_k(x, A + Du_l(y)) \, dy$$

$$\leq \sum_{j=1}^{k} |f_k(x, A_j) - \psi_k(x, A_j)| + \frac{1}{|B|} \int_B \psi_k(x, A + Du_l(y)) \, dy$$

$$\leq \sum_{j=1}^{k} |f(x, A_j) - P_k^j(x)| + \frac{1}{|B|} \int_B \psi_k(x, A + Du_l(y)) \, dy.$$

Taking the limit in l, and then applying a symmetry argument, we obtain

$$|Qf_k(x, A) - \varphi_k(x, A)| \leq \sum_{j=1}^{k} |f(x, A_j) - P_k^j(x)|$$

so that

$$\lim_k \fint \sup_{|A| \leq y} |Qf_k(x, A) - \varphi_k(x, A)| \, dx \leq \lim_k \sum_{j=1}^{k} \fint |f(x, A_j) - P_k^j(x)| \, dx = 0.$$

Step 6: the sequence (φ_k) and the constant sequence f are equivalent.
In fact, this follows from Steps 4 and 5 noticing that equivalence is transitive.

Step 7: conclusion.
By Step 2 each φ_k satisfies the hypotheses of Theorem 15.3, and hence also (17.17). By Step 6 we can apply Theorem 17.5 and conclude the proof. $\square$

Remark 17.11 Contrary to the periodic case, in Theorem 17.10 the growth conditions (17.1) cannot be weakened. In fact, the function g in Example 17.8 is Besicovitch almost-periodic in the first variable (actually, $g(\cdot, z)$ is equivalent to the constant z^2 for fixed z), and satisfies all the conditions of Theorem 14.8 (except the periodicity). Nevertheless, Example 17.8 shows that the conclusions of Theorem 17.10 do not hold.

18

LOSS OF POLYCONVEXITY BY HOMOGENIZATION

In this chapter we discuss the structure of the 'homogenized integrands'. It is clear that convexity is preserved by the passage to the Γ-limit, and hence also after homogenization, while the homogenized energy density is always quasiconvex. We will also show here that polyconvexity properties, in contrast, are not maintained by homogenization.

18.1 An example

The example we are going to construct draws inspiration from Šverák's construction of a quasiconvex function which is not polyconvex as in Proposition 6.10.

We apply the homogenization procedure to a 'composite' function constructed using two other functions f_1, f_2 with different behaviours. We require that:

(i) $f_1 : \mathbf{M}^{2\times 2} \to [0, +\infty)$ is *convex* and satisfies a standard growth condition of order p, with $1 < p < 2$; more precisely, we assume that there exist two positive constants α, β such that for every $A \in \mathbf{M}^{2\times 2}$

$$|A|^p - \alpha \le f_1(A) \le \beta(1 + |A|^p),$$

and $f_1(I) = f_1(-I) = 0$ (and hence $f_1(tI) = 0$ for $-1 \le t \le 1$);

(ii) $f_2 : \mathbf{M}^{2\times 2} \to [0, +\infty)$ is *polyconvex* and satisfies a growth condition of order 2; more precisely, for every $A \in \mathbf{M}^{2\times 2}$

$$|A|^2 - \alpha \le f_2(A) \le \beta(1 + |A|^2),$$

and $f_2(A) = 0$ if and only if $A \in \{I, -I\}$;

(iii) there exists a constant $\gamma > 0$ such that for both functions f_1 and f_2 we have

$$f_1(A) \ge \gamma(|A_{11} - A_{22}|^p + |A_{21} + A_{12}|^p)$$
$$f_2(A) \ge \gamma(|A_{11} - A_{22}|^p + |A_{21} + A_{12}|^p)$$

$(1 < p < 2$ as above).

It is not difficult to write down explicitly two such functions; for example, we can define

$$f(A) = |A_{11} - A_{22}| + |A_{12}| + |A_{21}| + (|A_{11}| - 1)^+$$

$(t^+ = \max\{t, 0\}$ is the positive part of $t)$,

$$f_1(A) = (f(A))^p, \qquad f_2(A) = (f(A))^2 + (f(A))^p + |1 - \det A|.$$

For all $\delta \geq 0$ we consider the function $f^\delta : \mathbf{R}^2 \times \mathbf{M}^{2\times 2} \to [0, +\infty)$ defined on $[0,1)^2 \times \mathbf{M}^{2\times 2} \to [0, +\infty)$ by

$$f^\delta(x, A) = \begin{cases} f_1(A) + \delta|A|^2 & \text{if } x \in [0,1)^2 \setminus \left[\frac{1}{4}, \frac{3}{4}\right]^2 \\[2ex] f_2(A) & \text{if } x \in \left[\frac{1}{4}, \frac{3}{4}\right]^2, \end{cases} \tag{18.1}$$

and extended by periodicity to the whole $\mathbf{R}^2$. We denote

$$E = \left[\tfrac{1}{4}, \tfrac{3}{4}\right]^2 + \mathbf{Z}^2;$$

note that $\frac{1}{j}E = \{x \in \mathbf{R}^2 : f^\delta(jx, \cdot) = f_2\}$ for all $\delta \geq 0$, and $j > 0$.

If $\delta > 0$ then the function f^δ satisfies a standard growth condition of order 2, up to addition of a constant; hence we can apply Theorem 14.5 and consider the homogenized energy density f^δ_{hom} of f^δ, which can be described as in Remark 14.6 by

$$f^\delta_{\text{hom}}(A) = \inf_{j \in \mathbf{N}} \inf\left\{ \int_{(0,1)^2} f^\delta(jy, Du(y) + A)\, dy : u \in W^{1,2}_{\#}((0,1)^2; \mathbf{R}^2) \right\} \tag{18.2}$$

for all $A \in \mathbf{M}^{2\times 2}$.

Theorem 18.1 *The function f^δ_{hom} is not polyconvex for δ small enough.*

Proof Note that by the growth conditions from below on f_1 and f_2 we obtain immediately that

$$|A|^p - c \leq f^\delta_{\text{hom}}(A) \tag{18.3}$$

for all $A \in \mathbf{M}^{2\times 2}$, with c independent of $\delta > 0$. We also get an estimate from above, remarking that

$$f^\delta_{\text{hom}}(A) \leq \inf\left\{ \int_{(0,1)^2} f^\delta(y, Du(y))\, dy : u \in Ax + W^{1,2}_0((0,1)^2; \mathbf{R}^2) \right\}. \tag{18.4}$$

It suffices to choose a cut-off function $\varphi \in W^{1,\infty}_0((0,1)^2)$ such that $\varphi = 1$ on $[\frac{1}{4}, \frac{3}{4}]^2$, $0 \leq \varphi \leq 1$ and $|D\varphi| \leq 4$, and define $v(x) = x\varphi(x) + Ax(1 - \varphi(x))$. We obtain

$$\begin{aligned} f^\delta_{\text{hom}}(A) &\leq \int_{(0,1)^2} f^\delta(y, Dv(y))\, dy \\ &= \int_{(0,1)^2 \setminus [\frac{1}{4}, \frac{3}{4}]^2} f_1\Big((y - Ay)D\varphi(y) + \varphi(y)I + (1 - \varphi(y))A\Big)\, dy \\ &\quad + \delta \int_{(0,1)^2 \setminus [\frac{1}{4}, \frac{3}{4}]^2} |(y - Ay)D\varphi(y) + \varphi(y)I + (1 - \varphi(y))A|^2\, dy \end{aligned}$$

$$\leq \int_{(0,1)^2} \beta(1 + |(y - Ay)D\varphi(y) + \varphi(y)I + (1 - \varphi(y))A|^p)\, dy$$

$$+ \delta \int_{(0,1)^2} |(y - Ay)D\varphi(y) + \varphi(y)I + (1 - \varphi(y))A|^2\, dy,$$

so that

$$f_{\text{hom}}^\delta(A) \leq c_1(1 + |A|^p) + \delta c_2(1 + |A|^2) \tag{18.5}$$

for all $A \in \mathbf{M}^{2\times 2}$, with c_1 and c_2 independent of δ.

Define the function

$$f_0(A) = \inf_{\delta > 0} f_{\text{hom}}^\delta(A) = \lim_{\delta \to 0+} f_{\text{hom}}^\delta(A). \tag{18.6}$$

By the estimates above f_0 satisfies a standard growth condition of order p. Suppose that all functions f_{hom}^δ are polyconvex. Then also f_0 should be polyconvex. In fact, by Remark 5.8 we know that a function $V : \mathbf{M}^{2\times 2} \to [0, +\infty)$ is polyconvex if and only if we have

$$V\Big(\sum_{i=1}^{6} \lambda_i A_i\Big) \leq \sum_{i=1}^{6} \lambda_i V(A_i)$$

for all choices of $A_i \in \mathbf{M}^{2\times 2}$ and $\lambda_i > 0$ with $\sum_{i=1}^{6} \lambda_i = 1$ and

$$\det\Big(\sum_{i=1}^{6} \lambda_i A_i\Big) = \sum_{i=1}^{6} \lambda_i \det A_i.$$

If this condition is satisfied by $V = f_{\text{hom}}^\delta$ for all $\delta > 0$ then it is satisfied also by f_0, since it is maintained by the pointwise limit. As f_0 satisfies a growth condition of order $p < 2$, if it were polyconvex, then it should be *convex*, by Remark 6.9. In order to show that this is not the case, first we remark that $f_0(I) = f_0(-I) = 0$: in fact, it can be immediately checked that $f_{\text{hom}}^\delta(I) = f_{\text{hom}}^\delta(-I) \leq c\delta$, taking $u \equiv 0$ in (18.2). The proof is concluded if we show that $f_0(0) > 0$.

Since by Proposition 7.5 and Proposition 11.7

$$f_{\text{hom}}^\delta(0) = \inf\Big\{ \liminf_j \int_{(0,1)^2} f^\delta(jy, Du_j)\, dy :$$

$$u_j \in W_0^{1,2}((0,1)^2; \mathbf{R}^2),\ u_j \to 0 \text{ in } L^2((0,1)^2; \mathbf{R}^2)\Big\}$$

$$\geq \inf\Big\{ \liminf_j \int_{(0,1)^2} f^0(jy, Du_j)\, dy :$$

$$u_j \in W_0^{1,2}((0,1)^2; \mathbf{R}^2),\ u_j \to 0 \text{ in } L^p((0,1)^2; \mathbf{R}^2)\Big\}$$

for all $\delta > 0$, it suffices to show that the latter infimum is strictly positive. We argue by contradiction. Consider a sequence $(u_j) = (u_j^1, u_j^2)$ in $W_0^{1,2}((0,1)^2; \mathbf{R}^2)$ converging to 0 in $L^p((0,1)^2; \mathbf{R}^2)$ such that

$$0 = \lim_j \int_{(0,1)^2} f^0(jx, Du_j)\, dx. \tag{18.7}$$

We then have by hypothesis (iii)

$$\lim_j \int_{(0,1)^2} (|D_1 u_j^1 - D_2 u_j^2|^p + |D_1 u_j^2 + D_2 u_j^1|^p)\, dx = 0. \tag{18.8}$$

Repeating the argument of Proposition 6.10, we can write

$$\Delta u_j^1 = D_1(D_1 u_j^1 - D_2 u_j^2) + D_2(D_2 u_j^1 + D_1 u_j^2)$$
$$\Delta u_j^2 = D_1(D_2 u_j^1 + D_1 u_j^2) - D_2(D_1 u_j^1 - D_2 u_j^2)$$

so that by (18.8) we have that $\Delta u_j^k \to 0$ in $W^{-1,p}((0,1)^2)$ ($k = 1, 2$). Using the L^p estimates for the Laplace operator (see Appendix C), we obtain that Du_j converges to 0 strongly in $L^p((0,1)^2; \mathbf{M}^{2\times 2})$. In particular this implies that the measure of the set $\{x \in (0,1)^2 : |Du_j| \geq 1\}$ converges to 0. Notice now that since f_2 is continuous and vanishes only at I and $-I$, we have $c_3 = \inf\{f_2(A) : |A| \leq 1\} > 0$, so that

$$\begin{aligned}
\lim_j \int_{(0,1)^2} f^0(jx, Du_j)\, dx &\geq \lim_j \int_{(0,1)^2 \cap \frac{1}{j}E} f_2(Du_j)\, dx \\
&\geq \lim_j \int_{(0,1)^2 \cap \frac{1}{j}E \cap \{|Du_j|<1\}} f_2(Du_j)\, dx \\
&\geq \lim_j c_3 |(0,1)^2 \cap \tfrac{1}{j}E \cap \{|Du_j| < 1\}| \\
&= \frac{1}{4} c_3 > 0,
\end{aligned}$$

and we reach a contradiction. $\qquad\square$

Remark 18.2 (Open problem) Theorem 18.1 shows that the family of polyconvex functions is in a sense not closed by homogenization. It is not clear whether all quasiconvex functions can be obtained from polyconvex functions via homogenization, or, more generally, which is the 'closure by homogenization' of polyconvex functions.

Part IV

Finer homogenization results

19

HOMOGENIZATION OF CONNECTED MEDIA

In this chapter we study the asymptotic behaviour of integral functionals that describe the physical properties of porous media having a periodic microstructure. More precisely, given a bounded open set Ω of $\mathbf{R}^n$, we compute the limit of Neumann minimum problems of the type

$$\inf\left\{\int_{\Omega\cap\varepsilon E} f\left(\frac{x}{\varepsilon}, Du\right)\,dx \;+\; \int_{\Omega\cap\varepsilon E} |g_\varepsilon - u|^p\,dx : u \in \mathrm{W}^{1,p}(\Omega\cap\varepsilon E; \mathbf{R}^m)\right\},$$

and of Dirichlet boundary value problems

$$\inf\left\{\int_{\Omega\cap\varepsilon E} f\left(\frac{x}{\varepsilon}, Du\right)\,dx : u \in \mathrm{W}^{1,p}(\Omega\cap\varepsilon E; \mathbf{R}^m),\; u = g \text{ on } \partial\Omega\right\}$$

as $\varepsilon \to 0$. We assume that the set E is connected, Lipschitz and periodic, and that the function $f = f(x, A)$ is periodic in x and satisfies a standard growth condition of order $p > 1$. Under these hypotheses, we show that the infima and (proper extensions to the whole Ω of) the minimizers of the problems above converge to the minimum and the minimizers, respectively, of the corresponding problems related to the functional

$$\int_\Omega f_{\mathrm{hom}}(Du(x))\,dx,$$

whose integrand f_{hom} still satisfies a growth condition of order p, and is given by the formula

$$f_{\mathrm{hom}}(A) = \lim_{t\to+\infty} \inf\left\{\frac{1}{t^n}\int_{(0,t)^n\cap E} f(x, Du + A)\,dx : \; u \in \mathrm{W}^{1,p}_0((0,t)^n; \mathbf{R}^m)\right\}.$$

The essential tool in dealing with such problems is a good extension property for the minimizers of the integrals above from $\Omega\cap\varepsilon E$ to the whole Ω.

19.1   A homogenization theorem on periodic connected domains

We say that a set $E \subset \mathbf{R}^n$ is 1-*periodic* if χ_E is 1-periodic; i.e., if $E + e_i = E$ for every $i = 1,\ldots,n$. We denote by $\mathcal{A}_n$ the family of all bounded open subsets of $\mathbf{R}^n$. In this section we use the notation $U(\lambda)$ for the *retracted subset* $\{x \in U : \mathrm{dist}\,(x, \partial U) > \lambda\}$ of the set $U \subset \mathbf{R}^n$.

In the sequel we consider a fixed 1-periodic, connected, open set $E \subset \mathbf{R}^n$, with Lipschitz boundary (i.e. ∂E is locally a graph of a Lipschitz function, see Definition B.1), and we study the limit behaviour of the functionals

$$F_\varepsilon(u, \Omega) = \int_{\Omega \cap \varepsilon E} f\left(\frac{x}{\varepsilon}, Du\right) dx, \qquad (19.1)$$

where Ω is a bounded open subset of $\mathbf{R}^n$, $u \in \mathrm{W}^{1,p}(\Omega; \mathbf{R}^m)$, and the integrand $f : E \times \mathbf{M}^{m \times n} \to \mathbf{R}$ is a fixed Borel function, with $f(\cdot, A)$ 1-periodic for every $A \in \mathbf{M}^{m \times n}$, satisfying the standard growth condition of order $p > 1$

$$|A|^p \leq f(x, A) \leq C(1 + |A|^p) \qquad (19.2)$$

for all $(x, A) \in E \times \mathbf{M}^{m \times n}$ ($C > 1$ fixed constant). We will prove the following result.

Theorem 19.1 *Let E be a 1-periodic, connected, open subset of $\mathbf{R}^n$, with Lipschitz boundary. Given a Borel function $f : E \times \mathbf{M}^{m \times n} \to \mathbf{R}$ with $f(\cdot, A)$ 1-periodic for every $A \in \mathbf{M}^{m \times n}$, satisfying (19.2) with $1 < p < +\infty$, we define, for every $\varepsilon > 0$, the functional $F_\varepsilon : \mathrm{L}_{\mathrm{loc}}^p(\mathbf{R}^n; \mathbf{R}^m) \times \mathcal{A}_n \to [0, +\infty]$ as*

$$F_\varepsilon(u, U) = \begin{cases} \displaystyle\int_{U \cap \varepsilon E} f\left(\frac{x}{\varepsilon}, Du(x)\right) dx & \text{if } u_{|U \cap \varepsilon E} \in \mathrm{W}^{1,p}(U \cap \varepsilon E; \mathbf{R}^m) \\ +\infty & \text{otherwise.} \end{cases} \qquad (19.3)$$

Then there exist a constant $k_1 > 0$ depending only on E, n, p (see Theorem 19.3 below), and a quasiconvex function $f_{\mathrm{hom}} : \mathbf{M}^{m \times n} \to \mathbf{R}$ with

$$\frac{1}{k_1} |A|^p \leq f_{\mathrm{hom}}(A) \leq C|(0,1)^n \cap E|(1 + |A|^p)$$

for all $A \in \mathbf{M}^{m \times n}$, such that, defining the functional $F_{\mathrm{hom}} : \mathrm{L}_{\mathrm{loc}}^p(\mathbf{R}^n; \mathbf{R}^m) \times \mathcal{A}_n \to [0, +\infty]$ by

$$F_{\mathrm{hom}}(u, U) = \begin{cases} \displaystyle\int_U f_{\mathrm{hom}}(Du(x)) \, dx & \text{if } u \in \mathrm{W}^{1,p}(U; \mathbf{R}^m) \\ +\infty & \text{otherwise,} \end{cases} \qquad (19.4)$$

we have

$$F_{\mathrm{hom}}(u, U) = \Gamma(\mathrm{L}^p)\text{-}\lim_{\varepsilon \to 0} F_\varepsilon(u, U) \qquad (19.5)$$

for every $U \in \mathcal{A}_n$ and every $u \in \mathrm{L}_{\mathrm{loc}}^p(\mathbf{R}^n; \mathbf{R}^m)$. Moreover, the function f_{hom} satisfies

$$f_{\text{hom}}(A) = \lim_{t \to +\infty} \inf \Big\{ \frac{1}{t^n} \int_{(0,t)^n \cap E} f(x, Du(x) + A)\, dx$$

$$: u \in W_0^{1,p}((0,t)^n; \mathbf{R}^m) \Big\} \tag{19.6}$$

for all $A \in \mathbf{M}^{m \times n}$

The proof of Theorem 19.1 will follow from Propositions 19.4, 19.5 and 19.6 below.

Remark 19.2 Note that if f is Carathéodory then, since f satisfies (19.2), the infimum in (19.6) can be taken over the space $C_0^\infty((0,t)^n; \mathbf{R}^m)$. Moreover, as in Remark 14.6, we also have

$$f_{\text{hom}}(A) = \lim_{t \to +\infty} \inf \Big\{ \frac{1}{t^n} \int_{(0,t)^n \cap E} f(x, Du(x) + A)\, dx \ :$$

$$u \in W_\#^{1,p}((0,t)^n; \mathbf{R}^m) \Big\}. \tag{19.7}$$

Finally, (19.7) reduces to

$$f_{\text{hom}}(A) = \inf \Big\{ \int_{(0,1)^n \cap E} f(y, Du(y) + A)\, dy : u \in W_\#^{1,p}((0,1)^n; \mathbf{R}^m) \Big\} \tag{19.8}$$

when $f(x, \cdot)$ is convex, as in Theorem 14.7.

The main tool in the proof of the Homogenization Theorem 19.1 will be the following extension lemma, whose proof is given in Appendix B.

Theorem 19.3 (Extension lemma) *Let E be a 1-periodic, connected, open subset of $\mathbf{R}^n$, with Lipschitz boundary. Given a bounded open set $\Omega \subset \mathbf{R}^n$ and a real number $\varepsilon > 0$, there exist a linear and continuous extension operator $T_\varepsilon : W^{1,p}(\Omega \cap \varepsilon E; \mathbf{R}^m) \to W_{\text{loc}}^{1,p}(\Omega; \mathbf{R}^m)$ and two constants $k_0, k_1 > 0$ such that*

$$T_\varepsilon u = u \quad a.e. \ in \ \Omega \cap \varepsilon E, \tag{19.9}$$

$$\int_{\Omega(\varepsilon k_0)} |T_\varepsilon u|^p\, dx \leq k_1 \int_{\Omega \cap \varepsilon E} |u|^p\, dx, \tag{19.10}$$

$$\int_{\Omega(\varepsilon k_0)} |D(T_\varepsilon u)|^p\, dx \leq k_1 \int_{\Omega \cap \varepsilon E} |Du|^p\, dx \tag{19.11}$$

for every $u \in W^{1,p}(\Omega \cap \varepsilon E; \mathbf{R}^m)$. The constants k_0, k_1 depend on E, n, p, but are independent of ε, m and Ω.

We want to remark that the roles played by the hypotheses of periodicity on the function f and on the set E are completely independent. In order to highlight this fact we will in some sense weaken the hypothesis of periodicity on the function f, by considering almost-periodic integrands.

Since we consider the product $f(x, A)\chi_E(x)$, which is in general *not* periodic, unless f and E share the same periods, we will require on the function f a weaker condition than periodicity (already introduced in Section 15) that behaves better under algebraic operations and superposition. That is, we suppose that for all $\varepsilon > 0$ the set T_ε of the *ε-almost periods* of the function f, namely the set of all $\tau \in \mathbf{R}^n$ such that

$$|f(x + \tau, A) - f(x, A)| \leq \varepsilon(1 + |A|^p) \text{ for every } x \in \mathbf{R}^n, A \in \mathbf{M}^{m \times n}, \quad (19.12)$$

is relatively dense in $\mathbf{R}^n$, so that there exists an *inclusion length* $L_\varepsilon > 0$ such that $T_\varepsilon + [0, L_\varepsilon)^n = \mathbf{R}^n$.

Moreover, if f is not periodic, we have to require uniform continuity of f with respect to the first variable: for every $\varepsilon > 0$ there exists $\delta > 0$ such that

$$|f(x, A) - f(y, A)| \leq \varepsilon(1 + |A|^p) \quad \text{for every } A \in \mathbf{M}^{m \times n}, \quad (19.13)$$

whenever $|x - y| \leq \delta$.

Conditions (19.12) and (19.13) are satisfied for functions of the form

$$f(x, A) = a(x)\psi(A),$$

if a is a *uniformly almost-periodic* function (in the sense of Bohr, see Definition 15.1), and ψ satisfies the growth condition

$$0 \leq \psi(A) \leq C(1 + |A|^p).$$

In particular, the function a could be **k**-periodic for some $\mathbf{k} = (k_1, \ldots, k_n) \in \mathbf{R}^n$ (i.e. $a(x + k_i e_i) = a(x)$ for every $x \in \mathbf{R}^n$ and $i = 1, \ldots, n$); in this case, note that we make no assumption on the periods of a, which may differ from those of E.

Under conditions (19.12), (19.13), the function $f(x, A)\chi_E(x)$ has some properties of almost periodicity. In fact, fix $\varepsilon > 0$, let δ be as in (19.13), and consider the set $S_\delta = \{x \in \mathbf{R}^n : |x| < \delta \text{ modulo } \mathbf{Z}^n\}$, i.e. the set of all points x such that $\mathbf{z} \in \mathbf{Z}^n$ exists with $|x - \mathbf{z}| < \delta$. By the properties of the sets of almost periods, we obtain that $T_\varepsilon \cap S_\delta$ is relatively dense, and hence so also is the set

$$Z_\varepsilon = \{\mathbf{z} \in \mathbf{Z}^n : \text{dist}(\mathbf{z}, T_\varepsilon) < \delta\}.$$

If $\mathbf{z} \in Z_\varepsilon$, then there exists $\tau \in T_\varepsilon$ such that $|\tau - \mathbf{z}| < \delta$, and hence

$$\begin{aligned}
&|f(x + \mathbf{z}, A) - f(x, A)| \\
&\leq |f(x + \mathbf{z}, A) - f(x + \tau, A)| + |f(x + \tau, A) - f(x, A)| \\
&\leq 2\varepsilon(1 + |A|^p)
\end{aligned} \quad (19.14)$$

for every $x \in \mathbf{R}^n, A \in \mathbf{M}^{m \times n}$, i.e. $Z_\varepsilon \subset T_{2\varepsilon}$. Of course, this also means that

$$|f(x + \mathbf{z}, A)\chi_E(x + \mathbf{z}) - f(x, A)\chi_E(x)| \leq 2\varepsilon(1 + |A|^p)$$

for every $\mathbf{z} \in Z_\varepsilon$, $x \in \mathbf{R}^n$, $A \in \mathbf{M}^{m \times n}$. All these properties are trivial if f is 1-periodic in the first variable, without assuming (19.13).

We first prove a compactness result.

Proposition 19.4 *Let E be as above, let f be a Borel function, 1-periodic in the first variable or satisfying (19.12) and (19.13), and let F_ε be defined as in (19.3). For every sequence (ε_j) of positive real numbers converging to 0, there exists a subsequence (ε_{j_k}) such that for every $U \in \mathcal{A}_n$ and every $u \in \mathrm{L}^p_{\mathrm{loc}}(\mathbf{R}^n; \mathbf{R}^m)$, there exists the Γ-limit*

$$\Gamma(\mathrm{L}^p)\text{-}\lim_k F_{\varepsilon_{j_k}}(u, U) = F_0(u, U). \tag{19.15}$$

Moreover, there exists a continuous function $\varphi : \mathbf{M}^{m \times n} \to \mathbf{R}$, such that

$$F_0(u, U) = \begin{cases} \int_U \varphi(Du(x))\, dx & \text{if } u \in \mathrm{W}^{1,p}(U; \mathbf{R}^m) \\ +\infty & \text{otherwise.} \end{cases} \tag{19.16}$$

Proof First of all, we remark that if we define

$$g(x, A) = \begin{cases} |A|^p & \text{if } x \in E \\ 0 & \text{otherwise,} \end{cases}$$

then we obtain, if $u \in \mathrm{W}^{1,p}(U \cap \varepsilon E; \mathbf{R}^m)$,

$$\int_U g\left(\frac{x}{\varepsilon}, Du\right) dx \leq F_\varepsilon(u, U) \leq C \int_U \left(1 + g\left(\frac{x}{\varepsilon}, Du\right)\right) dx.$$

Hence, by Remark 12.4, we obtain that the limit in (19.15) exists for every $U \in \mathcal{A}_n$, $u \in \mathrm{W}^{1,p}(U; \mathbf{R}^m)$, and that for every fixed $u \in \mathrm{W}^{1,p}(\mathbf{R}^n; \mathbf{R}^m)$ the set function $F_0(u, \cdot)$ is the restriction of a regular Borel measure to $\mathcal{A}_n$. Moreover, we have:

(i) for every $U \in \mathcal{A}_n$

$$0 \leq F_0(u, U) \leq C \int_U (1 + |Du|^p)\, dx$$

for every $u \in \mathrm{W}^{1,p}_{\mathrm{loc}}(\mathbf{R}^n; \mathbf{R}^m)$;

(ii) F_0 is local, i.e. $F_0(u, U) = F_0(v, A)$, whenever $A \in \mathcal{A}_n$ and $u = v$ a.e. on A (thanks to the local character of Γ-convergence);

(iii) for every $U \in \mathcal{A}_n$ the functional $u \mapsto F_0(u, U)$ is L^p-lower semicontinuous;

(iv) for every $U \in \mathcal{A}_n$, $u \in \mathrm{W}^{1,p}_{\mathrm{loc}}(\mathbf{R}^n; \mathbf{R}^m)$, $a \in \mathbf{R}^m$,

$$F_0(u + a, U) = F_0(u, U),$$

since all the functionals F_ε satisfy this condition;

(v) $F_0(u, U) = +\infty$ if $u \in \mathrm{L}^p_{\mathrm{loc}}(\mathbf{R}^n; \mathbf{R}^m) \setminus \mathrm{W}^{1,p}(U; \mathbf{R}^m)$. In fact, let (u_j) be a sequence in $\mathrm{L}^p_{\mathrm{loc}}(\mathbf{R}^n; \mathbf{R}^m)$ converging to $u \in \mathrm{L}^p_{\mathrm{loc}}(\mathbf{R}^n; \mathbf{R}^m)$ in $\mathrm{L}^p(U; \mathbf{R}^m)$, such that $\liminf_k F_{\varepsilon_{j_k}}(u_k, U) < +\infty$. Then, by (19.2), we have that

$$\liminf_k \int_{U \cap \varepsilon_{j_k} E} |Du_k|^p \, dx \leq \liminf_k F_{\varepsilon_{j_k}}(u_k, U),$$

and hence that, up to a subsequence,

$$\int_{U \cap \varepsilon_{j_k} E} |Du_k|^p \, dx \leq c.$$

For every k consider the function $T_{\varepsilon_{j_k}} u_k \in \mathrm{W}^{1,p}_{\mathrm{loc}}(U; \mathbf{R}^m)$ (the extension operator is considered on U). By Theorem 19.3 we have that $T_{\varepsilon_{j_k}} u_k = u_k$ a.e. in $U \cap \varepsilon_{j_k} E$, and

$$\|T_{\varepsilon_{j_k}} u_k\|_{\mathrm{W}^{1,p}(U'; \mathbf{R}^m)} \leq c$$

for every open set $U' \subset U$ with $\mathrm{dist}(U', \partial U) > \varepsilon_{j_k} k_0$. If U' is smooth, then, by Rellich's Theorem, $(T_{\varepsilon_{j_k}} u_k)$ converges to a function $v \in \mathrm{W}^{1,p}_{\mathrm{loc}}(U; \mathbf{R}^m)$ strongly in $\mathrm{L}^p(U'; \mathbf{R}^m)$, and weakly in $\mathrm{W}^{1,p}(U'; \mathbf{R}^m)$. Considering an invading sequence of smooth open subsets of U, by a diagonal argument we can extract a subsequence of $(T_{\varepsilon_{j_k}} u_k)$, still denoted by $(T_{\varepsilon_{j_k}} u_k)$, that converges to a function $v \in \mathrm{W}^{1,p}_{\mathrm{loc}}(U; \mathbf{R}^m)$ strongly in $\mathrm{L}^p_{\mathrm{loc}}(U; \mathbf{R}^m)$, and weakly in $\mathrm{W}^{1,p}_{\mathrm{loc}}(U; \mathbf{R}^m)$. For every open set $U' \subset U$ we have

$$
\begin{aligned}
\int_{U'} \chi_{\varepsilon_{j_k} E} |u - v|^p \, dx &= \int_{U' \cap \varepsilon_{j_k} E} |u - v|^p \, dx \\
&\leq c \int_{U' \cap \varepsilon_{j_k} E} |u - u_k|^p \, dx + c \int_{U' \cap \varepsilon_{j_k} E} |u_k - v|^p \, dx \\
&= c \int_{U' \cap \varepsilon_{j_k} E} |u - u_k|^p \, dx + c \int_{U' \cap \varepsilon_{j_k} E} |T_{\varepsilon_{j_k}} u_k - v|^p \, dx,
\end{aligned}
$$

from which, taking the limit as $k \to +\infty$,

$$|(0,1)^n \cap E| \int_{U'} |u - v|^p \, dx \leq 0.$$

By the arbitrariness of U' we get $v = u$ a.e. on U. Now, since

$$\|v\|_{\mathrm{W}^{1,p}(U'; \mathbf{R}^m)} \leq \liminf_k \|T_{\varepsilon_{j_k}} u_k\|_{\mathrm{W}^{1,p}(U'; \mathbf{R}^m)} \leq c$$

for every $U' \subset U$, it follows that $u \in \mathrm{W}^{1,p}(U; \mathbf{R}^m)$.

By (i)–(iv) and the fact that $F_0(u, \cdot)$ is a regular Borel measure, we can apply Theorem 9.1, to obtain the existence of a Carathéodory function $\varphi : \mathbf{R}^n \times \mathbf{M}^{m \times n} \to \mathbf{R}$ such that

$$F_0(u, U) = \int_U \varphi(x, Du(x))\, dx,$$

for every $U \in \mathcal{A}_n$ and $u \in W^{1,p}(U; \mathbf{R}^m)$. In order to complete the proof we have to show that the function φ can be chosen independent of the variable x. This is easily proven, as in Proposition 14.3 in the periodic case, as in Proposition 15.4 under hypotheses (19.12) and (19.13), using (19.14). $\qquad\square$

We can easily describe some properties of the function φ obtained in Proposition 19.4.

Proposition 19.5 *We have*

(i) (growth condition) *for every* $A \in \mathbf{M}^{m \times n}$

$$\frac{1}{k_1}|A|^p \le \varphi(A) \le C|(0,1)^n \cap E|(1 + |A|^p), \tag{19.17}$$

where k_1 *is the constant in Theorem 19.3;*

(ii) (quasiconvexity) *for every* $A \in \mathbf{M}^{m \times n}$,

$$|U|\varphi(A) \le \int_U \varphi(A + Du)\, dx \tag{19.18}$$

for every $U \in \mathcal{A}_n$ *and* $u \in W_0^{1,p}(U; \mathbf{R}^m)$.

Proof (i) From condition (7.3), and (19.2), we immediately obtain that

$$\begin{aligned}
\varphi(A) = F_0(Ax, (0,1)^n) &\le \liminf_k F_{\varepsilon_{j_k}}(Ax, (0,1)^n) \\
&\le C|(0,1)^n \cap E|(1 + |A|^p).
\end{aligned}$$

On the other hand, with fixed $A \in \mathbf{M}^{m \times n}$, by Proposition 11.7 there exists a sequence of functions $u_k \in W_0^{1,p}((0,1)^n; \mathbf{R}^m)$ such that $u_k \to 0$ in $L^p((0,1)^n; \mathbf{R}^m)$ and

$$\varphi(A) = F_0(Ax, (0,1)^n) = \lim_k F_{\varepsilon_{j_k}}(Ax + u_k, (0,1)^n).$$

If $U \in \mathcal{A}_n$ and $(0,1)^n \subset\subset U$, we can apply Theorem 19.3 to u_k (extended to 0 outside $(0,1)^n$), obtaining, for k sufficiently large so that $(0,1)^n \subset U(k_0\varepsilon_{j_k})$,

$$\begin{aligned}
\|T_{\varepsilon_{j_k}} u_k\|^p_{L^p((0,1)^n; \mathbf{R}^m)} &\le k_1 \|u_k\|^p_{L^p(U \cap \varepsilon_{j_k} E; \mathbf{R}^m)} \\
&= k_1 \|u_k\|^p_{L^p((0,1)^n \cap \varepsilon_{j_k} E; \mathbf{R}^m)} \to 0,
\end{aligned}$$

$$\begin{aligned}
\|D(T_{\varepsilon_{j_k}} u_k)\|^p_{L^p((0,1)^n; \mathbf{R}^m)} &\le k_1 \|Du_k\|^p_{L^p(U \cap \varepsilon_{j_k} E; \mathbf{R}^m)} \\
&= k_1 \|Du_k\|^p_{L^p((0,1)^n \cap \varepsilon_{j_k} E; \mathbf{R}^m)} \le c.
\end{aligned}$$

Hence we have $T_{\varepsilon_{j_k}} u_k \rightharpoonup 0$ weakly in $W^{1,p}((0,1)^n; \mathbf{R}^m)$. We then get, by (19.2),

$$|A|^p = \int_{(0,1)^n} |A|^p \leq \liminf_k \|A + D(T_{\varepsilon_{j_k}} u_k)\|^p_{L^p((0,1)^n;\mathbf{R}^m)}$$

$$\leq \liminf_k k_1 \|A + Du_k\|^p_{L^p((0,1)^n \cap \varepsilon_{j_k} E;\mathbf{R}^m)} \leq k_1 \varphi(A).$$

(ii) Inequality (19.18) follows immediately from the $W^{1,p}$-sequential lower semicontinuity of F_0, and the growth condition (19.2). $\qquad\square$

The proof of Theorem 19.1 will be completed if we show that the function φ is independent of the sequence (ε_{j_k}). For this purpose we introduce the function f_{hom} as in the following proposition.

Proposition 19.6 *Let E be as above, and let f be a Borel function satisfying (19.14) and the growth condition (19.2). Then the limit*

$$f_{\text{hom}}(A) = \lim_{t \to +\infty} \inf\left\{ \frac{1}{t^n} \int_{(0,t)^n \cap E} f(x, Du(x) + A)\, dx \right.$$

$$\left. : u \in W_0^{1,p}((0,t)^n; \mathbf{R}^m) \right\} \tag{19.19}$$

exists for every $A \in \mathbf{M}^{m \times n}$.

Proof For all $t > 0$ define

$$g_t(A) = \inf\left\{ \frac{1}{t^n} \int_{(0,t)^n \cap E} f(x, Du(x) + A)\, dx : u \in W_0^{1,p}((0,t)^n; \mathbf{R}^m) \right\},$$

so that we have to prove the existence of the limit $\lim_{t \to +\infty} g_t(A)$.

For $t > 0$ let $u_t \in W_0^{1,p}((0,t)^n; \mathbf{R}^m)$ be such that

$$\int_{(0,t)^n \cap E} f(x, Du_t(x) + A)\, dx \leq t^n g_t(A) + 1.$$

With fixed $t > 0$, for all $s > t$ we construct $u_s \in W_0^{1,p}((0,s)^n; \mathbf{R}^m)$ as follows. Consider $\varepsilon > 0$ and an inclusion length L_ε for the set of $\mathbf{z} \in \mathbf{Z}^n$ such that (19.14) holds. If $s \leq t + 1 + L_\varepsilon$ let $\mathbf{k} = (0, \ldots, 0)$. If $s > t + 1 + L_\varepsilon$, for each $\mathbf{k} \in \mathbf{Z}^n$ with

$$0 \leq k_j \leq \left[\frac{s}{t + 1 + L_\varepsilon}\right] - 1 \qquad j = 1, \ldots, n$$

choose $\mathbf{z_k} \in \mathbf{Z}^n \cap ([t + 1]\mathbf{k} + [0, L_\varepsilon)^n)$ such that

$$|f(x + \mathbf{z_k}, A) - f(x, A)| \leq 2\varepsilon(1 + |A|^p)$$

for every $x \in \mathbf{R}^n, A \in \mathbf{M}^{m \times n}$. Set

$$u_s(x) = \begin{cases} u_t(x - \mathbf{z_k}) & \text{if } x \in \mathbf{z_k} + (0, t)^n \\ 0 & \text{otherwise,} \end{cases}$$

and

$$Q'_s = (0, s)^n \setminus \bigcup_{\mathbf{k}} (\mathbf{z_k} + (0, t)^n).$$

We then have

$$s^n g_s(A) \leq \int_{E \cap (0,s)^n} f(x, Du_s + A) \, dx$$

$$= \sum_{\mathbf{k}} \int_{E \cap (0,t)^n} f(x + \mathbf{z_k}, Du_t + A) \, dx + \int_{Q'_s \cap E} f(x, A) \, dx$$

$$\leq 2 \sum_{\mathbf{k}} \int_{E \cap (0,t)^n} \varepsilon(1 + |Du_t + A|^p) \, dx$$

$$+ \sum_{\mathbf{k}} \int_{E \cap (0,t)^n} f(x, Du_t + A) \, dx + C \int_{Q'_s} (1 + |A|^p) \, dx$$

$$\leq 2\varepsilon \left(\frac{s}{t}\right)^n (t^n + t^n g_t(A) + 1) + s^n \left(g_t(A) + \frac{1}{t^n}\right)$$

$$+ Cs^n \left(1 - \left(\frac{t}{t + 1 + L_\varepsilon} - \frac{t}{s}\right)^n\right)(1 + |A|^p).$$

Dividing by s^n, and taking the limit first as $s \to +\infty$ and then as $t \to +\infty$, we obtain

$$\limsup_{s \to +\infty} g_s(A) \leq \liminf_{t \to +\infty} g_t(A)(1 + 2\varepsilon) + 2\varepsilon,$$

and hence the required equality, since ε is arbitrary. $\qquad\square$

In order to deal with limits of minimum problems with boundary data, we will need the following proposition.

Proposition 19.7 *Let U be a bounded open subset of $\mathbf{R}^N$, and let (B_j) be a sequence of Borel subsets of U such that $\chi_{B_j} \rightharpoonup c$ in $\mathrm{L}^\infty(U)$-weak*, where c is a strictly positive constant. Let $v_j \in \mathrm{L}^1(U)$, with $v_j = 0$ a.e. in B_j, and $v_j \rightharpoonup v$ weakly in $\mathrm{L}^1(U)$. Then $v = 0$ a.e. in U.*

Proof Let $\varphi \in \mathrm{L}^\infty(U)$. By assumption we have

$$\lim_j \int_U v_j \chi_{B_j} \varphi \, dx = c \int_U v\varphi \, dx,$$

so that $(v_j \chi_{B_j})$ converges weakly in $\mathrm{L}^1(U)$ to cv. Since $v_j \chi_{B_j} = 0$ a.e. on U we have $cv = 0$, and then $v = 0$ a.e. $\qquad\square$

Proof of Theorem 19.1. With fixed $A \in \mathbf{M}^{m \times n}$ let

$$g_\varepsilon(A) = \inf\left\{ \int_{(0,1)^n \cap \varepsilon E} f\left(\frac{x}{\varepsilon}, Du(x) + A\right) dx \; : \; u \in \mathrm{W}_0^{1,p}((0,1)^n; \mathbf{R}^m)\right\},$$

so that $f_{\mathrm{hom}}(A) = \lim_{\varepsilon \to 0} g_\varepsilon(A)$, and for every $\varepsilon > 0$ let $u_\varepsilon^A \in W_0^{1,p}((0,1)^n; \mathbf{R}^m)$ (extended to 0 outside $(0,1)^n$) be such that

$$\int_{(0,1)^n \cap \varepsilon E} f\left(\frac{x}{\varepsilon}, Du_\varepsilon^A + A\right) dx \le g_\varepsilon(A) + \varepsilon.$$

We have by (19.2)

$$\int_{(0,1)^n \cap \varepsilon E} |Du_\varepsilon^A + A|^p dx \le \int_{(0,1)^n \cap \varepsilon E} f\left(\frac{x}{\varepsilon}, Du_\varepsilon^A + A\right) dx \le C(1 + |A|^p) + \varepsilon;$$

hence

$$\int_{(0,1)^n \cap \varepsilon E} |Du_\varepsilon^A|^p dx \le c.$$

Let (ε_j) be a sequence of positive numbers converging to 0. If $U \in \mathcal{A}_n$ with $(0,1)^n \subset\subset U$, we can apply Theorem 19.3 to U and $u_{\varepsilon_j}^A$, obtaining extensions $T_{\varepsilon_j} u_{\varepsilon_j}^A \in W_{\mathrm{loc}}^{1,p}(U; \mathbf{R}^m)$. For j sufficiently large, we have $(0,1)^n \subset U(k_0\varepsilon_j)$, so that by (19.11) we get

$$\|D(T_{\varepsilon_j} u_{\varepsilon_j}^A)\|_{L^p((0,1)^n; \mathbf{R}^m)} \le \|D(T_{\varepsilon_j} u_{\varepsilon_j}^A)\|_{L^p(U(k_0\varepsilon_j); \mathbf{R}^m)} \le c.$$

Since $T_{\varepsilon_j} u_{\varepsilon_j}^A = 0$ in $(U \setminus (0,1)^n) \cap \varepsilon_j E$, by the Poincaré inequality, we also have that the sequence $(T_{\varepsilon_j} u_{\varepsilon_j}^A)$ is bounded in $W^{1,p}((0,1)^n; \mathbf{R}^m)$. We can suppose then that (up to subsequences) $T_{\varepsilon_j} u_{\varepsilon_j}^A \rightharpoonup u$ weakly in $W^{1,p}((0,1)^n; \mathbf{R}^m)$. Note that, since E is open and connected, there exists a constant $b > 0$, such that the intersection of $E \cap (0,1)^n$ with any hyperplane orthogonal to a coordinate axis has $(n-1)$-surface measure larger than b. Hence, for every sequence of positive numbers (ε_j) converging to 0, there exists a subsequence (which we still denote with (ε_j)) such that $\chi_{\varepsilon_j E} \rightharpoonup c$ in L^∞-weak* on each face of the cube $(0,1)^n$, where c is a positive constant, with $c \ge b$ (the value of the constant c possibly varying from face to face). We can then apply Proposition 19.7 with U a face of the cube $(0,1)^n$, and obtain $u \in W_0^{1,p}((0,1)^n; \mathbf{R}^m)$.

Suppose now that the sequence (ε_j) satisfies the thesis of Proposition 19.4. By (7.3), and (19.18), we obtain

$$\varphi(A) \le \int_{(0,1)^n} \varphi(A + Du) \, dx$$

$$\le \lim_j \int_{(0,1)^n \cap \varepsilon_j E} f\left(\frac{x}{\varepsilon_j}, Du_{\varepsilon_j}^A + A\right) dx = f_{\mathrm{hom}}(A). \tag{19.20}$$

On the other hand, by (7.5), there exists a sequence $(w_j) \subset L^p((0,1)^n; \mathbf{R}^m)$ with $w_j \to Ax$ in $L^p((0,1)^n; \mathbf{R}^m)$, such that

$$\varphi(A) = F_0(Ax, (0,1)^n) = \lim_j F_{\varepsilon_j}(w_j, (0,1)^n).$$

By Proposition 11.7, we can suppose that $w_j - Ax \in W_0^{1,p}((0,1)^n; \mathbf{R}^m)$, and obtain

$$\varphi(A) = \lim_j \int_{(0,1)^n \cap \varepsilon_j E} f\left(\frac{x}{\varepsilon_j}, Dw_j\right) dx$$

$$\geq \lim_j g_{\varepsilon_j}(A) = f_{\text{hom}}(A). \tag{19.21}$$

From (19.20) and (19.21) we obtain that $\varphi = f_{\text{hom}}$, and then that the Γ-limit in (19.15) is independent of the sequence (ε_{j_k}) and equal to the functional F_{hom} defined by (19.4) and (19.6), thus proving Theorem 19.1. $\qquad\square$

Remark 19.8 The Γ-convergence (but not the convergence of minima and minimum points) of functionals of the form $\int_\Omega f(\frac{x}{\varepsilon}, Du(x)) \, dx$ to the functional given by $\int_\Omega f_{\text{hom}}(Du(x)) \, dx$, where

$$f_{\text{hom}}(A) = \lim_{t \to +\infty} \inf\left\{\frac{1}{t^n} \int_{(0,t)^n} f(x, Du(x) + A) \, dx : u \in W_0^{1,p}((0,t)^n; \mathbf{R}^m)\right\},$$

can be obtained in more general situations than those described in (19.1), (19.2). Theorem 19.1 can be applied to the study of the asymptotic behaviour of $f_{\text{hom}}(A)$ as $|A| \to +\infty$, when f is of the type

$$f(x, A) = f_1(x, A)\chi_E(x) + f_2(x, A)\chi_{\mathbf{R}^n \setminus E}(x).$$

The function f may be interpreted as the energy density of a composite medium whose components may have very different behaviours (i.e. f_1, f_2 may satisfy different growth conditions).

When E is an open, connected and 1-periodic subset of $\mathbf{R}^n$ with Lipschitz boundary, and the functions f_1, f_2 satisfy

$$|A|^p \leq f_1(x, A) \leq C(1 + |A|^p), \qquad 0 \leq f_2(x, A) \leq C(1 + |A|^q),$$

for a.e. $x \in \mathbf{R}^n$ and for every $A \in \mathbf{R}^n$, with $1 < q \leq p < +\infty$, then by Theorem 19.1 we obtain that f_{hom} still satisfies a growth condition of order p. In fact, it suffices to notice that

$$\chi_E(x)|A|^p \leq f(x, A) \leq C(1 + |A|^p),$$

and to apply Theorem 19.1 to the function $|A|^p$ on the set E.

19.2   Convergence of Neumann boundary value problems

Remark 19.9 Given $\Omega \in \mathcal{A}_n$ fix a sequence $(g_\varepsilon) \subset L^p(\Omega; \mathbf{R}^m)$ converging to g_0 in $L^p(\Omega; \mathbf{R}^m)$, and let $G_\varepsilon, G_0 : L_{\text{loc}}^p(\Omega; \mathbf{R}^m) \times \mathcal{A}(\Omega) \to [0, +\infty]$ be the functionals defined by

$$G_\varepsilon(u, U) = F_\varepsilon(u, U) + \int_{U \cap \varepsilon E} |g_\varepsilon - u|^p \, dx$$

$$G_0(u, U) = F_{\mathrm{hom}}(u, U) + |(0,1)^n \cap E| \int_U |g_0 - u|^p \, dx$$

for every $u \in L^p_{\mathrm{loc}}(\Omega; \mathbf{R}^m)$ and every $U \in \mathcal{A}(\Omega)$. It is easy to show that

$$\Gamma(L^p_{\mathrm{loc}})\text{-}\lim_{\varepsilon \to 0} G_\varepsilon(u, U) = G_0(u, U)$$

for every $u \in L^p_{\mathrm{loc}}(\Omega; \mathbf{R}^m)$ and every $U \in \mathcal{A}_n$; that is, (7.3) and (7.5) hold when $u_j \to u$ in $L^p_{\mathrm{loc}}(U; \mathbf{R}^m)$. In fact, since

$$\lim_j \int_{U \cap \varepsilon_j E} |g_{\varepsilon_j} - u_j|^p \, dx = |(0,1)^n \cap E| \int_U |g_0 - u|^p \, dx$$

for all $\varepsilon_j \to 0+$, and $u_j \to u$ in $L^p(U, \mathbf{R}^m)$, we have

$$\Gamma(L^p)\text{-}\lim_{\varepsilon \to 0} G_\varepsilon(u, U) = G_0(u, U),$$

so that (7.5) holds trivially also in L^p_{loc}. To prove (7.3), we fix (ε_j), (u_j) and u in $L^p_{\mathrm{loc}}(U; \mathbf{R}^m)$ with $u_j \to u$ in $L^p_{\mathrm{loc}}(U; \mathbf{R}^m)$. If $U' \subset\subset U$ then, by the Γ-convergence above, we have

$$G_0(u, U') \leq \liminf_j G_{\varepsilon_j}(u_j, U') \leq \liminf_j G_{\varepsilon_j}(u_j, U),$$

and the conclusion follows by taking the supremum over $U' \subset\subset U$.

Proposition 19.10 *Under the hypotheses considered above on E, (g_ε), g_0 and f, the sequence (m_ε) defined by*

$$m_\varepsilon = \min_{u \in W^{1,p}(\Omega \cap \varepsilon E; \mathbf{R}^m)} \left\{ \int_{\Omega \cap \varepsilon E} f\left(\frac{x}{\varepsilon}, Du\right) dx + \int_{\Omega \cap \varepsilon E} |g_\varepsilon - u|^p \, dx \right\} \quad (19.22)$$

converges to the minimum value

$$m = \min_{u \in W^{1,p}(\Omega; \mathbf{R}^m)} \left\{ \int_\Omega f_{\mathrm{hom}}(Du) \, dx + |(0,1)^n \cap E| \int_\Omega |g_0 - u|^p \, dx \right\}, \quad (19.23)$$

as $\varepsilon \to 0$, with f_{hom} given by (19.6).

Proof Let $u_\varepsilon \in W^{1,p}(\Omega \cap \varepsilon E; \mathbf{R}^m)$ be a solution to problem (19.22) and let $T_\varepsilon : W^{1,p}(\Omega \cap \varepsilon E; \mathbf{R}^m) \to W^{1,p}_{\mathrm{loc}}(\Omega; \mathbf{R}^m)$ be the extension operator introduced in Theorem 19.3. By Theorem 7.2 and Remark 19.9 it is enough to prove that the sequence $(T_\varepsilon u_\varepsilon) \subset W^{1,p}_{\mathrm{loc}}(\Omega; \mathbf{R}^m)$ is compact for the L^p_{loc}-topology. To this end, note that, since $g_\varepsilon \to g_0$ in $L^p(\Omega; \mathbf{R}^m)$, by (19.2) and (19.22) the sequence $(\|u_\varepsilon\|_{W^{1,p}(\Omega \cap \varepsilon E; \mathbf{R}^m)})$ is bounded independently of ε, and hence, by (19.10) and (19.11),

$$\|T_\varepsilon u_\varepsilon\|_{W^{1,p}(U; \mathbf{R}^m)} \leq c$$

for every open set $U \subset\subset \Omega$ such that $\mathrm{dist}\,(U, \partial\Omega) > \varepsilon k_0$, with a constant c independent of ε and U. If U has Lipschitz boundary, by Rellich's Theorem there exists a subsequence $(T_{\varepsilon_j} u_{\varepsilon_j})$ that converges strongly in $L^p(U; \mathbf{R}^m)$ to a function $u \in W^{1,p}(U; \mathbf{R}^m)$. If we now consider an invading sequence of open subsets of Ω, using a diagonal process we can extract a subsequence of $(T_\varepsilon u_\varepsilon)$ that converges strongly in $L^p_{\mathrm{loc}}(\Omega; \mathbf{R}^m)$ to a function $u \in W^{1,p}_{\mathrm{loc}}(\Omega; \mathbf{R}^m)$. This means that the sequence $(T_\varepsilon u_\varepsilon)$ is compact for the L^p_{loc}-topology in $W^{1,p}_{\mathrm{loc}}(\Omega; \mathbf{R}^m)$, and concludes the proof. $\square$

19.3 Convergence of Dirichlet boundary value problems

We want to remark shortly how, using Proposition 19.7 and following the reasonings of the proof of Theorem 19.1, we can deal with the limits of Dirichlet boundary value problems.

Let Ω be a Lipschitz bounded open subset of $\mathbf{R}^n$, and let $g \in W^{1,p}(\Omega; \mathbf{R}^m)$. We can study the minimum problems related to the functionals F_ε with the boundary condition

$$u = g \text{ on } \partial\Omega.$$

Proposition 19.11 *Let E and f be as above, and consider the minimum problems*

$$m_\varepsilon = \inf\left\{ \int_{\Omega \cap \varepsilon E} f\left(\frac{x}{\varepsilon}, Du(x)\right) dx : u \in g + W^{1,p}_0(\Omega; \mathbf{R}^m) \right\}.$$

Then the minimum values m_ε converge to the minimum value

$$m = \min\left\{ \int_\Omega f_{\mathrm{hom}}(Du(x)) dx : u \in g + W^{1,p}_0(\Omega; \mathbf{R}^m) \right\},$$

and the T_ε-extensions of the restrictions to $\Omega \cap \varepsilon E$ of ε-minimizing sequences for the problems m_ε converge to minimizers for the problem m.

Proof Let Ω' be a bounded open subset of $\mathbf{R}^n$ such that $\Omega \subset\subset \Omega'$, and let $(u_\varepsilon) \subset g + W^{1,p}_0(\Omega; \mathbf{R}^m)$ be an ε-minimizing sequence for the problems m_ε, i.e.

$$\int_{\Omega \cap \varepsilon E} f\left(\frac{x}{\varepsilon}, Du_\varepsilon(x)\right) dx \le m_\varepsilon + \varepsilon.$$

Consider the functions

$$u_\varepsilon - g \in W^{1,p}_0(\Omega; \mathbf{R}^m) \subset W^{1,p}_0(\Omega'; \mathbf{R}^m)$$

extended to 0 in $\Omega' \setminus \Omega$. If we set

$$v_\varepsilon = (u_\varepsilon - g)_{|\Omega' \cap \varepsilon E}$$

then we have

$$\|v_\varepsilon\|_{W^{1,p}(\Omega' \cap \varepsilon E; \mathbf{R}^m)} = \|v_\varepsilon\|_{W^{1,p}(\Omega \cap \varepsilon E; \mathbf{R}^m)} \le c.$$

Take $\delta > 0$ such that $\Omega \subset\subset \Omega'(\delta)$; when ε is sufficiently small we have $\Omega'(\delta) \subset \Omega'(\varepsilon k_0)$, and hence, when we consider the extensions $T_\varepsilon v_\varepsilon$, we have by Theorem 19.3

$$\|T_\varepsilon v_\varepsilon\|_{\mathrm{W}^{1,p}(\Omega'(\delta);\mathbf{R}^m)} \leq c.$$

We can suppose (passing possibly to a subsequence) that

$$T_\varepsilon v_\varepsilon \rightharpoonup \bar{v} \text{ weakly in } \mathrm{W}^{1,p}(\Omega'(\delta);\mathbf{R}^m). \tag{19.24}$$

Let π be any hyperplane orthogonal to some coordinate axis; then by Proposition 19.7, as in the proof of Theorem 19.1, we see that $\bar{v} = 0$ in the space of traces $\mathrm{W}^{1-\frac{1}{p},p}((\Omega'(\delta) \setminus \overline{\Omega}) \cap \pi; \mathbf{R}^m)$, and hence

$$\bar{v}_{|\Omega} \in \mathrm{W}_0^{1,p}(\Omega; \mathbf{R}^m). \tag{19.25}$$

Define $\bar{u} = \bar{v} + g$; (19.25) shows, by (7.3) and (19.5), that

$$m \leq \int_\Omega f_{\mathrm{hom}}(D\bar{u}(x))\, dx \leq \liminf_{\varepsilon \to 0+} m_\varepsilon. \tag{19.26}$$

On the other hand, let $\tilde{u}$ be a minimum point for m; then Proposition 11.7 assures that, with a fixed sequence (ε_j) of positive real numbers converging to 0, we can find functions $u_j \in \mathrm{W}_0^{1,p}(\Omega; \mathbf{R}^m)$ such that $u_j \to 0$ strongly in $\mathrm{L}^p((\Omega; \mathbf{R}^m))$, and

$$m = F_{\mathrm{hom}}(\tilde{u}, \Omega) = \lim_j F_{\varepsilon_j}(\tilde{u} + u_j, \Omega) \geq \limsup_j m_{\varepsilon_j}.$$

By the arbitrariness of (ε_j), we have

$$m \geq \limsup_{\varepsilon \to 0+} m_\varepsilon. \tag{19.27}$$

The two inequalities (19.26) and (19.27) provide the required convergence of the minimum values, and hence (19.24) assures the weak convergence of minimum points. $\qquad\square$

$$20$$

HOMOGENIZATION WITH STIFF AND SOFT INCLUSIONS

We are going to improve the results of Chapter 19 and show the convergence to a non-degenerate homogeneous functional when the function f satisfies quite general growth assumptions, with large zones of various types of degeneracy. We require that the function f is periodic and that there exists $p > 1$ such that:

(i) there exists c such that the region of the points x where $f(x, A) \leq c(1 + |A|^p)$ does not hold for all matrices A is composed of well-separated sets;

(ii) the condition $f(x, A) \geq |A|^p$ for all matrices A is satisfied on a connected open set with Lipschitz boundary.

Condition (i) must be imposed to avoid a 'rigid behaviour' of the function f_{hom}, while (ii) prevents a 'weakening' of the structure in the limit. Note, however, that the latter condition can be substituted by weaker assumptions, invoking Theorem 14.8 (see Examples 14.9 and 14.10).

Conditions (i) and (ii) allow a variety of different behaviours for the function f. In particular, we may have the so-called 'stiff inclusions' modelled by a function of the form

$$f(x, A) = \begin{cases} 0 & \text{if } A = 0 \\ +\infty & \text{if } A \neq 0, \end{cases}$$

for x in a region composed of well-separated components, and at the same time we can have a 'soft' behaviour, or holes (in this case $f(x, A) \equiv 0$), provided that the complementary set of the 'void' region contains a connected periodic domain. The function f may also exhibit intermediate behaviours and satisfy globally a growth condition of the form

$$c_1 |A|^{p_1} \leq f(A) \leq c_2(1 + |A|^{p_2})$$

with $p_1 \leq p \leq p_2$. The main steps in the proof are Proposition 20.5 (the fundamental estimate in Definition 11.2 for the Γ-convergence, which we prove here in its full form), and Proposition 20.8 (the asymptotic formula for the homogenization).

20.1 Media with stiff and soft inclusions

The set K will be a fixed subset of $\mathbf{R}^n$. We suppose that $(i + K) \cap (j + K) = \emptyset$ for all $i, j \in \mathbf{Z}^n$, $i \neq j$; more precisely, we denote by $\eta > 0$ a positive number satisfying the following property:

$$\mathrm{dist.}(i + K, j + K) > 3\eta \quad \forall i, j \in \mathbf{Z}^n \ i \neq j. \tag{20.1}$$

We will consider the 1-periodic set $\mathcal{K} = \mathbf{Z}^n + K$; we will denote by K_η the set $\{x \in \mathbf{R}^n : \operatorname{dist}(x, K) < \eta\}$, and the set $\mathbf{Z}^n + K_\eta$ by $\mathcal{K}_\eta$.

Remark 20.1 It is easy to see that assumption (20.1) implies that the set K is contained in the union of a finite number of cubes of the type $i + [0, 1]^n$, $i \in \mathbf{Z}^n$, and in particular that its diameter is finite.

Consider now a Borel function $f : \mathbf{R}^n \times \mathbf{M}^{m \times n} \to [0, +\infty]$ and $1 < p < +\infty$ such that:

(i) $f(\cdot, A)$ is 1-periodic;

(ii) there exists a 1-periodic, connected open set E with Lipschitz boundary such that $|A|^p \leq f(x, A)$ for every $A \in \mathbf{M}^{m \times n}$ and $x \in E$;

(iii) $0 \leq f(x, A) \leq c(1 + |A|^p)$ for every $A \in \mathbf{M}^{m \times n}$ and $x \in [0, 1]^n \setminus \mathcal{K}$;

(iv) there exists a Borel function $a : \mathbf{R}^n \times \mathbf{M}^{m \times n} \to [0, +\infty]$, convex in the second variable, such that

$$a(x, A) \leq f(x, A) \leq c(1 + a(x, A))$$

for every $x \in \mathbf{R}^n$ and $A \in \mathbf{M}^{m \times n}$;

(v) there exists a positive constant α such that $a(x, 2A) \leq \alpha\, a(x, A)$ for every $A \in \mathbf{M}^{m \times n}$ and $x \in \mathbf{R}^n \setminus \mathcal{K}$;

(vi) there exists a constant $b \geq 0$ such that $f(x, 0) \leq b$ for every $x \in \mathcal{K}$.

For every bounded open subset Ω of $\mathbf{R}^n$, we will denote by $X(\Omega)$ a fixed subspace of $\mathrm{W}^{1,1}(\Omega; \mathbf{R}^m)$, with $C^1(\Omega; \mathbf{R}^m) \subseteq X(\Omega) \subseteq \mathrm{W}^{1,1}(\Omega; \mathbf{R}^m) \cap \mathrm{L}^p(\Omega; \mathbf{R}^m)$, such that for every $u \in X(\Omega)$ and $\varphi \in C_0^1(\Omega; \mathbf{R}^m)$ we have $u\varphi \in X(\Omega)$. It is understood that the meaning of the notation 'X' does not change with Ω, i.e. if $u \in X(\Omega)$, and $\Omega' \subset \Omega$, then the restriction $u_{|\Omega'}$ belongs to $X(\Omega')$, so that the notation $X_{\mathrm{loc}}(\mathbf{R}^n)$ makes sense. Moreover, we will consider the space $X_{\#}((0, 1)^n) = X_{\mathrm{loc}}(\mathbf{R}^n) \cap \mathrm{W}_{\#}^{1,1}((0, 1)^n; \mathbf{R}^m)$.

In the sequel we will study the limit behaviour of the following family of functionals $(F_\varepsilon)_{\varepsilon > 0}$ defined by

$$F_\varepsilon(u, \Omega) = \begin{cases} \displaystyle\int_\Omega f\left(\frac{x}{\varepsilon}, Du\right) dx & \text{if } u \in X(\Omega) \\ +\infty & \text{otherwise} \end{cases} \tag{20.2}$$

on $\mathrm{L}_{\mathrm{loc}}^p(\mathbf{R}^n; \mathbf{R}^m) \times \mathcal{A}_n$. We will sometimes use the same notation $F_\varepsilon(u, B)$ for $\int_B f(\frac{x}{\varepsilon}, Du)\, dx$ if $B \subset \Omega$ is a Borel set and $u \in X(\Omega)$.

In order to prove the L^p-fundamental estimate for $\varepsilon \to 0$ along the sequence (F_ε), it will be useful to consider, for every $\varepsilon > 0$ and $i \in \mathbf{Z}^n$, the operator R_i^ε defined on $\mathrm{W}_{\mathrm{loc}}^{1,\infty}(\mathbf{R}^n; \mathbf{R}^m)$ by

$$R_i^\varepsilon(u)(x) = \left(1 - \psi\left(\frac{x}{\varepsilon} - i\right)\right) u(x) + \psi\left(\frac{x}{\varepsilon} - i\right) \fint_{\varepsilon i + \varepsilon K_\eta} u(y)\, dy,$$

where ψ is a fixed cut-off function between K and K_η such that $|D\psi| \leq \frac{2}{\eta}$.

Remark 20.2 The operator R_i^ε acts over a function u in the following way:

$$R_i^\varepsilon(u)(x) = \fint_{\varepsilon i + \varepsilon K_\eta} u(y)\, dy \qquad \text{if } x \in \varepsilon i + \varepsilon K;$$

in particular $R_i^\varepsilon(u)$ is constant on $\varepsilon i + \varepsilon K$, and

$$R_i^\varepsilon(u)(x) = u(x) \qquad \text{if dist}\,(x, \varepsilon i + \varepsilon K) \geq \varepsilon\eta.$$

Thus, we define the operator

$$\mathcal{R}^\varepsilon(u)(x) = \begin{cases} R_i^\varepsilon(u)(x) & \text{if dist}\,(x, \varepsilon i + \varepsilon K) < \varepsilon\eta,\ i \in \mathbf{Z}^n \\[2ex] u(x) & \text{otherwise.} \end{cases}$$

Remark 20.3 If $u \in \mathrm{W}^{1,\infty}_{\mathrm{loc}}(\mathbf{R}^n)$ it is easy to see that for every bounded open set Ω

$$\|D\mathcal{R}^\varepsilon(u)\|_{\mathrm{L}^\infty(\Omega;\mathbf{R}^n)} \leq \frac{2}{\varepsilon\eta} \sup_{i \in \mathbf{Z}^n} \left\| u - \fint_{\varepsilon i + \varepsilon K_\eta} u(y)\, dy \right\|_{\mathrm{L}^\infty(\varepsilon i + \varepsilon K_\eta)}$$
$$+ \|Du\|_{\mathrm{L}^\infty(\Omega;\mathbf{R}^n)}.$$

Moreover, denoting $d = \mathrm{diam}\, K_\eta$, we have

$$\left\| u - \fint_{\varepsilon i + \varepsilon K_\eta} u(y)\, dy \right\|_{\mathrm{L}^\infty(\varepsilon i + \varepsilon K_\eta)} \leq \operatorname*{ess\text{-}sup}_{\varepsilon i + \varepsilon K_\eta} u - \operatorname*{ess\text{-}inf}_{\varepsilon i + \varepsilon K_\eta} u$$
$$\leq \varepsilon d \|Du\|_{\mathrm{L}^\infty(\Omega;\mathbf{R}^n)},$$

thus, if $u \in \mathrm{W}^{1,\infty}_{\mathrm{loc}}(\mathbf{R}^n; \mathbf{R}^m)$, we obtain

$$\|D\mathcal{R}^\varepsilon(u)\|_{\mathrm{L}^\infty(\Omega;\mathbf{M}^{m\times n})} \leq \left(\frac{2}{\eta}d + 1\right) \|Du\|_{\mathrm{L}^\infty(\Omega;\mathbf{M}^{m\times n})}.$$

In the sequel we will denote $\mathcal{R}^1$ and R_i^1 simply by $\mathcal{R}$ and R_i.

20.2 The Homogenization Theorem

We will prove the following homogenization result.

Theorem 20.4 *Let $\mathcal{K}$ be the set satisfying the assumption (20.1) defined above. Let $f : \mathbf{R}^n \times \mathbf{M}^{m\times n} \to [0, +\infty]$ be a Borel function satisfying (i)–(vi), and, for every $\varepsilon > 0$, let $F_\varepsilon : \mathrm{L}^p_{\mathrm{loc}}(\mathbf{R}^n; \mathbf{R}^m) \times \mathcal{A}_n \to [0, +\infty]$ be the functional defined* *in (20.2). Then there exist two positive constants c_1 and c_2, and a quasiconvex function $f_{\text{hom}} : \mathbf{M}^{m \times n} \to [0, +\infty)$, with*

$$c_1 |A|^p \leq f_{\text{hom}}(A) \leq c_2(1 + |A|^p) \tag{20.3}$$

for all $A \in \mathbf{M}^{m \times n}$, such that, defining $F_{\text{hom}} : \mathrm{L}^p_{\text{loc}}(\mathbf{R}^n; \mathbf{R}^m) \times \mathcal{A}_n \to [0, +\infty]$ by

$$F_{\text{hom}}(u, \Omega) = \begin{cases} \displaystyle\int_\Omega f_{\text{hom}}(Du(x))\, dx & \text{if } u \in \mathrm{W}^{1,p}(\Omega; \mathbf{R}^m) \\ +\infty & \text{otherwise,} \end{cases} \tag{20.4}$$

we have

$$F_{\text{hom}}(u, \Omega) = \Gamma(\mathrm{L}^p)\text{-}\lim_{\varepsilon \to 0} F_\varepsilon(u, \Omega)$$

for every $\Omega \in \mathcal{A}_n$ and every $u \in \mathrm{L}^p(\Omega; \mathbf{R}^m)$. Moreover, the function f_{hom} satisfies the formula

$$f_{\text{hom}}(A) = \lim_{T \to +\infty} \inf\left\{ \frac{1}{T^n} \int_{(0,T)^n} f(x, A + Du)\, dx \; : \; u \in X_A^T \right\},$$

where

$$X_A^T = \{u \in X((0,T)^n) \; : \; u = \mathcal{R}(Ax) - Ax \text{ in a neighbourhood of } \partial(0,T)^n\}.$$

To prove Theorem 20.4 it is not possible to apply directly the homogenization theorems of Chapters 15 and 19. Indeed the functionals F_ε do not satisfy standard growth conditions of order p in all the space. However, the region where the growth condition from above is violated, thanks to assumption (20.1), is composed of well-isolated domains.

Proposition 20.5 *The family $(F_\varepsilon)_{\varepsilon > 0}$ of functionals defined by (20.2) satisfies uniformly the L^p-fundamental estimate as $\varepsilon \to 0$.*

Proof We fix the open sets U, U' and V, with $U' \subset\subset U$, and $\sigma > 0$. Let U'' be an open set such that $U' \subset\subset U'' \subset\subset U$. Given $N \in \mathbf{N}$, to be fixed later, we define the open set

$$U_l = \left\{ x \; : \; \text{dist}\,(x, U') < \frac{\text{dist}\,(U', \partial U'')}{3N} l \right\}$$

for every $l = 1, \ldots, 3N$, and $U_0 = U'$. For every $k = 0, \ldots, N - 1$, let φ_k be a cut-off function between the sets U_{3k+1} and U_{3k+2} such that $|D\varphi_k| < 4N/\text{dist}\,(U', \partial U'')$. Now we can fix ε_σ such that

$$\varepsilon_\sigma d < \frac{1}{3N} \text{dist}\,(U', \partial U''),$$

where d denotes the diameter of the set K_η that, by assumption (20.1), is finite (see Remark 20.1). Since $\text{dist}\,(U_l, \partial U_{l+1}) = \text{dist}\,(U', \partial U'')/3N$, if, for some $i \in \mathbf{Z}^n$

and $\varepsilon < \varepsilon_\sigma$, $\varepsilon i + \varepsilon K$ intersects U_l, it does not intersect $\mathbf{R}^n \setminus U_{l+1}$. Then the functions $\phi_k = \mathcal{R}^\varepsilon(\varphi_k)$, $0 \le k \le N - 1$, are cut-off functions between the sets U_{3k} and $U_{3(k+1)}$.

We will show that, for every $u \in X(U)$ and $v \in X(V)$, it is possible to prove the fundamental estimate by choosing the cut-off function between this finite family $(\phi_k)_{k=0,\ldots,N-1}$. Thus, with fixed $u \in X(U)$ and $v \in X(V)$, for every $k = 0, \ldots, N - 1$ we have

$$
\begin{aligned}
F_\varepsilon(\phi_k u + (1 - \phi_k)v, U' \cup V) &= F_\varepsilon(u, (U' \cup V) \cap \overline{U}_{3k}) + F_\varepsilon(v, V \setminus U_{3(k+1)}) \\
&\quad + F_\varepsilon(\phi_k u + (1 - \phi_k)v, V \cap (U_{3(k+1)} \setminus \overline{U}_{3k})) \\
&\le F_\varepsilon(u, U) + F_\varepsilon(v, V) \\
&\quad + F_\varepsilon(\phi_k u + (1 - \phi_k)v, V \cap (U_{3(k+1)} \setminus \overline{U}_{3k})).
\end{aligned}
\tag{20.5}
$$

Now we want to estimate the last term, which we denote by I_k. Let $S_k = V \cap (U_{3(k+1)} \setminus \overline{U}_{3k})$. By construction we have $D\phi_k = 0$ on εK and by assumption (iv) we obtain

$$
\begin{aligned}
F_\varepsilon(\phi_k u + (1 - \phi_k)v, S_k \cap \varepsilon K) &= \int_{S_k \cap \varepsilon K} f\left(\frac{x}{\varepsilon}, \phi_k Du + (1 - \phi_k)Dv\right) dx \\
&\le c \int_{S_k \cap \varepsilon K} \left(1 + \phi_k a\left(\frac{x}{\varepsilon}, Du\right) + (1 - \phi_k)a\left(\frac{x}{\varepsilon}, Dv\right)\right) dx \\
&\le c|S_k \cap \varepsilon K| + cF_\varepsilon(u, S_k \cap \varepsilon K) + cF_\varepsilon(v, S_k \cap \varepsilon K).
\end{aligned}
\tag{20.6}
$$

Moreover, by (iv) and (v) we have

$$
\begin{aligned}
&F_\varepsilon(\phi_k u + (1 - \phi_k)v, S_k \setminus \varepsilon K) \\
&\le c \int_{S_k \setminus \varepsilon K} \left(1 + a\left(\frac{x}{\varepsilon}, D\phi_k(u - v) + \phi_k Du + (1 - \phi_k)Dv\right)\right) dx \\
&\le c|S_k \setminus \varepsilon K| \\
&\quad + c\frac{\alpha}{2} \int_{S_k \setminus \varepsilon K} \left(a\left(\frac{x}{\varepsilon}, D\phi_k(u - v)\right) + \phi_k a\left(\frac{x}{\varepsilon}, Du\right) + (1 - \phi_k)a\left(\frac{x}{\varepsilon}, Dv\right)\right) dx \\
&\le c\left(1 + c\frac{\alpha}{2}\right)|S_k \setminus \varepsilon K| + \frac{\alpha}{2}cF_\varepsilon(u, S_k \setminus \varepsilon K) + c\frac{\alpha}{2}F_\varepsilon(v, S_k \setminus \varepsilon K) \\
&\quad + c^2\frac{\alpha}{2}\|D\phi_k\|_{L^\infty}^p \int_{S_k \setminus \varepsilon K} |u - v|^p \, dx,
\end{aligned}
\tag{20.7}
$$

where in the last inequality we have used that, by (iii) and (iv), $a(x, A) \le c(1 + |A|^p)$ for every $A \in \mathbf{M}^{m \times n}$ and $x \in [0, 1]^n \setminus K$. Since $\|D\varphi_k\|_{L^\infty} \le 4N/\mathrm{dist}\,(U', \partial U'')$ by Remark 20.3 we obtain

$$\|D\phi_k\|_{L^\infty}^p \le \left(\frac{4N}{\mathrm{dist}\,(U',\partial U'')}\left(\frac{2}{\eta}d+1\right)\right)^p$$

Taking $M_\sigma = c^2\frac{\alpha}{2}\left((4N/\mathrm{dist}\,(U',\partial U''))((2/\eta)d+1)\right)^p$, by (20.6) and (20.7) we get

$$\min_{0\le k\le N-1} I_k \le \frac{1}{N}\sum_{k=0}^{N-1} I_k \le \frac{c}{N}\left(1+c\frac{\alpha}{2}\right)|U''\setminus\overline{U}'|$$
$$+ \left(1+\frac{\alpha}{2}\right)\frac{c}{N}(F_\varepsilon(u,U)+F_\varepsilon(v,V))$$
$$+ M_\sigma \int_{(U\cap V)\setminus(U'\cup\varepsilon\mathcal{K})}|u-v|^p\,dx. \qquad (20.8)$$

Thus, choosing N such that $\max\{c(1+c\frac{\alpha}{2})|U''\setminus\overline{U}'|,(1+\frac{\alpha}{2})c\} < \sigma N$, by (20.5) and (20.8) we obtain that there exists k, $0\le k\le N-1$, such that the fundamental estimate holds with ϕ_k as the cut-off function. $\qquad\square$

Now fix a sequence (ε_j), and denote F_{ε_j} by F_j. We will consider the $\Gamma(\mathrm{L}^p)$-lim sup and the $\Gamma(\mathrm{L}^p)$-lim inf of the sequence (F_j), which we denote by F'' and F', respectively. A first important step in order to prove the homogenization theorem is to check the growth conditions for F'' and F'.

Proposition 20.6 *There exist two constants $c_1 > 0$ and $c_2 > 0$ such that*

$$c_1\int_\Omega |Du|^p\,dx \le F'(u,\Omega) \le F''(u,\Omega) \le c_2\int_\Omega(1+|Du|^p)\,dx$$

for every bounded Lipschitz open set Ω and every $u \in \mathrm{W}^{1,p}(\Omega;\mathbf{R}^m)$.

Proof In order to estimate F' from below, consider the sequence of functionals $F_j^0 : \mathrm{L}_{\mathrm{loc}}^p(\mathbf{R}^n;\mathbf{R}^m)\times\mathcal{A}_n \to [0,+\infty]$ defined by

$$F_j^0(u,\Omega) = \begin{cases} \displaystyle\int_\Omega \widetilde{f}\left(\frac{x}{\varepsilon_j},Du\right)dx & \text{if } u\in \mathrm{W}^{1,1}(\Omega;\mathbf{R}^m)\cap\mathrm{L}^p(\Omega;\mathbf{R}^m) \\ +\infty & \text{otherwise,} \end{cases}$$

with $\widetilde{f}(x,A) = |A|^p$ for every $x\in E$ and $A\in\mathbf{M}^{m\times n}$, and $\widetilde{f}(x,A) = 0$ for every $x\in\mathbf{R}^n\setminus E$ and $A\in\mathbf{M}^{m\times n}$. Since by (ii) E is a 1-periodic connected open set with Lipschitz boundary, by Theorem 19.1 there exists a constant $c_1 = 1/k_1 > 0$ such that

$$c_1\int_\Omega |Du|^p\,dx \le \Gamma(\mathrm{L}^p)\text{-}\lim_j F_j^0(u,\Omega) \qquad (20.9)$$

for every $u\in\mathrm{W}^{1,p}(\Omega;\mathbf{R}^m)$. Since, by (ii), $F_j^0(u,\Omega) \le F_j(u,\Omega)$ for every open set Ω and every $u\in\mathrm{L}^p(\Omega;\mathbf{R}^m)$, we have that $\Gamma(\mathrm{L}^p)\text{-}\lim_j F_j^0 \le F'$ and then by (20.9) we obtain the estimate from below.

Now we prove the estimate from above. Fix a bounded open set Ω and consider a piecewise affine function u. We can write $u = \sum_{j=1}^{s}(A_j x + b_j)\chi_{U_j}$, where $A_j \in \mathbf{M}^{m \times n}$, $b_j \in \mathbf{R}^n$, (U_j) is a finite family of disjoint open subsets of Ω such that $|\Omega \setminus \bigcup_j U_j| = 0$. For every $k \in \mathbf{N}$ let $u_k = \mathcal{R}^{\varepsilon_k}(u)$. Since $Du_k = 0$ on $\varepsilon_k \mathcal{K}$ and $Du = A_j$ a.e. in U_j, by the growth condition (iii), by the boundedness condition (vi), and by Remark 20.3 we get

$$
\begin{aligned}
\int_\Omega f\left(\frac{x}{\varepsilon_k}, Du_k\right) dx &= \sum_j \int_{U_j} f\left(\frac{x}{\varepsilon_k}, Du_k\right) dx \\
&\leq c \sum_j \int_{U_j} (1 + |Du_k|^p)\, dx \\
&\leq c \sum_j |U_j|\left(1 + \|Du_k\|^p_{L^\infty(U_j; \mathbf{M}^{m \times n})}\right) \\
&\leq c \sum_j |U_j|\left(1 + \left(\frac{2}{\eta}d + 1\right)^p |A_j|^p\right) \\
&\leq c \left(\frac{2}{\eta}d + 1\right)^p \sum_j |U_j|(1 + |A_j|^p) \\
&\leq c \left(\frac{2}{\eta}d + 1\right)^p \int_\Omega (1 + |Du|^p)\, dx.
\end{aligned}
\tag{20.10}
$$

Then taking the $\limsup$ as $k \to \infty$ we obtain

$$
F''(u, \Omega) \leq \limsup_k \int_\Omega f\left(\frac{x}{\varepsilon_k}, Du_k\right) dx \leq c_2 \int_\Omega (1 + |Du|^p)\, dx,
$$

where $c_2 = c\left(\left(\frac{2d}{\eta}\right) + 1\right)^p$, so that we have proved the growth condition from above for every piecewise affine function u and every open bounded set Ω. Finally, if Ω is a Lipschitz bounded open subset of $\mathbf{R}^n$ and $u \in W^{1,p}(\Omega; \mathbf{R}^m)$ then there exists a sequence (u_j) of piecewise affine functions that converges strongly to u in $W^{1,p}(\Omega; \mathbf{R}^m)$; hence, by the lower semicontinuity of F'' we have

$$
\begin{aligned}
F''(u, \Omega) &\leq \liminf_j F''(u_j, \Omega) \\
&\leq \liminf_j c_2 \int_\Omega (1 + |Du_j|^p)\, dx = c_2 \int_\Omega (1 + |Du|^p)\, dx.
\end{aligned}
$$

This concludes the proof. $\qquad\square$

Theorem 20.7 *For every sequence (ε_j) of positive real numbers converging to 0 there exists a subsequence (ε_{j_k}) such that, for every Lipschitz bounded open set Ω, there exists the $\Gamma(L^p)$-limit $F_0(\cdot, \Omega)$ of $F_{j_k}(\cdot, \Omega)$. Moreover, there exists a quasiconvex function $\varphi: \mathbf{M}^{m \times n} \to \mathbf{R}$ such that*

$$F_0(u, \Omega) = \begin{cases} \displaystyle\int_\Omega \varphi(Du)\, dx & \text{if } u \in W^{1,p}(\Omega; \mathbf{R}^m) \\ +\infty & \text{otherwise.} \end{cases}$$

Proof The proof follows word by word the proof of Proposition 12.3 and Theorem 12.5, taking into account Proposition 20.6. $\square$

As usual, the proof of Theorem 20.4 will be completed if we show that the function φ is independent of the sequence (ε_{j_k}). To this end we introduce the function f_{hom} as in the following proposition.

Proposition 20.8 *For every $A \in \mathbf{M}^{m \times n}$ there exists the limit*

$$f_{\mathrm{hom}}(A) = \lim_{T \to +\infty} \inf\left\{ \frac{1}{T^n} \int_{(0,T)^n} f(x, A + Du)\, dx \ : \ u \in X_A^T \right\}, \qquad (20.11)$$

where

$$X_A^T = \left\{ v \in X((0,T)^n) \ : \ v = \mathcal{R}(Ax) - Ax \ \text{ in a neighbourhood of } \partial(0,T)^n \right\}$$

and $f_{\mathrm{hom}}(A) = \varphi(A)$ for every $A \in \mathbf{M}^{m \times n}$.

Proof Fix $A \in \mathbf{M}^{m \times n}$ and define for $T > 0$

$$g_T(A) = \inf\left\{ \int_{(0,T)^n} f(x, A + Du)\, dx \ : \ u \in X_A^T \right\}.$$

Let $u_A^T \in X_A^T$ be a function that satisfies

$$\int_{(0,T)^n} f(x, A + Du_A^T)\, dx \le g_T(A) + \frac{1}{T}. \qquad (20.12)$$

We extend u_A^T to the function v_A^T defined in $[0, [T+1]]^n$ by

$$v_A^T(x) = \begin{cases} \mathcal{R}(Ax) - Ax & \text{if } x \in [0, [T+1]]^n \setminus [0, T]^n \\ u_A^T(x) & \text{if } x \in [0, T]^n, \end{cases}$$

and then we extend it to a function defined on all $\mathbf{R}^n$, which we still denote v_A^T, by $[T+1]$-periodicity. Since the function $v_k(x) = \varepsilon_{j_k} v_A^T(x/\varepsilon_{j_k})$ converges to zero as $k \to \infty$ in $L^p_{\mathrm{loc}}(\mathbf{R}^n; \mathbf{R}^m)$, by L^p-lower semicontinuity and by periodicity, we get, using Example 2.7,

$$F_0(Ax, (0,1)^n) \le \liminf_k F_{j_k}(v_k + Ax, (0,1)^n)$$
$$= \frac{1}{[T+1]^n} \int_{(0,[T+1])^n} f(x, A + Dv_A^T)\, dx.$$

By definition of v_A^T, by (20.12) and by the growth conditions (iii) and (vi) for f, taking into account Remark 20.3 we have that

$$F_0(Ax, (0,1)^n) \leq \frac{1}{T^n}\left(\frac{T^n}{[T+1]^n}\left(g_T(A) + \frac{1}{T}\right)\right) + c(1 + |A|^p)\frac{[T+1]^n - T^n}{[T+1]^n}.$$

Taking the limit as $T \to +\infty$ we get

$$\varphi(A) = F_0(Ax, (0,1)^n) \leq \liminf_{T \to +\infty}\frac{1}{T^n}g_T(A). \tag{20.13}$$

Conversely, by the definition of $\Gamma(L^p)$-convergence, there exists a sequence (u_k) of functions of $X((0,1)^n)$ converging to Ax in $L^p((0,1)^n; \mathbf{R}^m)$, such that

$$F_0(Ax, (0,1)^n) = \lim_{k \to \infty}F_{j_k}(u_k, (0,1)^n). \tag{20.14}$$

We can use the fundamental estimate to modify the functions u_k on a neighbourhood of $\partial(0,1)^n$. Fix a compact set C of $(0,1)^n$ and an open set U' such that $C \subset U' \subset\subset (0,1)^n$, $\sigma > 0$, and choose in the fundamental estimate $V = (0,1)^n \setminus C$, $U = (0,1)^n$, $u = u_k$, and $v = v_k = \mathcal{R}^{\varepsilon_{j_k}}(Ax)$. Then for every k large enough there exists a cut-off function $\phi_k \in C_0^1((0,1)^n)$ between U' and U, such that, if we define $w_k = \phi_k u_k + (1 - \phi_k)v_k$, we have

$$F_{j_k}(w_k, (0,1)^n) \leq (1 + \sigma)(F_{j_k}(u_k, (0,1)^n) + F_{j_k}(v_k, (0,1)^n \setminus C))$$

$$+ M_\sigma \int_{C \setminus U'} |u_k - v_k|^p \, dx + \sigma.$$

Denote $T_k = 1/\varepsilon_{j_k}$. Since $(u_k - v_k)$ converges to zero in $L^p((0,1)^n)$, taking the limit as $k \to +\infty$, using the definition of g_T and v_k, by (20.14) and (20.10), we get

$$F_0(Ax, (0,1)^n) \geq \frac{1}{1 + \sigma}\limsup_k F_{j_k}(w_k, (0,1)^n)$$

$$- \limsup_k F_{j_k}(v_k, (0,1)^n \setminus C) - \frac{\sigma}{1 + \sigma}$$

$$\geq \frac{1}{1 + \sigma}\limsup_k \frac{1}{T_k^n}g_{T_k}(A)$$

$$- c_2|(0,1)^n \setminus C|(1 + |A|^p) - \frac{\sigma}{1 + \sigma}.$$

By the arbitrariness of C and σ we obtain

$$\varphi(A) \geq \limsup_k \frac{1}{T_k^n}g_{T_k}(A).$$

Then, by (20.13), we have that the limit $\lim_k \frac{1}{T_k^n}g_{T_k}(A)$ exists, and

$$\lim_k \frac{1}{T_k^n} g_{T_k}(A) = \varphi(A) = \liminf_{T \to +\infty} \frac{1}{T^n} g_T(A). \qquad (20.15)$$

Since the last term of (20.15) does not depend on the choice of the sequence (ε_{j_k}), we conclude that the limit $\lim_{T \to +\infty} \frac{1}{T^n} g_T(A)$ exists and coincides with $\varphi(A)$. $\qquad \square$

20.3 Convergence of minima

Let Ω be a Lipschitz bounded open subset of $\mathbf{R}^n$. Let (g_ε) be a sequence of functions of $L^p(\Omega; \mathbf{R}^m)$ converging strongly in $L^p(\Omega; \mathbf{R}^m)$ to a function g. Let $G_0, G_\varepsilon : L^p(\Omega; \mathbf{R}^m) \times \mathcal{A}(\Omega) \to [0, +\infty]$ be the functionals defined by

$$G_\varepsilon(u, U) = F_\varepsilon(u, U) + \int_U |g_\varepsilon - u|^p \, dx$$

and

$$G_0(u, U) = F_0(u, U) + \int_U |g - u|^p \, dx$$

for every $U \in \mathcal{A}(\Omega)$ and $u \in L^p(\Omega; \mathbf{R}^m)$. Since for every sequence (ε_j) and for every sequence (u_j) of functions in $L^p(\Omega; \mathbf{R}^m)$ converging strongly in $L^p(\Omega; \mathbf{R}^m)$ to a function u we have

$$\int_U |g_{\varepsilon_j} - u_j|^p \, dx \to \int_U |g - u|^p \, dx \qquad \forall U \in \mathcal{A}(\Omega),$$

by the $\Gamma(L^p)$-convergence of F_ε to F_0 we get

$$G_0(\cdot, U) = \Gamma(L^p)\text{-}\lim_{\varepsilon \to 0} G_\varepsilon(\cdot, U)$$

for every $U \in \mathcal{A}(\Omega)$.

Suppose that there exists $1 \le q \le +\infty$ such that

$$|A|^q \le a(x, A) \qquad \text{for all } x \in \mathbf{R}^n \setminus E \text{ and } A \in \mathbf{M}^{m \times n}, \qquad (20.16)$$

where a is the function given in the definition of f (condition (iv)). In this case, by conditions (ii), (iii) and (iv), for every $u \in W^{1,1}(\Omega; \mathbf{R}^m) \cap L^p(\Omega; \mathbf{R}^m)$, we obtain

$$G_\varepsilon(u, \Omega) = \int_{\Omega \setminus \varepsilon E} f\left(\frac{x}{\varepsilon}, Du\right) dx + \int_{\Omega \cap \varepsilon E} f\left(\frac{x}{\varepsilon}, Du\right) dx + \int_\Omega |g_\varepsilon - u|^p \, dx$$

$$\ge \int_{\Omega \setminus \varepsilon E} |Du|^q \, dx + \int_{\Omega \cap \varepsilon E} |Du|^p \, dx + \int_\Omega |g_\varepsilon - u|^p \, dx.$$

Since (g_ε) converges in $L^p(\Omega; \mathbf{R}^m)$, we get

$$\|u\|_{W^{1,s}(\Omega;\mathbf{R}^m)} \leq C(G_\varepsilon(u,\Omega) + 1), \qquad (20.17)$$

where $s = p \wedge q$ and C is a positive constant independent of ε and u.

Proposition 20.9 *If* (20.16) *holds and either* $q < n$ *with* $q^* = qn/(n-q) > p$ *or* $q \geq n$, *then*

$$\lim_{\varepsilon \to 0} \inf_{X(\Omega)} G_\varepsilon(\cdot,\Omega) = \min_{W^{1,p}(\Omega;\mathbf{R}^m)} G_0(\cdot,\Omega)$$

and the ε-minimizing sequences for G_ε converge to the minimizers for G_0.

Proof By definition of Γ-convergence it follows immediately that

$$\limsup_{\varepsilon \to 0} \inf_{X(\Omega)} G_\varepsilon(\cdot,\Omega) \leq \min_{W^{1,p}(\Omega;\mathbf{R}^m)} G_0(\cdot,\Omega). \qquad (20.18)$$

We now prove the opposite inequality for the $\liminf$. Since $\min G_0(\cdot,\Omega) < +\infty$, by (20.18) we may assume that there exists a positive constant M such that

$$\inf_{X(\Omega)} G_\varepsilon(\cdot,\Omega) < M \qquad (20.19)$$

for every $\varepsilon > 0$. Let $u_\varepsilon \in X(\Omega)$ be such that $G_\varepsilon(u_\varepsilon,\Omega) \leq \inf_{X(\Omega)} G_\varepsilon(\cdot,\Omega) + \varepsilon$. By (20.17) and (20.19) we obtain that u_ε is uniformly bounded in $W^{1,s}(\Omega;\mathbf{R}^m)$, where $s = p \wedge q$. Let (u_{ε_j}) be a sequence contained in (u_ε). Thanks to our assumption on q, by Rellich's Theorem, we get that, up to a subsequence, (u_{ε_j}) converges strongly in $L^p(\Omega;\mathbf{R}^m)$ to some function u_0. Then, by definition of $\Gamma(L^p)$-limit, we have

$$G_0(u_0,\Omega) \leq \liminf_j G_{\varepsilon_j}(u_{\varepsilon_j},\Omega) = \liminf_j \inf_{X(\Omega)} G_{\varepsilon_j}(\cdot,\Omega).$$

By (20.19) $u_0 \in W^{1,p}(\Omega;\mathbf{R}^m)$, and by (20.18) we can conclude that

$$\min_{W^{1,p}(\Omega;\mathbf{R}^m)} G_0(\cdot,\Omega) = G_0(u_0,\Omega) = \lim_{\varepsilon \to 0} \inf_{X(\Omega)} G_\varepsilon(\cdot,\Omega),$$

as required. $\qquad\square$

Similarly we will show how to study the limit of Dirichlet boundary value problems, when (20.16) is satisfied.

Let $\varphi \in W^{1,p}_{\mathrm{loc}}(\mathbf{R}^n;\mathbf{R}^m)$ and let (φ_ε) be a sequence of $W^{1,p}_{\mathrm{loc}}(\mathbf{R}^n;\mathbf{R}^m) \cap X_{\mathrm{loc}}(\mathbf{R}^n)$ such that

(α) $\varphi_\varepsilon \to \varphi$ in $L^p_{\mathrm{loc}}(\mathbf{R}^n;\mathbf{R}^m)$, as $\varepsilon \to 0$;

(β) there exists a function $\omega(\rho)$ with $\omega(\rho) \to 0$ as $\rho \to 0$, such that

$$\limsup_{\varepsilon \to 0} F_\varepsilon(\varphi_\varepsilon, U) \leq \omega(|U|)$$

for every open set U.

Remark 20.10 If $\varphi \in W^{1,\infty}_{\mathrm{loc}}(\mathbf{R}^n; \mathbf{R}^m) \cap X_{\mathrm{loc}}(\mathbf{R}^n)$, a sequence φ_ε that satisfies (α) and (β) is given, for instance, by $\varphi_\varepsilon = \mathcal{R}^\varepsilon(\varphi)$.

Let Ω be a bounded open set. Suppose that K is connected and with Lipschitz boundary, and let $\varphi \in W^{1,p}_{\mathrm{loc}}(\mathbf{R}^n; \mathbf{R}^m) \cap X_{\mathrm{loc}}(\mathbf{R}^n)$. If we set $I_\varepsilon = \{i \in \mathbf{Z}^n : (\varepsilon i + \varepsilon K_\eta) \cap \Omega \neq \emptyset\}$ and $\Omega_t = \{x \in \mathbf{R}^n : \ \mathrm{dist}(x, \Omega) < t\}$ for $t > 0$, using Poincaré's inequality, we obtain

$$\int_\Omega |\mathcal{R}^\varepsilon(\varphi) - \varphi|^p \, dx \le \sum_{i \in I_\varepsilon} \int_{\varepsilon i + \varepsilon K_\eta} \left| \fint_{\varepsilon i + \varepsilon K_\eta} \varphi(y)dy - \varphi(x) \right|^p dx$$

$$\le \varepsilon^p C \sum_{i \in I_\varepsilon} \int_{\varepsilon i + \varepsilon K_\eta} |D\varphi|^p \, dx \le \varepsilon^p C \|D\varphi\|^p_{\mathrm{L}^p(\Omega_{\varepsilon d}; \mathbf{R}^m)},$$

where $\mathcal{R}^\varepsilon(\varphi)$ is defined as for $\varphi \in W^{1,\infty}_{\mathrm{loc}}(\mathbf{R}^n, \mathbf{R}^m)$ and C is a positive constant independent of φ, ε and Ω. Thus the sequence $\varphi_\varepsilon = \mathcal{R}^\varepsilon(\varphi)$ satisfies assumption (α). Moreover, since $D\mathcal{R}^\varepsilon(\varphi)(x) = D\varphi(x)$ if $x \notin \bigcup_i(\varepsilon i + \varepsilon K_\eta)$,

$$D\mathcal{R}^\varepsilon(\varphi)(x) = \frac{1}{\varepsilon} D\psi \left(\fint_{\varepsilon i + \varepsilon K_\eta} \varphi(y)dy - \varphi(x) \right) + \psi \, D\varphi$$

if $x \in \varepsilon i + \varepsilon(K_\eta \setminus K)$, and $D\mathcal{R}^\varepsilon \varphi(x) = 0$ if $x \in \varepsilon i + \varepsilon K$, by Poincaré inequality we always get

$$\int_\Omega |D\mathcal{R}^\varepsilon(\varphi)(x)|^p \, dx \le 2^{p-1} \left(\frac{2^p}{\eta^p} + 1 \right) C \int_{\Omega_{\varepsilon d}} |D\varphi|^p \, dx,$$

and thus the sequence $\varphi_\varepsilon = \mathcal{R}^\varepsilon(\varphi)$ also satisfies (β).

Let $\Phi_0, \Phi_\varepsilon : \mathrm{L}^p_{\mathrm{loc}}(\mathbf{R}^n; \mathbf{R}^m) \times \mathcal{A}_n \to [0, +\infty]$ be the functionals defined by

$$\Phi_\varepsilon(u, \Omega) = \begin{cases} F_\varepsilon(u, \Omega) & \text{if } u - \varphi_\varepsilon \in W^{1,p}_0(\Omega; \mathbf{R}^m) \\ +\infty & \text{otherwise}, \end{cases}$$

and

$$\Phi_0(u, \Omega) = \begin{cases} F_0(u, \Omega) & \text{if } u - \varphi \in W^{1,p}_0(\Omega; \mathbf{R}^m) \\ +\infty & \text{otherwise}. \end{cases}$$

Proposition 20.11 *If (20.16) holds and either $q < n$ with $q^* = qn/(n-q) > p$ or $q \ge n$, then*

$$\lim_{\varepsilon \to 0} \inf_{W^{1,p}(\Omega; \mathbf{R}^m)} \Phi_\varepsilon(\cdot, \Omega) = \min_{W^{1,p}(\Omega; \mathbf{R}^m)} \Phi_0(\cdot, \Omega)$$

and the ε-minimizing sequences for Φ_ε converge to the minimizers for Φ_0.

Proof First we prove that

$$\limsup_{\varepsilon\to 0} \inf_{W^{1,p}(\Omega;\mathbf{R}^m)} \Phi_\varepsilon(\cdot,\Omega) \le \min_{W^{1,p}(\Omega;\mathbf{R}^m)} \Phi_0(\cdot,\Omega). \qquad (20.20)$$

Let $u_0 \in W^{1,p}(\Omega;\mathbf{R}^m)$ such that $u_0 - \varphi \in W_0^{1,p}(\Omega;\mathbf{R}^m)$ and $\Phi_0(u_0,\Omega) = \min \Phi_0(\cdot,\Omega)$. Then $\Phi_0(u_0,\Omega) = F_0(u_0,\Omega)$ and there exists a sequence (u_ε) in $X(\Omega)$ converging to u_0 in $L^p(\Omega;\mathbf{R}^m)$ such that $F_0(u_0,\Omega) = \lim_{\varepsilon\to 0} F_\varepsilon(u_\varepsilon,\Omega)$. By the fundamental estimate, as in the proof of Proposition 20.8, with $u = u_\varepsilon$ and $v = \varphi_\varepsilon$, for every compact subset C of Ω we can find a function $w_\varepsilon \in X(\Omega)$ such that $w_\varepsilon - \varphi_\varepsilon \in W_0^{1,p}(\Omega;\mathbf{R}^m)$ and

$$\begin{aligned}
\Phi_0(u_0,\Omega) &= \lim_{\varepsilon\to 0} F_\varepsilon(u_\varepsilon,\Omega) \\
&\ge \frac{1}{1+\sigma} \limsup_{\varepsilon\to 0} F_\varepsilon(w_\varepsilon,\Omega) - \limsup_{\varepsilon\to 0} F_\varepsilon(\varphi_\varepsilon,\Omega \setminus C) \\
&\quad - \frac{1}{1+\sigma} M_\sigma \int_{\Omega\setminus C} |u - \varphi|^p\, dx - \frac{\sigma}{1+\sigma}.
\end{aligned}$$

Since $F_\varepsilon(w_\varepsilon,\Omega) = \Phi_\varepsilon(w_\varepsilon,\Omega)$, by condition (β) we get

$$\Phi_0(u_0,\Omega) \ge \frac{1}{1+\sigma} \limsup_{\varepsilon\to 0} \inf_{W^{1,p}(\Omega;\mathbf{R}^m)} \Phi_\varepsilon(\cdot,\Omega) - \widetilde{\omega}(|\Omega \setminus C|) - \frac{\sigma}{1+\sigma},$$

where $\widetilde{\omega}$ is a positive function with $\widetilde{\omega}(\rho) \to 0$ as $\rho \to 0$, and by the arbitrariness of C and σ we obtain (20.20). It remains to prove that, for every sequence (ε_j) converging to 0, we have

$$\liminf_{\varepsilon_j\to 0} \inf_{W^{1,p}(\Omega;\mathbf{R}^m)} \Phi_{\varepsilon_j}(\cdot,\Omega) \ge \min_{W^{1,p}(\Omega;\mathbf{R}^m)} \Phi_0(\cdot,\Omega). \qquad (20.21)$$

The estimate (20.21) follows using the same method as the proof of Proposition 20.9, and remarking that, since the boundary values of the ε-minimizing sequence are the restrictions of a converging sequence in $W^{1,p}(\Omega;\mathbf{R}^m)$, to apply Rellich's Theorem it is sufficient to estimate uniformly the norm $L^s(\Omega;\mathbf{R}^m)$, with $s = p \wedge q$, of the gradient of the ε-minimizing sequence. $\qquad\square$

20.4 A Lavrentiev phenomenon

In the following example we show that the homogenized functional may depend on the space $X(\Omega)$, even though conditions (i)–(vi) guarantee a growth condition for f_{hom} independent of the choice of $X(\Omega)$.

Let $0 < r < \frac{1}{2}$ and $K = K_r = (r,\frac{1}{2})^n \cup (\frac{1}{2}, 1 - r)^n$, $p < n$. We will show that, at least if we choose r small enough, the homogenization formula gives two different results if in the definition of the functional $F(\cdot,\Omega)$ we take the space $X(\Omega)$ equal to $C^1(\Omega;\mathbf{R}^m)$ or $W^{1,p}(\Omega;\mathbf{R}^m)$. Let $f : \mathbf{R}^n \times \mathbf{M}^{m\times n} \to [0, +\infty]$ be a function 1-periodic in the first variable such that

$$f(x, A) = \begin{cases} |A|^p & \text{if } x \in (0,1)^n \setminus K_r, \\ 0 & \text{if } x \in K_r,\ A = 0, \\ +\infty & \text{if } x \in K_r,\ A \neq 0. \end{cases}$$

Since f is convex in the second variable, by Theorem 20.4, the homogenization formula reduces to

$$f_{\text{hom}}(A) = \inf\left\{ \int_{(0,1)^n} f(y, Du(y) + A)\, dy : u \in X_{\#}((0,1)^n) \right\},$$

which, in the two cases (one with $X(\Omega) = C^1(\Omega; \mathbf{R}^m)$ and the other one with $X(\Omega) = \mathrm{W}^{1,p}(\Omega; \mathbf{R}^m)$), gives, respectively,

$$f_{\text{hom}}^1(A) = \inf\left\{ \int_{(0,1)^n} |A + Du(y)|^p\, dy \right.$$
$$\left. : u \in C_{\#}^1((0,1)^n; \mathbf{R}^m),\ Du = -A \text{ on } K_r \right\}$$

and

$$f_{\text{hom}}^2(A) = \inf\left\{ \int_{(0,1)^n} |A + Du(y)|^p\, dy \right.$$
$$\left. : u \in \mathrm{W}_{\#}^{1,p}((0,1)^n; \mathbf{R}^m),\ Du = -A \text{ on } K_r \right\}.$$

We want to choose $r > 0$ and $A \in \mathbf{M}^{m \times n}$, such that $f_{\text{hom}}^2(A) < f_{\text{hom}}^1(A)$. To this end consider the functionals F_r^1 and F_r^2 defined by

$$F_r^1(u, (0,1)^n) = \begin{cases} \displaystyle\int_{(0,1)^n} |A + Du(x)|^p\, dx & \text{if } u \in C_{\#}^1((0,1)^n; \mathbf{R}^m) \\[2mm] & \text{and } A = -Du \text{ on } K_r, \\[4mm] +\infty & \text{otherwise,} \end{cases}$$

and

$$F_r^2(u, (0,1)^n) = \begin{cases} \displaystyle\int_{(0,1)^n} |A + Du(x)|^p\, dx & \text{if } u \in \mathrm{W}_{\#}^{1,p}((0,1)^n; \mathbf{R}^m) \\[2mm] & \text{and } A = -Du \text{ on } K_r, \\[4mm] +\infty & \text{otherwise.} \end{cases}$$

Since F_r^1 and F_r^2 are decreasing with r, it is easy to see (Remark 7.4(ii)) that F_r^1 $\Gamma(\mathrm{L}^p)$-converge to

$$F_0^1(u, (0,1)^n) = \begin{cases} \displaystyle\int_{(0,1)^n} |A + Du(x)|^p \, dx & \text{if } u \in W_\#^{1,p}((0,1)^n; \mathbf{R}^m), \\ & \text{and } u + Ax \text{ is equal to} \\ & \text{a constant a.e. on } K_0, \\ \\ +\infty & \text{otherwise,} \end{cases}$$

and that F_r^2 $\Gamma(L^p)$-converge to

$$F_0^2(u, (0,1)^n) = \begin{cases} \displaystyle\int_{(0,1)^n} |A + Du(x)|^p \, dx & \text{if } u \in W_\#^{1,p}((0,1)^n; \mathbf{R}^m) \\ & \text{and } A = -Du \text{ on } K_0, \\ \\ +\infty & \text{otherwise.} \end{cases}$$

We consider the case where $n = 2$, $m = 1$ and $A = (-1, 0)$.

Let $u_0 \in W_\#^{1,p}((0,1)^2)$ be the minimum point of $F_0^1(\cdot, (0,1)^2)$. Let u_1 be a function in $W^{1,p}((0, \frac{1}{2}) \times (\frac{1}{2}, 1))$ such that

$$\int_{(0,\frac{1}{2})\times(\frac{1}{2},1)} |A + Du_1|^p \, dx$$

$$= \min\left\{ \int_{(0,\frac{1}{2})\times(\frac{1}{2},1)} |A + Dv|^p \, dx : \ v = u_0 \text{ on } \partial((0,\tfrac{1}{2}) \times (\tfrac{1}{2}, 1)) \right\}.$$

Consider the function

$$v_1(x_1, x_2) = \tfrac{1}{2}u_1(x_1, x_2) - \tfrac{1}{2}u_1(\tfrac{1}{2} - x_1, \tfrac{3}{2} - x_2).$$

Since $A = (-1, 0)$ and $Du_0 = (1, 0)$ on K_0, we can assume, by periodicity of u_0 and by translation invariance of the functional F_0^1, that $u_0(x_1, x_2) = x_1$ if $x_1 \in (0, \frac{1}{2})$ and $x_2 = \frac{1}{2}$ or $x_2 = 1$, $u_0(0, x_2) = 1$ if $x_2 \in (\frac{1}{2}, 1)$, and $u_0(\frac{1}{2}, x_2) = \frac{1}{2}$ if $x_2 \in (\frac{1}{2}, 1)$, so that $v_1(x_1, x_2) = x_1 - \frac{1}{4}$ if $x_1 \in (0, \frac{1}{2})$ and $x_2 = \frac{1}{2}$ or $x_2 = 1$, $v_1(0, x_2) = \frac{1}{4}$ if $x_2 \in (\frac{1}{2}, 1)$, and $v_1(\frac{1}{2}, x_2) = -\frac{1}{4}$ if $x_2 \in (\frac{1}{2}, 1)$. Moreover, since $A \mapsto |A|^p$ is strictly convex and $v_1 \neq u_1$, we have

$$\int_{(0,\frac{1}{2})\times(\frac{1}{2},1)} |A + Dv_1|^p \, dx < \frac{1}{2} \int_{(0,\frac{1}{2})\times(\frac{1}{2},1)} |A + Du_1(x_1, x_2)|^p \, dx$$

$$+ \frac{1}{2} \int_{(0,\frac{1}{2})\times(\frac{1}{2},1)} |A + Du_1(\tfrac{1}{2}-x_1, \tfrac{3}{2}-x_2)|^p \, dx$$

$$= \int_{(0,\frac{1}{2})\times(\frac{1}{2},1)} |A + Du_1|^p \, dx$$

$$\leq \int_{(0,\frac{1}{2})\times(\frac{1}{2},1)} |A + Du_0|^p \, dx. \tag{20.22}$$

With a similar construction we can find a function $v_2 \in W^{1,p}((\frac{1}{2}, 1) \times (0, \frac{1}{2}))$ such that $v_2(x_1, x_2) = x_1 - \frac{3}{4}$ if $x_1 \in (\frac{1}{2}, 1)$ and $x_2 = 0$ or $x_2 = \frac{1}{2}$, $v_2(\frac{1}{2}, x_2) = -\frac{1}{4}$ if $x_2 \in (0, \frac{1}{2})$, $v_2(1, x_2) = \frac{1}{4}$ if $x_2 \in (0, \frac{1}{2})$, and

$$\int_{(\frac{1}{2},1)\times(0,\frac{1}{2})} |A + Dv_2|^p \, dx < \int_{(\frac{1}{2},1)\times(0,\frac{1}{2})} |A + Du_0|^p \, dx. \qquad (20.23)$$

Then if we consider the function

$$v = \begin{cases} v_1 & \text{in } (0, \frac{1}{2}) \times (\frac{1}{2}, 1), \\[2mm] v_2 & \text{in } (\frac{1}{2}, 1) \times (0, \frac{1}{2}), \\[2mm] x_1 - \frac{1}{4} & \text{in } (0, \frac{1}{2})^2, \\[2mm] x_1 - \frac{3}{4} & \text{in } (\frac{1}{2}, 1)^2 \end{cases}$$

we have that $v \in \dot{W}^{1,p}_{\#}((0,1)^2)$, $Dv = -A$ on K_0, and, by (20.22) and (20.23),

$$F_0^2(v, (0,1)^2) < F_0^1(u_0, (0,1)^2) = \min_u F_0^1(u, (0,1)^2);$$

so that $\min F_0^2 < \min F_0^1$. Since, by Theorem 7.2,

$$\liminf_{r\to 0} F_r^1 = \min F_0^1 \qquad \text{and} \qquad \liminf_{r\to 0} F_r^2 = \min F_0^2,$$

for r small enough, we have

$$f_{\text{hom}}^2(A) = \inf F_r^2 < \inf F_r^1 = f_{\text{hom}}^1(A),$$

and this concludes our example.

20.5 Loss of polyconvexity after homogenization

The homogenization of media with stiff inclusions allows us to present simple examples underlining the phenomenon of loss of polyconvexity induced by the process of averaging. In particular, we can modify the example in Chapter 18 to avoid the introduction of the parameter δ in (18.1).

We take $n = m = 2$ and we define the two functions f_1, f_2 as follows:

$$f_1(A) = |A_{11} - A_{22}|^p + |A_{12}|^p + |A_{21}|^p + \left((|A_{11}| - 1)^+\right)^p$$

$$f_2(A) = \begin{cases} (1 - \det A) & \text{if } A = tI \text{ with } -1 \leq t \leq 1, \\[2mm] +\infty & \text{otherwise.} \end{cases}$$

The function f_1 is convex (hence polyconvex); the function f_2 is polyconvex: it can be written as $f_2(A) = h(A, \det A)$, where $h : \mathbf{R}^4 \times \mathbf{R} \to \mathbf{R}$ is the convex function

$$h(A, r) = \begin{cases} 1 - r & \text{if } A = tI, \text{ with } -1 \leq t \leq 1 \\ +\infty & \text{otherwise.} \end{cases}$$

If we take K as the ball of centre $(\frac{1}{2}, \frac{1}{2})$ and radius $\frac{1}{4}$, then we can define the function $f : \mathbf{R}^2 \times \mathbf{M}^{2 \times 2} \to [0, +\infty]$ by

$$f(x, A) = \begin{cases} f_2(A) & \text{if } x \in K + \mathbf{Z}^2 \\ f_1(A) & \text{otherwise.} \end{cases} \tag{20.24}$$

By Theorem 20.4 we obtain the following proposition.

Proposition 20.12 *Let $p > 1$, let f and F_ε be defined by (20.24) and (20.2), respectively. Then there exists a function f_{hom} satisfying*

$$c_1 |A|^p - c_2 \leq f_{\mathrm{hom}}(A) \leq c_3(1 + |A|^p)$$

for suitable positive constants c_i, for all $A \in \mathbf{M}^{2 \times 2}$, such that, if F_{hom} is the functional defined by (20.4), then $F_{\mathrm{hom}} = \Gamma(\mathrm{L}^p)\text{-}\lim_{\varepsilon \to 0} F_\varepsilon$.

Proof It suffices to check that the hypotheses (i)–(vi) of Theorem 20.4 are satisfied by the function $\widetilde{f}(x, A) = f(x, A) + C$, where C is a suitable constant added to satisfy condition (ii) in the form $\widetilde{f}(x, A) \geq \widetilde{c}|A|^p$ ($\widetilde{c}$ a suitable positive constant). This is easily done by taking $\mathcal{K} = K + \mathbf{Z}^2$, $b = 1 + C$, and defining the function $a : \mathbf{R}^2 \times \mathbf{M}^{2 \times 2} \to [0, +\infty]$ in (iv) by

$$a(x, A) = \begin{cases} 0 & \text{if } A = tI, \text{ with } -1 \leq t \leq 1, \\ +\infty & \text{otherwise} \end{cases} \qquad \forall x \in \mathcal{K},$$

and $a(x, A) = \widetilde{c}|A|^p$ if $x \in E = \mathbf{R}^n \setminus \mathcal{K}$. $\qquad \square$

Proposition 20.13 *Let $1 < p < 2$. Then f_{hom} is not polyconvex.*

Proof We can repeat the argument of Chapter 18: if f_{hom} were polyconvex then it should be convex by Remark 6.9. Hence we have to show that f_{hom} is not convex. Since $f_{\mathrm{hom}}(I) = f_{\mathrm{hom}}(-I) = 0$, it suffices to prove that $f_{\mathrm{hom}}(0) > 0$, i.e. $F_0(0, (0, 1)^2) > 0$. Suppose otherwise; then, by the definition of Γ-limit we can find a sequence (ε_j) of positive numbers converging to 0, and a sequence (u_j) of functions in $X((0, 1)^2)$ converging to 0 in $\mathrm{L}^p((0, 1)^2; \mathbf{R}^2)$, such that $\lim_j F_{\varepsilon_j}(u_j, (0, 1)^2) = F_0(0, (0, 1)^2) = 0$. Since $a(x, A) \geq c(|A_{11} - A_{22}|^p + |A_{12} + A_{21}|^p)$ we deduce that Δu_j converge strongly to zero in $\mathrm{W}^{-1,p}((0, 1)^2; \mathbf{R}^2)$ (see

Chapter 18). By elliptic regularity we must have $u_j \to 0$ in $W^{1,p}_{\text{loc}}((0,1)^2; \mathbf{R}^2)$; in particular $|\{x \in (0,1)^2 : |Du_j| > 1\}| \to 0$. This convergence implies that

$$\lim_j F_{\varepsilon_j}(u_j, (0,1)^2) \geq \liminf_j F_{\varepsilon_j}(u_j, (0,1)^2 \cap \{|Du_j| \leq 1\})$$
$$\geq \liminf_j F_{\varepsilon_j}(u_j, (0,1)^2 \cap \{|Du_j| \leq 1\} \cap \varepsilon_j \mathcal{K})$$
$$\geq \frac{1}{2}\lim_j |(\varepsilon_j \mathcal{K}) \cap [0,1]^2| = \frac{1}{2}|K| > 0,$$

and the proof is concluded by contradiction. $\qquad\qquad\qquad\qquad\qquad\qquad\qquad\qquad$ $\square$

21

HOMOGENIZATION WITH NON-STANDARD GROWTH CONDITIONS

In this chapter we prove a homogenization result for functionals whose integrands satisfy some 'non-standard growth conditions', thanks to an accurate use of Rellich's Theorem and a careful generalization of the direct methods of Γ-convergence.

21.1 A class of non-standard integrals

Let $p \leq q < p^*$ and let $f : \Omega \times \mathbf{M}^{m \times n} \longrightarrow [0, +\infty)$ be a Borel function satisfying the non-standard growth condition

$$\alpha |A|^p \leq f(x, A) \leq \beta(1 + |A|^q). \tag{21.1}$$

We will consider the functional $F : L^p(\Omega; \mathbf{R}^m) \times \mathcal{A}(\Omega) \to [0, +\infty]$ defined as follows

$$F(u, U) = \begin{cases} \displaystyle\int_U f(x, Du(x))\, dx & \text{if } u \in W^{1,p}(\Omega; \mathbf{R}^m) \\[2mm] +\infty & \text{otherwise,} \end{cases}$$

and denote by $\mathcal{F}(\alpha, \beta, p, q)$ the class of functionals which admit a representation of this type for some such f.

We want to show that the same compactness and integral representation theorems as in Chapter 12 can be proved for these functionals, although the growth condition taken into account is weaker. To this end we will restate all the necessary results, but in the proofs we will describe in detail only those changes due to the different hypothesis.

Proposition 21.1 *Let (F_ε) be a family of functionals in $\mathcal{F}(\alpha, \beta, p, q)$ satisfying the L^q-fundamental estimate as $\varepsilon \to 0$, and let (ε_j) be a sequence of positive numbers converging to 0. Denote*

$$F'(u, U) = \Gamma(L^p)\text{-}\liminf_j F_{\varepsilon_j}(u, U) \tag{21.2}$$

$$F''(u, U) = \Gamma(L^p)\text{-}\limsup_j F_{\varepsilon_j}(u, U) \tag{21.3}$$

for every $u \in L^p(\Omega; \mathbf{R}^m)$ and $U \in \mathcal{A}(\Omega)$. Then we have

$$F'(u, U' \cup V) \leq F'(u, U) + F''(u, V) \tag{21.4}$$

$$F''(u, U' \cup V) \leq F''(u, U) + F''(u, V) \tag{21.5}$$

for all $u \in L^p(\Omega; \mathbf{R}^m)$ *and* $U, U', V \in \mathcal{A}(\Omega)$ *with* $U' \subset\subset U$ *and* U, V *with Lipschitz boundary.*

Proof The inequality in (21.4) is trivial if $F'(u, U)$ or $F''(u, V)$ are $+\infty$, so let them both be finite. Let (u_j) be a sequence of functions converging to u in $L^p(U; \mathbf{R}^m)$ such that $F'(u, U) = \liminf_j F_{\varepsilon_j}(u_j, U)$, and let (v_j) be a sequence converging to u in $L^p(V; \mathbf{R}^m)$ such that $F''(u, V) = \limsup_j F_{\varepsilon_j}(v_j, V)$. We can extract a subsequence of (ε_j), still denoted by (ε_j), such that both quantities are limits. From (21.1) it follows that $|Du_j|^p \leq f_{\varepsilon_j}(x, Du_j)$; since the sequence $f_{\varepsilon_j}(x, Du_j)$ is equibounded in $L^1(U; \mathbf{R}^m)$, so is the sequence (Du_j) in $L^p(U; \mathbf{R}^m)$. Thus (u_j) is equibounded in $W^{1,p}(U; \mathbf{R}^m)$. By Rellich's Theorem (u_j) converges strongly to u in $L^q(U; \mathbf{R}^m)$ (at this point we need the set U to have a regular boundary). The same holds for the sequence (v_j) in $L^q(V; \mathbf{R}^m)$. This allows us to use the L^q-fundamental estimate: for each fixed j and σ, there exists $w_j = \varphi_j u_j + (1 - \varphi_j) v_j$, where φ_j are suitable cut-off functions between U' and U, such that

$$F_{\varepsilon_j}(w_j, U' \cup V) \leq (1 + \sigma)(F_{\varepsilon_j}(u_j, U) + F_{\varepsilon_j}(v_j, V)) + M_\sigma \int_{U \cap V} |u_j - v_j|^q \, dx + \sigma.$$

Since $w_j \to u$ in $L^p(U' \cup V; \mathbf{R}^m)$ and (u_j) and (v_j) converge in $L^q(U \cap V; \mathbf{R}^m)$ to the same limit, taking into account the definition of Γ-$\liminf$ and letting $j \to +\infty$, by the arbitrariness of σ we obtain (21.4).

The inequality (21.5) is obtained by a similar argument. $\qquad\square$

Since (21.4) and (21.5) are proved only for sets with Lipschitz boundary, we need to modify Proposition 11.6 slightly to obtain the inner regularity of $F'(u, \cdot)$ and $F''(u, \cdot)$.

Proposition 21.2 *Let* (F_ε), F' *and* F'' *be as in Proposition 21.1. Then for all* $u \in W^{1,q}(\Omega; \mathbf{R}^m)$ $F'(u, \cdot)$ *and* $F''(u, \cdot)$ *are inner regular increasing set functions on* $\mathcal{A}(\Omega)$.

Proof Fix $W \in \mathcal{A}(\Omega)$. Let K be a compact subset of W, choose $U', U \in \mathcal{A}(\Omega)$ with ∂U Lipschitz such that $K \subset U' \subset\subset U \subset\subset W$, and define $V = W \setminus K$. Let (u_j) be a sequence of functions converging to u in $L^p(U; \mathbf{R}^m)$ such that

$$F'(u, U) = \liminf_j F_{\varepsilon_j}(u_j, U).$$

As in the previous proof the sequence converges to u also in $L^q(U; \mathbf{R}^m)$. Using the fundamental estimate with u fixed in the set V we have

$$F_{\varepsilon_j}(\varphi u_j + (1 - \varphi)u, U' \cup V)$$
$$\leq (1 + \sigma)(F_{\varepsilon_j}(u_j, U) + F_{\varepsilon_j}(u, V)) + M_\sigma \int_{U \cap V} |u_j - u|^q \, dx + \sigma.$$

From (21.1) we obtain

$$F_{\varepsilon_j}(\varphi u_j + (1 - \varphi)u, W)$$
$$\leq (1 + \sigma)\Big(F_{\varepsilon_j}(u_j, U) + \beta \int_V (1 + |Du|^q)\, dx\Big) + M_\sigma \int_{U \cap V} |u_j - u|^q\, dx + \sigma.$$

Passing to the lower limit, by the arbitrariness of σ, we get

$$F'(u, W) \leq F'(u, U) + c \int_{W \setminus K} (1 + |Du|^q)\, dx.$$

The conclusion follows as in the proof of Proposition 11.6. $\qquad\square$

With these tools we can now prove a compactness theorem which is the analogue of Proposition 12.3.

Theorem 21.3 *Let (F_ε) be a family of functionals in $\mathcal{F}(\alpha, \beta, p, q)$ satisfying the L^q-fundamental estimate as $\varepsilon \to 0$, and let (ε_j) be a sequence of positive numbers converging to 0. Then there exists a subsequence (ε_{j_k}) such that the Γ-limit*

$$F(u, U) = \Gamma(L^p)\text{-}\lim_k F_{\varepsilon_{j_k}}(u, U) \tag{21.6}$$

exists for all $u \in W^{1,q}(\Omega; \mathbf{R}^m)$ and $U \in \mathcal{A}(\Omega)$. Moreover, $F(u, \cdot)$ is the restriction of a Borel measure to $\mathcal{A}(\Omega)$.

Proof Proposition 21.2 ensures the inner regularity of F' and F'', so that Theorem 10.3 can be applied to obtain the existence of the Γ-limit. The proof of the measure property for $F(u, \cdot)$ when u is in $W^{1,q}(\Omega; \mathbf{R}^m)$ is exactly the same as in Proposition 12.3. $\qquad\square$

Theorem 12.4 can be restated as follows.

Theorem 21.4 *Let (F_ε) be a family of functionals in $\mathcal{F}(\alpha, \beta, p, q)$ satisfying the L^q-fundamental estimate as $\varepsilon \to 0$. Then for every sequence (ε_j) of positive numbers converging to 0 there exist a subsequence (ε_{j_k}) and a Carathéodory function $\psi : \Omega \times \mathbf{M}^{m \times n} \to [0, +\infty)$, satisfying the growth condition (21.1), such that if we define*

$$F_0(u, U) = \int_U \psi(x, Du)\, dx \tag{21.7}$$

then we have

$$F_0(u, U) = \Gamma(L^p)\text{-}\lim_k F_{\varepsilon_{j_k}}(u, U)$$

for all $u \in W^{1,q}(\Omega; \mathbf{R}^m)$ and $U \in \mathcal{A}(\Omega)$.

Proof We need only to remark that the Integral Representation Theorem relies only on the estimates from above. $\qquad\square$

The analogue of Proposition 11.7 can be directly proved only for functions in $W^{1,q}(\Omega; \mathbf{R}^m)$.

Proposition 21.5 *Let (F_ε) be a family of functionals in $\mathcal{F}(\alpha,\beta,p,q)$ satisfying the L^q-fundamental estimate as $\varepsilon \to 0$, and let (ε_j) be a sequence of positive numbers converging to 0. If we take $\phi \in \mathrm{W}^{1,q}(\Omega; \mathbf{R}^m)$, and we define*

$$G^\phi_{\varepsilon_j}(u,\Omega) = \begin{cases} F_{\varepsilon_j}(u,\Omega) & \text{if } u - \phi \in \mathrm{W}_0^{1,q}(\Omega; \mathbf{R}^m) \\ +\infty & \text{otherwise} \end{cases}$$

then we have

$$F'(u,\Omega) = G'(u,\Omega) \tag{21.8}$$

$$F''(u,\Omega) = G''(u,\Omega) \tag{21.9}$$

for all $u \in \mathrm{W}^{1,q}(\Omega; \mathbf{R}^m)$ such that $u - \phi \in \mathrm{W}_0^{1,q}(\Omega; \mathbf{R}^m)$, where G' and G'' have the obvious meaning as F' and F''.

Proof The proof runs in the same way as that of Proposition 11.7, with the same modifications made in the proof of Proposition 21.1 with respect to Proposition 11.5. $\qquad\qquad\square$

21.2   Convex homogenization

Now we restrict our attention to the case of homogenization, which particularly concerns the integral representation. First we consider the case when the function f is periodic and convex, and using this we will then deal with the non-convex case.

Let $f : \mathbf{R}^n \times \mathbf{M}^{m \times n} \to [0, +\infty)$ be a Borel function, satisfying (21.1), such that

$$f(\cdot, A) \text{ is 1-periodic for all } A \in \mathbf{M}^{m \times n}, \tag{21.10}$$

and

$$f(x, \cdot) \text{ is convex for all } x \in \mathbf{R}^n. \tag{21.11}$$

Since in the proofs of Chapter 14 the growth condition is not directly used, but only used in the results corresponding to Propositions 21.1–21.5 and Theorems 21.3 and 21.4, we can apply Theorems 14.5 and 14.7. We know that if U is a bounded open subset of $\mathbf{R}^n$ and we set, for all $\varepsilon > 0$,

$$F_\varepsilon(u,U) = \int_U f\left(\frac{x}{\varepsilon}, Du(x)\right) dx$$

for $u \in \mathrm{W}^{1,p}(U; \mathbf{R}^m)$ (and $+\infty$ for $u \in \mathrm{L}^p(U; \mathbf{R}^m) \setminus \mathrm{W}^{1,p}(U; \mathbf{R}^m)$), then we have

$$\Gamma(\mathrm{L}^p)\text{-}\lim_{\varepsilon \to 0} F_\varepsilon(u,U) = F_0(u,U)$$

$$= \int_U \psi(Du(x))\, dx \text{ for all } u \in \mathrm{W}^{1,q}(U; \mathbf{R}^m), \tag{21.12}$$

where $\psi : \mathbf{M}^{m \times n} \to [0, +\infty)$ is a convex function satisfying

$$\alpha|A|^p \le \psi(A) \le \beta(1 + |A|^q).$$

From these estimates it follows that, if we set

$$W^{1,\psi}(U; \mathbf{R}^m) = \left\{ w \in W^{1,p}(U; \mathbf{R}^m) : \int_U \psi(Dw)\, dx < +\infty \right\}$$

for every bounded open subset U of $\mathbf{R}^n$, then

$$W^{1,q}(U; \mathbf{R}^m) \subset W^{1,\psi}(U; \mathbf{R}^m) \subset W^{1,p}(U; \mathbf{R}^m).$$

The integral representation (21.12) holds for functions in $W^{1,q}$; by repeating the proof of Theorem 14.8 this representation can be extended to $W^{1,p}$, F_0 being finite only in $W^{1,\psi}$

Theorem 21.6 *Let $f : \mathbf{R}^n \times \mathbf{M}^{m\times n} \to [0, +\infty)$ be a Borel function satisfying (21.1), (21.10) and (21.11), and let $\Omega \subset \mathbf{R}^n$ be a bounded open set with Lipschitz boundary. Then there exists the Γ-limit*

$$\Gamma(\mathrm{L}^p)\text{-}\lim_{\varepsilon \to 0} \int_U f\left(\frac{x}{\varepsilon}, Du(x)\right)\, dx = \int_U \psi(Du(x))\, dx$$

for all $U \in \mathcal{A}(\Omega)$ and all $u \in W^{1,p}(\Omega; \mathbf{R}^m)$. The function ψ satisfies the formula

$$\psi(A) = \inf \left\{ \int_{(0,1)^n} f(y, A + Du(y))\, dy \; : \; u \in W^{1,p}_{\#}((0,1)^n; \mathbf{R}^m) \right\} \qquad (21.13)$$

for all $A \in \mathbf{M}^{m\times n}$.

Proof We can repeat word for word Steps 1 and 3 of the proof of Theorem 14.8, with $W^{1,p}(\Omega; \mathbf{R}^m)$ and $W^{1,q}(\Omega; \mathbf{R}^m)$ in place of $W^{1,1}(\Omega; \mathbf{R}^m)$ and $W^{1,p}(\Omega; \mathbf{R}^m)$, respectively, and ψ instead of f_{hom}, noticing that only the convexity of f_{hom} is used therein.

To conclude, it remains to note that the proof of formula (21.13) is the same as in Theorem 14.7. $\qquad\qquad\square$

21.3 Non-convex homogenization

We now consider the family $\mathcal{G}(\alpha, \beta, p, q)$ of integral functionals $G : \mathrm{L}^p(\Omega; \mathbf{R}^m) \times \mathcal{A}(\Omega) \to [0, +\infty]$ such that there exist two functions $f, g : \Omega \times \mathbf{M}^{m\times n} \to [0, +\infty)$, with f satisfying (21.1), (21.11) and

$$f(x, 2A) \le \beta(1 + f(x, A)), \qquad\qquad (21.14)$$

such that

$$f(x, A) \le g(x, A) \le \beta(1 + f(x, A)), \qquad\qquad (21.15)$$

and G is given by

$$G(u, U) = \begin{cases} \displaystyle\int_U g(x, Du(x))\, dx & \text{if } u \in W^{1,p}(\Omega; \mathbf{R}^m) \\ +\infty & \text{otherwise.} \end{cases} \tag{21.16}$$

We want to show that the results proved above for the functional F hold also with these more general hypotheses, apart from the analogue of Proposition 21.5. First of all we prove the following proposition.

Proposition 21.7 *The family* $\mathcal{G}(\alpha, \beta, p, q)$ *uniformly satisfies the* L^q*-fundamental estimate.*

Proof Let $U, U', V \in \mathcal{A}(\Omega)$, with $U' \subset\subset U$, and set $\delta = \mathrm{dist}(U', \partial U)$. Let $\eta > 0$, $0 < r < \delta - \eta$, and let φ be a cut-off function between $\{x \in U : \mathrm{dist}\,(x, U') < r\}$ and $\{x \in U : \mathrm{dist}\,(x, U') < r + \eta\}$, with $|D\varphi| \leq 2/\eta$. If $u, v \in W^{1,p}(\Omega; \mathbf{R}^m)$ and $V_r^\eta = \{x \in V : r < \mathrm{dist}(x, U') < r + \eta\}$ then we have

$$G(\varphi u + (1 - \varphi)v, U' \cup V)$$
$$= \int_{U' \cup V} g(x, \varphi Du + (1 - \varphi)Dv + (u - v)D\varphi)\, dx$$
$$\leq G(u, U) + G(v, V)$$
$$\qquad + \beta \int_{V_r^\eta} (1 + f(x, \varphi Du + (1 - \varphi)Dv + (u - v)D\varphi))\, dx$$
$$\leq G(u, U) + G(v, V)$$
$$\qquad + \beta \int_{V_r^\eta} \left(1 + f\left(x, 2\left(\frac{1}{2}(\varphi Du + (1 - \varphi)Dv) + \frac{1}{2}(u - v)D\varphi\right)\right)\right) dx$$
$$\leq G(u, U) + G(v, V)$$
$$\qquad + \beta \int_{V_r^\eta} \left(1 + \beta + \beta\left(\frac{1}{2}f(x, \varphi Du + (1 - \varphi)Dv) + \frac{1}{2}f(x, (u - v)D\varphi)\right)\right) dx$$
$$\leq G(u, U) + G(v, V)$$
$$\qquad + \beta \int_{V_r^\eta} \left(1 + \beta + \frac{\beta^2}{2} + \frac{\beta}{2}|u - v|^q |D\varphi|^q\right) dx$$
$$\qquad + \frac{\beta^2}{2} \int_{V_r^\eta} (\varphi f(x, Du) + (1 - \varphi)f(x, Dv))\, dx$$
$$\leq G(u, U) + G(v, V) + \frac{\beta^2 2^q}{2\eta^q} \int_{(U \cap V) \setminus U'} |u - v|^q\, dx$$
$$\qquad + K \int_{V_r^\eta} (1 + f(x, Du) + f(x, Dv))\, dx,$$

with $K = \beta(1 + \beta + (\beta^2/2))$. Now consider the measure

$$\mu(E) = K \int_E (1 + f(x, Du) + f(x, Dv))\, dx$$

and note that, by the growth condition (21.15), we have

$$\mu(U \cap V) \leq K(|U \cap V| + G(u, U) + G(v, V)).$$

Moreover, for every $N = 1, 2, \ldots$

$$\mu(U \cap V) \geq \sum_{k=1}^{N} \mu\Big(\Big\{x \in V : \delta\frac{k-1}{N} < \mathrm{dist}(x, U') < \delta\frac{k}{N}\Big\}\Big).$$

Hence for every $N = 1, 2, \ldots$ there exists $\overline{k} \in \{1, \ldots, N\}$ such that

$$\mu\Big(\Big\{x \in V : \delta\frac{\overline{k}-1}{N} < \mathrm{dist}(x, U') < \delta\frac{\overline{k}}{N}\Big\}\Big) \leq \frac{1}{N}\mu(U \cap V).$$

Now, with fixed $\sigma > 0$, if we set

$$N \geq \frac{K}{\sigma}\max\{|U \cap V|, 1\}, \quad \eta = \frac{\delta}{N}, \quad \text{and} \quad r = \frac{\overline{k}-1}{N}\delta,$$

we obtain

$$K \int_{V_r^\eta} (1 + f(x, Du) + f(x, Dv))\, dx \leq \frac{K}{N}\Big(|U \cap V| + G(u, U) + G(v, V)\Big)$$

$$\leq \sigma + \sigma(G(u, U) + G(v, V)),$$

and eventually, taking $M_\sigma = 2^{(q-1)}N^q\beta^2\delta^{-q}$,

$$G(\varphi u + (1 - \varphi)v, U' \cup V)$$

$$\leq (1 + \sigma)\,(G(u, U) + G(v, V)) + M_\sigma \int_{(U \cap V)\setminus U'} |u - v|^q\, dx + \sigma.$$

Since the constant M_σ depends only on $U, U', V, \alpha, \beta, p, q$, the estimate is uniform on $\mathcal{G}(\alpha, \beta, p, q)$. $\qquad\square$

The following is the analogue of Proposition 21.1.

Proposition 21.8 *Let (G_ε) be a family of functionals in $\mathcal{G}(\alpha, \beta, p, q)$, and let (ε_j) be a sequence of positive numbers converging to 0. Denote*

$$G'(u, U) = \Gamma(\mathrm{L}^p)\text{-}\liminf_j G_{\varepsilon_j}(u, U)$$

$$G''(u, U) = \Gamma(\mathrm{L}^p)\text{-}\limsup_j G_{\varepsilon_j}(u, U)$$

for every $u \in \mathrm{L}^p(\Omega; \mathbf{R}^m)$ and $U \in \mathcal{A}(\Omega)$. Then we have

$$G'(u, U' \cup V) \leq G'(u, U) + G''(u, V)$$

$$G''(u, U' \cup V) \leq G''(u, U) + G''(u, V)$$

for all $u \in \mathrm{L}^p(\Omega; \mathbf{R}^m)$ and $U, U', V \in \mathcal{A}(\Omega)$ with $U' \subset\subset U$ and U and V with Lipschitz boundary.

Proof The proof is exactly the same as that of Proposition 21.1, as $g_\varepsilon(x, A) \leq c(1 + |A|^q)$, and since the family (G_ε) uniformly satisfies the L^q-fundamental estimate. $\qquad\square$

We now restrict our attention to the case of homogenization, and to this end we suppose that both functions f and g are defined on all of $\mathbf{R}^n$ and are 1-periodic in the first variable. The analogue of Proposition 11.6 holds as well, with a slight difference.

Proposition 21.9 *Let Ω have a Lipschitz boundary. Let G_ε be defined for every $U \in \mathcal{A}(\Omega)$ and $u \in \mathrm{W}^{1,p}(\Omega; \mathbf{R}^m)$ by*

$$G_\varepsilon(u, U) = \int_U g\left(\frac{x}{\varepsilon}, Du\right) dx,$$

and let G' and G'' be as in the previous proposition. Then $G'(u, \cdot)$ and $G''(u, \cdot)$ are inner regular increasing set functions, for each $u \in \mathrm{W}^{1,\psi}(\Omega; \mathbf{R}^m)$.

Proof The proof follows the same argument as in Proposition 11.6, since (11.10) can be replaced by an analogous condition on G'' of the form

$$G''(u, U) \leq \beta \int_U (1 + \psi(Du))\, dx$$

for all $U \in \mathcal{A}(\Omega)$ and all $u \in \mathrm{W}^{1,\psi}(U; \mathbf{R}^m)$. In fact, let w_j be a sequence such that $\lim_j \int_U f(\frac{x}{\varepsilon_j}, Dw_j)\, dx = \int_U \psi(Du)\, dx$; then

$$G''(u, U) \leq \limsup_j G_{\varepsilon_j}(w_j, U) = \limsup_j \int_U g\left(\frac{x}{\varepsilon_j}, Dw_j\right) dx$$

$$\leq \limsup_j \int_U \beta\left(1 + f\left(\frac{x}{\varepsilon_j}, Dw_j\right)\right) dx = \int_U \beta(1 + \psi(Du))\, dx$$

as required. $\qquad\square$

Now the compactness and integral representation theorems are simple restatements of Theorem 21.3 and Theorem 21.4 which we sum up in the following theorem.

Theorem 21.10 *Let G_ε be as in Proposition 21.9, and let Ω have a Lipschitz boundary. For every sequence (ε_j) of positive numbers converging to 0 there exist a subsequence (ε_{j_k}) and a Carathéodory function $\gamma : \mathbf{R}^n \times \mathbf{M}^{m \times n} \to [0, +\infty)$ such that the Γ-limit*

$$G_0(u, U) = \Gamma(L^p)\text{-}\lim_k G_{\varepsilon_{j_k}}(u, U)$$

exists and

$$G_0(u, U) = \int_U \gamma(x, Du)\, dx$$

for all $U \in \mathcal{A}(\Omega)$ and all $u \in \mathrm{W}^{1,\psi}(U; \mathbf{R}^m)$.

So far, f takes the role played by $|\cdot|^p$ in the theory of Chapter 12, with no major changes in statements and proofs. We see that it is not possible to prove the analogue of Proposition 11.7, since $g(\frac{x}{\varepsilon}, Du)$ does not satisfy an estimate independent of ε. This result was used in the proofs of Proposition 14.3 and Theorem 14.5. We will see that it is possible to show the same results without Proposition 11.7.

Proposition 21.11 *The function γ can be chosen to be independent of the first variable.*

Proof By Proposition 9.2 it suffices to show that for all $A \in \mathbf{M}^{m \times n}$, y, $z \in \mathbf{R}^n$ and $\rho > 0$ such that $B_\rho(y) \cup B_\rho(z) \subset \Omega$

$$\Gamma(\mathrm{L}^p)\text{-}\lim_k G_{\varepsilon_{j_k}}(Ax, B_\rho(y)) = \Gamma(\mathrm{L}^p)\text{-}\lim_k G_{\varepsilon_{j_k}}(Ax, B_\rho(z)). \qquad (21.17)$$

Let $u_k \in \mathrm{W}^{1,q}(B_\rho(y); \mathbf{R}^m)$ be a sequence, converging to zero in $\mathrm{L}^p(B_\rho(y); \mathbf{R}^m)$ such that

$$\lim_k G_{\varepsilon_{j_k}}(Ax + u_k, B_\rho(y)) = \Gamma(\mathrm{L}^p)\text{-}\lim_k G_{\varepsilon_{j_k}}(Ax, B_\rho(y)),$$

and define the vector $\tau_k \in \mathbf{R}^n$ with components

$$(\tau_k)_i = \varepsilon_{j_k} \left[\frac{z_i - y_i}{\varepsilon_{j_k}} \right]$$

which is a period for $g(\cdot/\varepsilon_{j_k}, A)$. Note that $\tau_k \to z - y$ as k tends to $+\infty$.

Now let $r > 0$ be such that $B_{r\rho}(z) \subset B_\rho(y + \tau_k) \cap B_\rho(z)$. Since the restrictions $u_k(x - \tau_k)_{|B_{r\rho}(z)}$ belong to $\mathrm{W}^{1,q}(B_{r\rho}(z); \mathbf{R}^m)$, and they tend to 0 in $\mathrm{L}^p(B_{r\rho}(z); \mathbf{R}^m)$, we have

$$
\begin{aligned}
G_0(Ax, B_{r\rho}(z)) &= \Gamma(\mathrm{L}^p)\text{-}\lim_k G_{\varepsilon_{j_k}}(Ax, B_{r\rho}(z)) \\
&\le \liminf_k G_{\varepsilon_{j_k}}(Ax + u_k(x - \tau_k), B_{r\rho}(z)) \\
&\le \liminf_k G_{\varepsilon_{j_k}}(Ax + u_k(x - \tau_k), B_\rho(y + \tau_k)) \\
&= \liminf_k G_{\varepsilon_{j_k}}(Ax + u_k(x), B_\rho(y)) \\
&= \Gamma(\mathrm{L}^p)\text{-}\lim_k G_{\varepsilon_{j_k}}(Ax, B_\rho(y)).
\end{aligned}
$$

Letting r tend to 1, by inner regularity of the Γ-limit, as in Proposition 21.9, we obtain one inequality of (21.17). The other one is obtained by a symmetry argument. $\qquad \square$

The last result we are going to prove is the representation of γ in terms of g.

Proposition 21.12 *Let γ be as in Theorem 21.10. Then*

$$\gamma(A) = \lim_{T \to +\infty} \inf \left\{ \frac{1}{T^n} \int_{(0,T)^n} g(x, A + Du(x))\, dx \ : \ u \in \mathrm{W}_0^{1,p}((0,T)^n; \mathbf{R}^m) \right\}$$

for all $A \in \mathbf{M}^{m \times n}$.

Proof Let us fix $A \in \mathbf{M}^{m \times n}$ and for $T > 0$ set

$$h_T(A) = \inf \left\{ \frac{1}{T^n} \int_{(0,T)^n} g(x, A + Du(x)) \, dx \ : \ u \in \mathrm{W}_0^{1,p}((0,T)^n; \mathbf{R}^m) \right\}.$$

We will prove our thesis by showing first that $\gamma(A) \leq \liminf_{T \to +\infty} h_T(A)$ and then that $\limsup_{T \to +\infty} h_T(A) \leq \gamma(A)$.

For the first part, let $u^T \in \mathrm{W}_0^{1,p}((0,T)^n; \mathbf{R}^m)$ be such that

$$\frac{1}{T^n} \int_{(0,T)^n} g(x, A + Du^T(x)) \, dx \leq h_T(A) + \frac{1}{T};$$

extend u^T trivially to $(0, [T+1])^n$, then to all of $\mathbf{R}^n$ by periodicity and set $u_k^T(x) = \varepsilon_{j_k} u^T(x/\varepsilon_{j_k})$. Note that $u_k^T \to 0$ in $\mathrm{L}_{\mathrm{loc}}^p(\mathbf{R}^n; \mathbf{R}^m)$ as $k \to +\infty$.

Taking into account that the number of squares of side $[T+1]$ intersecting the square of side $1/\varepsilon_{j_k}$ is

$$\left(\left[\frac{(\varepsilon_{j_k})^{-1}}{[T+1]} \right] + 1 \right)^n \leq \left(\frac{1}{\varepsilon_{j_k} T} + 1 \right)^n,$$

we have

$$\int_{(0,1)^n} \gamma(A) \, dx \leq \liminf_k G_{\varepsilon_{j_k}}(Ax + u_k^T, (0,1)^n)$$

$$\leq \liminf_k \int_{(0,1)^n} g\left(\frac{x}{\varepsilon_{j_k}}, A + Du_k^T \right) dx$$

$$= \liminf_k \int_{(0,1/\varepsilon_{j_k})^n} \varepsilon_{j_k}^n g(x, A + Du^T) \, dx$$

$$\leq \liminf_k \varepsilon_{j_k}^n \left(\frac{1}{\varepsilon_{j_k} T} + 1 \right)^n \frac{1}{\varepsilon_{j_k}^n T^n} \int_{(0,T)^n} g(x, A + Du^T) \, dx$$

$$+ c([T+1]^n - T^n)\left(\frac{1}{T} + \varepsilon_{j_k} \right)^n (1 + |A|^q).$$

Hence, by taking the limit as $T \to +\infty$, we get

$$\gamma(A) \leq \liminf_{T \to +\infty} h_T(A).$$

For the second part, we need to use the L^q-fundamental estimate. Let (u_k) be a sequence in $\mathrm{W}^{1,p}((0,1)^n; \mathbf{R}^m)$ such that $u_k \to 0$ in $\mathrm{L}^p((0,1)^n; \mathbf{R}^m)$ and

$$\gamma(A) = \lim_k \int_{(0,1)^n} g\left(\frac{x}{\varepsilon_{j_k}}, A + Du_k \right) dx.$$

This implies by (21.15) and (21.1) that (u_k) is bounded in $\mathrm{W}^{1,p}((0,1)^n; \mathbf{R}^m)$. Note that, since $Q = (0,1)^n$ is Lipschitz, we then have, by Rellich's Theorem, that $u_k \to 0$ also in $\mathrm{L}^q((0,1)^n; \mathbf{R}^m)$. Fix $U' \subset\subset U \subset\subset Q$ and let $V = Q \setminus U'$. Then for every $\sigma > 0$, there exists M_σ and, with fixed k, for the functions u_k and $v = 0$ there exists $w_k = \varphi_k u_k$ (where φ_k is a cut-off function between U' and U, so that $w_k \in \mathrm{W}_0^{1,p}(Q; \mathbf{R}^m)$) such that

$$\int_U g\left(\frac{x}{\varepsilon_{j_k}}, A + Du_k\right) dx \geq -\int_V g\left(\frac{x}{\varepsilon_{j_k}}, A + 0\right) dx + \frac{1}{1+\sigma} \int_Q g\left(\frac{x}{\varepsilon_{j_k}}, A + Dw_k\right) dx$$
$$- \frac{M_\sigma}{1+\sigma} \int_{Q\setminus U'} |u_k|^q \, dx - \frac{\sigma}{1+\sigma}.$$

Thus we get, since $w_k \in \mathrm{W}_0^{1,p}(Q; \mathbf{R}^m)$,

$$\int_Q g\left(\frac{x}{\varepsilon_k}, A + Du_k\right) dx \geq \int_U g\left(\frac{x}{\varepsilon_k}, A + Du_k\right) dx$$
$$\geq \frac{1}{1+\sigma} h_{(1/\varepsilon_{j_k})}(A) - c|Q \setminus U'|$$
$$- \frac{M_\sigma}{1+\sigma} \int_{Q\setminus U'} |u_k|^q \, dx - \frac{\sigma}{1+\sigma}.$$

Taking limits on both sides and using the fact that $u_k \to 0$ in $\mathrm{L}^q((0,1)^n; \mathbf{R}^m)$, we obtain

$$\gamma(A) \geq \frac{1}{1+\sigma} \limsup_k h_{(1/\varepsilon_k)}(A) - c|Q \setminus U'| - \frac{\sigma}{1+\sigma},$$

for all $\sigma > 0$ and $U' \subset\subset Q$. Let $\sigma \to 0$ and $U' \to Q$; summing up the previous step we get

$$\limsup_{k\to+\infty} h_{(1/\varepsilon_{j_k})}(A) \leq \gamma(A) \leq \liminf_{T\to+\infty} h_T(A) \leq \liminf_{k\to+\infty} h_{(1/\varepsilon_{j_k})}(A),$$

and then

$$\gamma(A) = \lim_{k\to+\infty} h_{(1/\varepsilon_{j_k})}(A) = \liminf_{T\to+\infty} h_T(A).$$

This equality gives the thesis, as it proves that γ is independent of (ε_{j_k}), and that the limit $\lim_{T\to+\infty} h_T(A)$ exists by a compactness argument. $\square$

Considering Theorem 21.10 together with Propositions 21.11 and 21.12, we can now state our main theorem.

Theorem 21.13 *Let $\Omega \subset \mathbf{R}^n$ have a Lipschitz boundary. Let $g : \mathbf{R}^n \times \mathbf{M}^{m\times n} \to [0, +\infty)$ be a Borel function, 1-periodic in the first variable and such that*

$$f(x, A) \leq g(x, A) \leq \beta(1 + f(x, A)),$$

for some Borel function f, convex in A, 1-periodic in x and satisfying

$$\alpha|A|^p \leq f(x, A) \leq \beta(1 + |A|^q) \qquad p \leq q < p^*$$

and

$$f(x, 2A) \leq \beta(1 + f(x, A))$$

for all $A \in \mathbf{M}^{m \times n}$. Then the limit

$$\Gamma(\mathrm{L}^p)\text{-}\lim_{\varepsilon \to 0} \int_\Omega g\left(\frac{x}{\varepsilon}, Du\right) dx = \int_\Omega \gamma(Du)\, dx$$

exists for each $u \in \mathrm{W}^{1,p}(\Omega; \mathbf{R}^m)$, where γ is given by

$$\gamma(A) = \lim_{T \to +\infty} \inf\left\{\frac{1}{T^n} \int_{(0,T)^n} g(x, A + Du(x))\, dx \ : \ u \in \mathrm{W}_0^{1,p}((0,T)^n; \mathbf{R}^m)\right\}$$

for all $A \in \mathbf{M}^{m \times n}$.

Remark 21.14 The theorem above can be generalized, repeating the same proof, by requiring that f satisfy

$$\alpha|A|^p \leq f(x, A) \leq \beta(a(x) + |A|^q) \qquad p \leq q < p^* \tag{21.18}$$

with $a \in L^1_{\mathrm{loc}}(\mathbf{R}^n)$ and 1-periodic. Moreover, we can define, for fixed $r \geq p$,

$$G_\varepsilon(u) = \begin{cases} \displaystyle\int_\Omega g\left(\frac{x}{\varepsilon}, Du\right) dx & \text{if } u \in \mathrm{W}^{1,r}(\Omega; \mathbf{R}^m) \\ +\infty & \text{otherwise.} \end{cases}$$

In this case we have that

$$\Gamma(\mathrm{L}^p)\text{-}\lim_{\varepsilon \to 0} G_\varepsilon(u, \Omega) = \int_\Omega \gamma(Du)\, dx,$$

exists for all $u \in \mathrm{W}^{1,p}(\Omega; \mathbf{R}^m)$, where γ now depends on r and is given by

$$\gamma(A) = \lim_{T \to +\infty} \inf\left\{\frac{1}{T^n} \int_{(0,T)^n} g(x, A + Du(x))\, dx \ : \ u \in \mathrm{W}_0^{1,r}((0,T)^n; \mathbf{R}^m)\right\}$$

for all $A \in \mathbf{M}^{m \times n}$. Note that if g is a Carathéodory function and $r \geq q$ then we can take $C^\infty(\Omega; \mathbf{R}^m)$ as well in place of $\mathrm{W}^{1,r}(\Omega; \mathbf{R}^m)$ by the strong continuity of $u \mapsto \int_\Omega g(\frac{x}{\varepsilon}, Du)\, dx$ in $\mathrm{W}^{1,q}(\Omega; \mathbf{R}^m)$.

Example 21.15 The function γ in Remark 21.14 may depend on r, even in the convex case, when it is given by

$$\gamma(A) = \inf\left\{ \int_{(0,1)^n} g(x, A + Du(x))\, dx \ : \ u \in W^{1,r}_\#((0,1)^n; \mathbf{R}^m) \right\}. \qquad (21.19)$$

As an example we can take $n = 2$, and the function φ defined on $(-\frac{1}{2}, \frac{1}{2})^2$ as

$$\varphi(x_1, x_2) = \frac{x_1}{|x_1| \vee |x_2|},$$

extended to $\mathbf{R}^2$ in such a way that $x \mapsto \varphi(x) - x_1$ is 1-periodic. The function φ belongs to $W^{1,s}_{\mathrm{loc}}(\mathbf{R}^2)$ for all $s < 2$. Let $p \in (1,2)$, and let $g : \mathbf{R}^2 \times \mathbf{R}^2 \to [0, +\infty)$ be given by

$$g(x, z) = |\det[D\varphi(x), z]| + |z|^p$$

(we denote by $[y, z]$ the matrix whose columns are given by the vectors y and z). The function g is a Carathéodory function satisfying

$$|z|^p \le g(x, z) \le |D\varphi(x)|\,|z| + |z|^p$$
$$\le \frac{1}{s}|D\varphi(x)|^s + \frac{(s-1)}{s}|z|^{s/(s-1)} + |z|^p$$

so that (21.18) and (21.14) are satisfied with $f = g$ and any $2 < q < p^*$. Note that if $r > 2$ then by the polyconvexity of the determinant, for all $u \in W^{1,r}_\#((0,1)^2)$ we have

$$\int_{(0,1)^2} |\det[D\varphi(x), z + Du]|\, dx \ge |z_2|. \qquad (21.20)$$

Hence, by (21.19) and (21.20) we get

$$\gamma(z) \ge \inf\left\{ \int_{(0,1)^2} |Du + z|^p\, dx \ : \ u \in W^{1,r}_\#((0,1)^2) \right\} + |z_2| = |z|^p + |z_2| \qquad (21.21)$$

for all $z \in \mathbf{R}^2$. On the other hand, if $1 \le r \le 2$ then

$$\gamma(z) \le c_p |z|^p \qquad (21.22)$$

for all $z \in \mathbf{R}^2$, where c can be chosen to be dependent only on p. In fact, it suffices to consider the function

$$\varphi_z(x_1, x_2) = \frac{\langle x, z \rangle}{|x_1| \vee |x_2|},$$

extended to $\mathbf{R}^2$ in such a way that $x \mapsto \varphi_z(x) - \langle x, z \rangle$ is 1-periodic. The function φ_z belongs to $W^{1,r}_{\mathrm{loc}}(\mathbf{R}^2)$, and

$$\det[D\varphi, D\varphi_z] = 0 \qquad \text{a.e. on } \mathbf{R}^2.$$

We can check formula (21.19) on this function to get

$$\gamma(z) \le \int_{(0,1)^2} |D\varphi_z|^p \, dx \le c_p |z|^p,$$

which proves (21.22). Equations (21.21) and (21.22) prove that the values of $\gamma(z)$ are different on the set $\{z \in \mathbf{R}^2 : |z_2| + |z|^p > c_p |z|^p\}$ in the cases $r < 2$ and $r > 2$.

Remark 21.16 Theorem 21.13 can be applied to integrands $f(x, A) = |A|^{P(x)}$, with P periodic. An example by S. Kozlov (1989) shows that even when the function γ is isotropic it may not satisfy a standard growth condition of any order. More precisely, Kozlov exhibits functions P such that

$$c_1(|A|^p \log(1 + |A|) - 1) \le \gamma(A) \le c_2(|A|^p \log(1 + |A|) + 1)$$

(see also Exercise 21.2).

21.4 Exercises

Exercise 21.1 Let $n = 2$, $m = 1$ and $f(x, A) = |A|^{P(x_1)}$, where P is 1-periodic and satisfies

$$P(t) = \begin{cases} p & \text{if } 0 < t \le \frac{1}{2} \\ q & \text{if } \frac{1}{2} < t \le 1, \end{cases}$$

with $1 < p < q < p^*$. Prove that $\psi(s, 0) \le c|s|^p$ and that $\psi(0, s) \ge c'|s|^q$. Deduce that $W^{1,\psi}(\Omega)$ is not a Sobolev space.

Solution: as f is independent of x_2, the formula for $\psi(s, 0)$ reduces to a 1-dimensional minimum problem, namely

$$\psi(s, 0) = \inf_v \left\{ \int_0^1 |v'(t)|^{P(t)} \, dt : v(0) = 0, v(1) = s \right\}.$$

Taking $v(t) = 2ts$ if $0 \le t \le \frac{1}{2}$, and $v(t) = s$ if $\frac{1}{2} < t \le 1$, we obtain $\psi(s, 0) \le 2^{p-1}|s|^p$. Conversely, we have

$$\psi(0, s) = \inf_v \left\{ \int_0^1 \int_0^1 |v'(t) + s|^{P(\tau)} \, dt \, d\tau : v(0) = v(1) = 0 \right\}$$

$$= \int_0^1 |s|^{P(\tau)} d\tau = \frac{1}{2}|s|^p + \frac{1}{2}|s|^q.$$

Since ψ satisfies a growth condition of order p along the x_1-axis, and of order $q > p$ along the x_2-axis, $W^{1,\psi}(\Omega)$ is not a Sobolev space.

Exercise 21.2 Let $n = 2$, $m = 1$ and $f(x, A) = |A|^{3+\sin x_1}$. Prove that $\psi(0, s)$ satisfies a growth condition of the form

$$c_1(|s|^4(\log(1 + |s|))^{-1/2} - 1) \leq \psi(0, s) \leq c_2(|s|^4(\log(1 + |s|))^{-1/2} + 1).$$

Hint: since f does not depend on x_2, we have $\psi(0, s) = \frac{1}{2\pi} \int_0^{2\pi} |s|^{3+\sin t} dt$. Calculate

$$\lim_{s \to +\infty} \frac{1}{s^4} \sqrt{|\log s|} \int_0^{2\pi} |s|^{3+\sin t} dt = \lim_{s \to +\infty} \sqrt{|\log s|} \int_0^{2\pi} |s|^{\sin t - 1} dt.$$

22

ITERATED HOMOGENIZATION

We now consider the case of media with 'multiple scales of homogenization', whose microscopic behaviour can be studied by means of functionals

$$\int_\Omega f\left(\frac{x}{\varepsilon}, \frac{x}{\varepsilon^2}, \ldots, \frac{x}{\varepsilon^k}, Du\right) dx.$$

We will see that these functionals, as $\varepsilon \to 0$, behave as a homogeneous integral functional, whose representation can be obtained from the function f by iteration of the homogenization formula.

22.1 Statement of the Iterated Homogenization Theorem

For the sake of simplicity, we deal in detail only with the case of two scales of homogenization, i.e. $k = 2$ in the notation above. The general case can be easily obtained by induction (see Remark 22.8).

We consider a fixed function $f : \mathbf{R}^n \times \mathbf{R}^n \times \mathbf{M}^{m \times n} \to [0, +\infty)$ with the following properties:

$$f(x, \cdot, A) \text{ is a measurable function for all } (x, A) \in \mathbf{R}^n \times \mathbf{M}^{m \times n}; \quad (22.1)$$

$$f(x, y, \cdot) \text{ is a continuous function for all } (x, y) \in \mathbf{R}^n \times \mathbf{R}^n; \quad (22.2)$$

$$|f(x, y, A) - f(x', y, A)| \le \omega(|x - x'|)(1 + f(x, y, A)) \quad (22.3)$$

for all x, x', $y \in \mathbf{R}^n$, $A \in \mathbf{M}^{m \times n}$, where $\omega : [0, +\infty) \to \mathbf{R}$ is a continuous increasing function with $\omega(0) = 0$; the functions $f(\cdot, y, A)$ and $f(x, \cdot, A)$ are 1-periodic, and

$$|A|^p \le f(x, y, A) \le C(1 + |A|^p) \quad (22.4)$$

for all $x, y \in \mathbf{R}^n$ and $A \in \mathbf{M}^{m \times n}$.

Using the integrand f we define for all Ω bounded open subsets of $\mathbf{R}^n$ the following functionals:

$$F_\varepsilon(u, \Omega) = \begin{cases} \displaystyle\int_\Omega f\left(\frac{x}{\varepsilon}, \frac{x}{\varepsilon^2}, Du(x)\right) dx & \text{if } u \in W^{1,p}(\Omega; \mathbf{R}^m) \\ +\infty & \text{otherwise} \end{cases} \quad (22.5)$$

on $L^p(\Omega; \mathbf{R}^m)$. These functionals are well defined and finite on $W^{1,p}(\Omega; \mathbf{R}^m)$ thanks to (22.1)–(22.4) and Proposition 3.3. We will discuss the Γ-convergence of the family (F_ε). More precisely, we will prove the following result.

Theorem 22.1 (Iterated homogenization) *Let (ε_j) be a sequence of positive numbers converging to 0. Then the limit $F_{\mathrm{hom}}(u, \Omega) = \Gamma(\mathrm{L}^p)\text{-}\lim_j F_{\varepsilon_j}(u, \Omega)$ exists for every Ω bounded open subset of $\mathbf{R}^n$ and $u \in \mathrm{L}^p(\Omega; \mathbf{R}^m)$. We have, moreover,*

$$F_{\mathrm{hom}}(u, \Omega) = \begin{cases} \displaystyle\int_\Omega \varphi(Du(x))\, dx & \text{if } u \in \mathrm{W}^{1,p}(\Omega; \mathbf{R}^m) \\ +\infty & \text{otherwise} \end{cases}$$

for all $u \in \mathrm{W}^{1,p}(\Omega; \mathbf{R}^m)$, where the function $\varphi : \mathbf{M}^{m \times n} \to [0, +\infty)$ is quasiconvex and satisfies

$$\varphi(A) = \lim_{T \to +\infty} \inf\left\{ \frac{1}{T^n} \int_{(0,T)^n} Pf(y, Du(y) + A)\, dy :\ u \in \mathrm{W}_0^{1,p}((0,T)^n; \mathbf{R}^m) \right\},$$
$$(22.6)$$

where Pf is a Carathéodory function given by

$$Pf(x, A) = \lim_{T \to +\infty} \inf\left\{ \frac{1}{T^n} \int_{(0,T)^n} f(x, y, Du(y) + A)\, dy : u \in \mathrm{W}_0^{1,p}((0,T)^n; \mathbf{R}^m) \right\}$$
$$(22.7)$$

for all $x \in \mathbf{R}^n$, $A \in \mathbf{M}^{m \times n}$.

22.2 Proof of the Iterated Homogenization Theorem

It is clear that $F_{\mathrm{hom}}(u, \Omega) = +\infty$ if $u \in \mathrm{L}^p(\Omega; \mathbf{R}^m) \setminus \mathrm{W}^{1,p}(\Omega; \mathbf{R}^m)$; hence, we will focus on the existence and representation of the Γ-limit on $\mathrm{W}^{1,p}(\Omega; \mathbf{R}^m)$. By the Compactness Theorem 12.5 we have the following proposition.

Proposition 22.2 *Let (ε_j) be a sequence of positive numbers converging to 0. Then there exists a subsequence (not relabelled) and a Carathéodory function $\psi : \mathbf{R}^n \times \mathbf{M}^{m \times n} \to [0, +\infty)$, satisfying*

$$|A|^p \leq \psi(x, A) \leq C(1 + |A|^p),$$

such that the limit

$$F(u, \Omega) = \Gamma(\mathrm{L}^p)\text{-}\lim_j F_{\varepsilon_j}(u, \Omega) = \int_\Omega \psi(x, Du(x))\, dx \qquad (22.8)$$

exists for every bounded open subset Ω of $\mathbf{R}^n$, and $u \in \mathrm{W}^{1,p}(\Omega; \mathbf{R}^m)$.

In order to have the function Pf well defined, we recall the following proposition, whose proof is as in Exercise 14.6.

Proposition 22.3 (Parameterized homogenization) *Let f be as in Section 22.1. For every Ω bounded open subset of $\mathbf{R}^n$ and $\varepsilon > 0$ let*

$$\tilde{F}_\varepsilon(u, \Omega) = \begin{cases} \displaystyle\int_\Omega f\left(x, \frac{x}{\varepsilon}, Du(x)\right) dx & \text{if } u \in \mathrm{W}^{1,p}(\Omega; \mathbf{R}^m) \\ +\infty & \text{otherwise} \end{cases}$$

on $L^p(\Omega; \mathbf{R}^m)$. Then we have $\Gamma(L^p)\text{-}\lim_{\varepsilon \to 0} \tilde{F}_\varepsilon(u, \Omega) = F^1(u, \Omega)$, where

$$F^1(u, \Omega) = \begin{cases} \displaystyle\int_\Omega Pf(x, Du(x)) \, dx & \text{if } u \in W^{1,p}(\Omega; \mathbf{R}^m) \\ +\infty & \text{otherwise,} \end{cases}$$

and the function Pf is defined in (22.7).

In order to estimate from below the limit of F_{ε_j} in terms of F^1, we will need the following proposition.

Proposition 22.4 *Let $U \subset \mathbf{R}^n$ be a Lipschitz set. For all $M, \delta > 0$ there exists $\varepsilon_0 > 0$ such that for all $\varepsilon < \varepsilon_0$ and $u \in W^{1,p}(U; \mathbf{R}^m)$ with $\|Du\|_{L^p(U; \mathbf{M}^{m \times n})} \le M$ there exists $v \in W^{1,p}(U; \mathbf{R}^m)$ such that*

$$\int_U f\left(x, \frac{x}{\varepsilon}, Du\right) dx \ge \int_U Pf(x, Du + Dv) \, dx - \delta \tag{22.9}$$

and $\|v\|_{L^p(U; \mathbf{R}^m)} \le \delta$. The choice of ε_0 can be made uniformly for all U' translations of U.

Proof By contradiction suppose that there exist $M, \delta > 0$, a sequence (ε_j) of positive numbers converging to 0, and a sequence (u_j) in $W^{1,p}(U; \mathbf{R}^m)$ with $\|Du_j\|_{L^p(U; \mathbf{M}^{m \times n})} \le M$ such that

$$\int_U f\left(x, \frac{x}{\varepsilon_j}, Du_j\right) dx$$
$$< \inf\left\{ \int_U Pf(x, Du_j + Dv) \, dx : \ \|v\|_{L^p(U; \mathbf{R}^m)} \le \delta \right\} - \delta. \tag{22.10}$$

Up to passing to a subsequence and up to a translation argument we can suppose that (u_j) converges weakly in $W^{1,p}(U; \mathbf{R}^m)$ to a function u, and that

$$\|u_j - u\|_{L^p(U; \mathbf{R}^m)} \le \frac{\delta}{2}$$

for all j. Note that the last inequality implies that

$$\inf\left\{ \int_U Pf(x, Du_j + Dv) \, dx : \ \|v\|_{L^p(U; \mathbf{R}^m)} \le \delta \right\}$$
$$\le \inf\left\{ \int_U Pf(x, Dw) \, dx : \ \|w - u\|_{L^p(U; \mathbf{R}^m)} \le t \right\}$$

for all $t \le \delta/2$. Hence, by Definition 7.1(i), (22.10) and Remark 7.3(i), we get

$$\int_U Pf(x, Du) \, dx \le \liminf_j \int_U f\left(x, \frac{x}{\varepsilon_j}, Du_j\right) dx$$

$$< \sup_{t>0} \inf\left\{ \int_U Pf(x, Dw)\, dx \ : \ \|w - u\|_{L^p(U;\mathbf{R}^m)} \le t \right\} - \delta$$

$$= \liminf_{w \to u} \int_U Pf(x, Dw)\, dx - \delta$$

$$= \int_U Pf(x, Du)\, dx - \delta$$

which is a contradiction.

By the continuity and periodicity of f in the first variable, it is clear that, in order to prove that the thesis of the theorem holds uniformly, it suffices to consider all the sets $z + U$ with $z \in [0, 1]^n$. In this case the proof above still works with an additional compactness argument for z. $\qquad\square$

Proposition 22.5 *For all $A \in \mathbf{M}^{m \times n}$ and a.e. $x \in \mathbf{R}^n$ we have $\psi(x, A) \ge \varphi(A)$.*

Proof Fix $A \in \mathbf{M}^{m \times n}$. To prove the thesis it is sufficient to show that for every $y \in \mathbf{R}^n$ and $t > 0$ we have $F(Ax, y + (0, t)^n) \ge t^n \varphi(A)$. For the sake of simplicity, we perform the proof only in the case $y = 0$ and $t = 1$.

Let $w_j \to 0$ in $L^p((0, 1)^n; \mathbf{R}^m)$ and $F(Ax, (0, 1)^n) = \lim_j F_{\varepsilon_j}(Ax + w_j, (0, 1)^n)$. Using the L^p-fundamental estimate we can suppose that we have $w_j = 0$ if $\mathrm{dist}(x, \partial(0, 1)^n) \le 2\varepsilon_j$. Moreover, by Remark C.6, up to passing to a subsequence, we can suppose that the sequence $(|Dw_j|^p)$ is equi-integrable. We set

$$u_j(x) = \frac{1}{\varepsilon_j} w_j(\varepsilon_j x), \qquad T_j = \frac{1}{\varepsilon_j}.$$

Note that

$$F(Ax, (0, 1)^n) = \lim_j \frac{1}{T_j^n} \int_{(0, T_j)^n} f\left(x, \frac{x}{\varepsilon_j}, A + Du_j\right) dx$$

and that $u_j = 0$ if $\mathrm{dist}(x, \partial(0, T_j)^n) \le 2$. By the equi-integrability of $(|Dw_j|^p)$, and after a change of variables, we get that there exists a positive increasing continuous function ζ with $\zeta(0) = 0$ such that

$$\int_E (1 + |A + Du_j|^p)\, dx \le T_j^n \zeta(|\varepsilon_j E|)$$

for all j and for all measurable sets $E \subset (0, T_j)^n$.

Define

$$I_j = \{i \in \mathbf{Z}^n \ : \ 1 \le i_1, \dots, i_n \le T_j - 2\}, \qquad U = (0, 1)^n.$$

Since $\bigcup_{i \in I_j}(i + U) = (1, [T_j - 1])^n$, we have

$$u_j = 0 \text{ on } C_j = (0, T_j)^n \setminus \bigcup_{i \in I_j}(i + U).$$

We can write

$$\int_{(0,T_j)^n} f\left(x, \frac{x}{\varepsilon_j}, A + Du_j\right) dx$$

$$= \sum_{i \in I_j} \int_{(i+U)} f\left(x, \frac{x}{\varepsilon_j}, A + Du_j\right) dx + \int_{C_j} f\left(x, \frac{x}{\varepsilon_j}, A\right) dx. \qquad (22.11)$$

Let $M > 0$ be fixed, and let

$$I_j^1 = \{i \in I_j : \|Du_j\|_{L^p(i+U;\mathbf{M}^{m \times n})} \le M\}, \qquad I_j^2 = I_j \setminus I_j^1.$$

Note that since

$$\int_{(0,T_j)^n} |Du_j|^p \, dx \le c \, T_j^n$$

we have

$$\#(I_j^2) \le c \, M^{-p} T_j^n. \qquad (22.12)$$

For all $i \in I_j^1$ we can apply Proposition 22.4, and obtain that for j large enough (independent of i) there exist functions $v_j^i \in W^{1,p}(i + U; \mathbf{R}^m)$ such that

$$\int_{i+U} f\left(x, \frac{x}{\varepsilon_j}, A + Du_j\right) dx \ge \int_{i+U} Pf\left(x, A + Du_j + Dv_j^i\right) dx - \delta \qquad (22.13)$$

and

$$\int_{i+U} |v_j^i|^p \, dx \le \delta^p. \qquad (22.14)$$

Using the L^p-fundamental estimate, with fixed $0 < t < 1$ and $\sigma > 0$, we can find a cut-off function φ_j^i between $i + (t, 1 - t)^n$ and $i + U$ such that, if we set $w_j^i = u_j + \varphi_j^i v_j^i$ and $U_t = (0, 1)^n \setminus (t, 1 - t)^n$, we have

$$\int_{i+U} Pf(x, A + Dw_j^i) \, dx$$

$$\le (1 + \sigma)\left(\int_{i+U} Pf(x, A + Du_j + Dv_j^i) \, dx + \int_{i+U_t} Pf(x, A + Du_j) \, dx\right)$$

$$+ c(t) \int_{i+U_t} |v_j^i|^p \, dx + \sigma$$

$$\le (1 + \sigma)\left(\int_{i+U} Pf(x, A + Du_j + Dv_j^i) \, dx\right.$$

$$\left. + \beta \int_{i+U_t} (1 + |A + Du_j|^p) \, dx\right) + c(t)\delta^p + \sigma. \qquad (22.15)$$

If we define

$$\tilde{u}_j = \begin{cases} w_j^i & \text{on } i + U, \, i \in I_j^1 \\ u_j & \text{otherwise,} \end{cases}$$

then we have $\tilde{u}_j \in W_0^{1,p}((0,T_j)^n; \mathbf{R}^m)$ and

$$\int_{(0,T_j)^n} Pf(x, A + D\tilde{u}_j)\, dx \leq \sum_{i \in I_j^1} \int_{i+U} Pf(x, A + Dw_j^i)\, dx$$

$$+ \sum_{i \in I_j^2} \int_{i+U} Pf(x, A + Du_j)\, dx + \int_{C_j} Pf(x, A)\, dx$$

$$\leq \sum_{i \in I_j^1} \int_{i+U} Pf(x, A + Dw_j^i)\, dx$$

$$+ C \int_{D_j} (1 + |A + Du_j|^p)\, dx + 2n\, T_j^{n-1} C(1 + |A|^p), \qquad (22.16)$$

where $D_j = \bigcup_{i \in I_j^2}(i + U)$. Note that $|D_j| = \#(I_j^2) \leq cM^{-p}T_j^n$.

By (22.13)–(22.16) we get

$$\int_{(0,T_j)^n} Pf(x, A + D\tilde{u}_j)\, dx$$

$$\leq \sum_{i \in I_j^1} \Bigg((1+\sigma)\Big(\int_{i+U} Pf(x, A + Du_j + Dv_j^i)\, dx$$

$$+ C \int_{i+U_\ell} (1 + |A + Du_j|^p)\, dx \Big) + c(t)\delta^p + \sigma \Bigg)$$

$$+ C \int_{D_j} (1 + |A + Du_j|^p)\, dx + 2n\, T_j^{n-1} C(1 + |A|^p)$$

$$\leq \sum_{i \in I_j^1} \Bigg((1+\sigma)\Big(\int_{i+U} f\Big(x, \frac{x}{\varepsilon_j}, A + Du_j\Big)\, dx + \delta$$

$$+ C \int_{i+U_\ell} (1 + |A + Du_j|^p)\, dx \Big) + c(t)\delta^p + \sigma \Bigg)$$

$$+ C \int_{D_j} (1 + |A + Du_j|^p)\, dx + 2n\, T_j^{n-1} C(1 + |A|^p)$$

$$\leq (1+\sigma) \int_{(0,T_j)^n} f\Big(x, \frac{x}{\varepsilon_j}, A + Du_j\Big)\, dx$$

$$+ (1+\sigma)CT_j^n(\zeta(|\varepsilon_j(D_j \cup E_j)|) + \delta)$$

$$+ (c(t)\delta^p + \sigma)T_j^n + 2n\, T_j^{n-1} C(1 + |A|^p), \qquad (22.17)$$

where $E_j = \bigcup_{i \in I_j^1}(i + U_t)$. Note that $|E_j| \leq 2ntT_j^n$.

Changing back variables we obtain

$$\frac{1}{T_j^n} \int_{(0,T_j)^n} Pf(x, A + Dw_j)\, dx \le (1 + \sigma) \int_{(0,1)^n} f\left(\frac{x}{\varepsilon_j}, \frac{x}{\varepsilon_j^2}, A + Dw_j\right) dx$$
$$+ (1 + \sigma)C(\zeta(cM^{-p} + 2nt) + \delta) + c(t)\delta^p + \sigma + 2n\varepsilon_j C(1 + |A|^p).$$

Letting $j \to +\infty$ we get, by (22.6),

$$\varphi(A) \le \liminf_j \frac{1}{T_j^n} \int_{(0,T_j)^n} Pf(x, A + Dw_j)\, dx$$
$$\le (1 + \sigma)F(Ax, (0,1)^n) + (1 + \sigma)(\zeta(cM^{-p} + 2nt) + \delta) + c(t)\delta^p + \sigma.$$

Letting, in the following order, $\delta \to 0$, $M \to +\infty$, $t \to 0$, and $\sigma \to 0$, we obtain the desired inequality. $\qquad\square$

Proof of Theorem 22.1 By the previous proposition, it suffices to show that for all $A \in \mathbf{M}^{m \times n}$ and a.a. $x \in \mathbf{R}^n$ we have $\varphi(A) \ge \psi(x, A)$. In order to prove this, it is enough to check that $\varphi(A)|Q| \ge F(Ax, Q)$ for all cubes Q. By a translation argument, it is clearly sufficient to perform the proof only for $A = 0$.

Fixed $\eta > 0$, we can choose $T > 0$ and $u \in \mathrm{W}_0^{1,p}((0,T)^n; \mathbf{R}^m)$ such that

$$\frac{1}{T^n} \int_{(0,T)^n} Pf(x, Du)\, dx \le \varphi(0) + \eta. \tag{22.18}$$

By Proposition 22.3 we can find a sequence $v_j \in \mathrm{W}_0^{1,p}((0,T)^n; \mathbf{R}^m)$ such that $v_j \to 0$ in $\mathrm{L}^p((0,T)^n; \mathbf{R}^m)$ and

$$\lim_j \int_{(0,T)^n} f\left(x, \frac{x}{\varepsilon_j}, Dv_j + Du\right) dx = \int_{(0,T)^n} Pf(x, Du)\, dx.$$

We define then $u_j = u + v_j$, so that $u_j \in \mathrm{W}_0^{1,p}((0,T)^n; \mathbf{R}^m)$ and

$$\lim_j \int_{(0,T)^n} f\left(x, \frac{x}{\varepsilon_j}, Du_j\right) dx \le (\varphi(0) + \eta)T^n.$$

We extend now u_j to the whole $\mathbf{R}^n$ by a patchwork almost-periodic procedure. Note that we cannot proceed directly by periodicity since $x \mapsto f(x, x/\varepsilon_j, A)$ is in general not-periodic. For all $i \in \mathbf{Z}^n$ let $x_i^j \in \varepsilon_j \mathbf{Z}^n \cap (i[T+1] + [0, \varepsilon_j)^n)$ (which is thus uniquely defined). We set

$$\tilde{u}_j(x) = \begin{cases} u_j(x - x_i^j) & \text{if } x \in x_i^j + (0,T)^n, \ i \in \mathbf{Z}^n \\ \\ 0 & \text{otherwise.} \end{cases}$$

For each $i \in \mathbf{Z}^n$ we have

$$\int_{i[T+1]+(0,[T+1])^n} f\left(x, \frac{x}{\varepsilon_j}, D\tilde{u}_j\right) dx$$

$$= \int_{x_i^j+(0,T)^n} f\left(x,\frac{x}{\varepsilon_j},D\tilde{u}_j\right)dx + \int_{(i[T+1]+(0,[T+1])^n)\backslash(x_i^j+(0,T)^n)} f\left(x,\frac{x}{\varepsilon_j},0\right)dx$$

$$\leq \int_{(0,T)^n} f\left(x+x_i^j,\frac{x+x_i^j}{\varepsilon_j},Du_j\right)dx + c([T+1]^n - T^n)$$

$$= \int_{(0,T)^n} f\left(x+(x_i^j - i[T+1]),\frac{x}{\varepsilon_j},Du_j\right)dx + c([T+1]^n - T^n)$$

$$\leq \int_{(0,T)^n} f\left(x,\frac{x}{\varepsilon_j},Du_j\right)dx$$

$$+\omega(\sqrt{n}\,\varepsilon_j)\int_{(0,T)^n}\left(1+f\left(x,\frac{x}{\varepsilon_j},Du_j\right)\right)dx + c([T+1]^n - T^n)$$

$$= (1+\omega(\sqrt{n}\,\varepsilon_j))\int_{(0,T)^n} f\left(x,\frac{x}{\varepsilon_j},Du_j\right)dx + \omega(\sqrt{n}\,\varepsilon_j)T^n + c([T+1]^n - T^n)$$

$$\leq (1+\omega(\sqrt{n}\,\varepsilon_j))T^n(\varphi(0)+\eta+o(1)) + \omega(\sqrt{n}\,\varepsilon_j)T^n + c([T+1]^n - T^n),$$

where we have used (22.4) and the fact that $x_i^j - i[T+1] \in [0,\varepsilon_j)^n$.

We can set now

$$w_j(x) = \varepsilon_j \tilde{u}_j\left(\frac{x}{\varepsilon_j}\right).$$

Let Q be any open cube. We have $w_j \to 0$ in $L^p(Q;\mathbf{R}^m)$, so that we can estimate

$$F(0,Q) \leq \liminf_j F_{\varepsilon_j}(w_j,Q)$$

$$\leq |Q|\left((\varphi(0)+\eta) + c\left(\frac{[T+1]^n}{T^n} - 1\right)\right).$$

Since we can let $T \to +\infty$ and then $\eta \to 0$, the proof is concluded. $\qquad\square$

Remark 22.6 If f is convex in the last variable, then the function $\varphi : \mathbf{M}^{m\times n} \to [0,+\infty)$ is convex and satisfies

$$\varphi(A) = \min\left\{\int_{(0,1)^n} Pf(y,Du(y)+A)\,dy : \ u \in W_\#^{1,p}((0,1)^n;\mathbf{R}^m)\right\}, \quad (22.19)$$

where Pf is then given by

$$Pf(x,A) = \min\left\{\int_{(0,1)^n} f(x,y,Du(y)+A)\,dy : u \in W_\#^{1,p}((0,1)^n;\mathbf{R}^m)\right\} \quad (22.20)$$

for all $x \in \mathbf{R}^n$, $A \in \mathbf{M}^{m\times n}$.

Remark 22.7 Theorem 22.1 can be extended to functions $f = f(x,y,A)$ satisfying (22.1), (22.2), (22.4) and weaker assumptions on the dependence on the variable x. For example, we can require only piecewise continuity of the type (22.3). The proof follows in the same way, remarking that continuity everywhere is not necessary for the definitions of the functionals F_ε and for the proof of Proposition 22.3.

Remark 22.8 Proceeding exactly as in the proof of Exercise 14.6, using Theorem 22.1, it is easy to prove the Γ-convergence of functionals of the form

$$\int_\Omega f\left(x, \frac{x}{\varepsilon}, \frac{x}{\varepsilon^2}, Du(x)\right) dx$$

as $\varepsilon \to 0^+$ to the integral functional

$$\int_\Omega \varphi(x, Du(x))\, dx, \tag{22.21}$$

where

$$\varphi(x, A) = \lim_{T \to +\infty} \inf\left\{ \frac{1}{T^n} \int_{(0,T)^n} Pf(x, y, Du(y)+A)\, dy : u \in W_0^{1,p}((0,T)^n; \mathbf{R}^m) \right\} \tag{22.22}$$

and

$$Pf(x, w, A) = \lim_{T \to +\infty} \inf\left\{ \frac{1}{T^n} \int_{(0,1)^n} f(x, w, y, Du(y) + A)\, dy : \right.$$
$$\left. u \in W_0^{1,p}((0,T)^n; \mathbf{R}^m) \right\}.$$

In the same way we can prove by induction that the Γ-limit of the functionals

$$\int_\Omega f\left(x, \frac{x}{\varepsilon}, \frac{x}{\varepsilon^2}, \ldots, \frac{x}{\varepsilon^k}, Du(x)\right) dx$$

as $\varepsilon \to 0^+$ is a functional given by (22.21), where φ satisfies formula (22.22), $Pf(x, \ldots)$ being the integrand of the Γ-limit of the functionals

$$\int_\Omega f\left(x, y, \frac{y}{\varepsilon}, \frac{y}{\varepsilon^2}, \ldots, \frac{y}{\varepsilon^{k-1}}, Du(y)\right) dy$$

as $\varepsilon \to 0^+$.

22.3 Exercises

In Exercises 22.1 and 22.2 $n = m = 1$, and the functionals are tacitly supposed to take the value $+\infty$ outside $W^{1,2}(\Omega)$.

Exercise 22.1 Let G be defined as in Exercise 13.2. Compute the homogenized functional

$$F(u, \Omega) = \Gamma\text{-}\lim_{\varepsilon \to 0^+} \int_\Omega G\left(\frac{t}{\varepsilon}\right) G\left(\frac{t}{\varepsilon^2}\right) |u'|^2\, dt.$$

Hint: use Theorem 22.1 and Remark 22.6 with $f(x, y, z) = G(x)G(y)z^2$, and Remark 22.7, applying Exercise 13.2 twice to obtain

$$Pf(x,z) = G(x)\min\left\{\int_0^1 G(y)|u'(y)|^2\,dy : u \in \mathrm{W}_0^{1,2}(0,1)\right\} = G(x)\varphi_G(z),$$

with φ_G as in Exercise 13.2. Deduce that

$$F(u,\Omega) = \frac{4\alpha^2\beta^2}{(\alpha+\beta)^2}\int_\Omega |u'|^2\,dt.$$

Exercise 22.2 Compute the homogenized functional

$$F(u,\Omega) = \Gamma\text{-}\lim_{\varepsilon\to 0^+}\int_\Omega\left(G\Big(\frac{t}{\varepsilon}\Big) + G\Big(\frac{t}{\varepsilon^2}\Big)\right)|u'|^2\,dt.$$

Note that in this case the integrand of the functional F equals the limit as $k \to +\infty$ of the integrands of the Γ-limits as $\varepsilon \to 0^+$ of

$$\int_\Omega\left(G\Big(\frac{t}{\varepsilon}\Big) + G\Big(\frac{tk}{\varepsilon}\Big)\right)|u'|^2\,dt$$

(see Exercise 13.3).

Hint: apply Theorem 22.1, Remark 22.6 and Remark 22.7 with $f(x,y,z) = (G(x) + G(y))z^2$, and obtain

$$Pf(x,z) = \begin{cases} \min\left\{\displaystyle\int_0^1 (\alpha + G(y))|u'(y)|^2\,dy : u \in \mathrm{W}_0^{1,2}(0,1)\right\} \\ \qquad\qquad\qquad\qquad \text{if } n < x \le n+\tfrac{1}{2},\, n \in \mathbf{N} \\[2ex] \min\left\{\displaystyle\int_0^1 (\beta + G(y))|u'(y)|^2\,dy : u \in \mathrm{W}_0^{1,2}(0,1)\right\} \\ \qquad\qquad\qquad\qquad \text{if } n - \tfrac{1}{2} < x \le n,\, n \in \mathbf{N}; \end{cases}$$

that is, using the formula for φ_G in Exercise 13.2,

$$Pf(x,z) = \begin{cases} \dfrac{4\alpha(\alpha+\beta)}{3\alpha+\beta}z^2 & \text{if } n < x \le n+\tfrac{1}{2},\, n \in \mathbf{N} \\[2ex] \dfrac{4\beta(\alpha+\beta)}{\alpha+3\beta}z^2 & \text{if } n - \tfrac{1}{2} < x \le n,\, n \in \mathbf{N}. \end{cases}$$

Deduce that

$$F(u,\Omega) = \frac{8(\alpha\beta)(\alpha+\beta)}{\alpha^2+\beta^2+6\alpha\beta}\int_\Omega |u'|^2\,dt.$$

In the next exercise we model a 2-dimensional composite consisting of a fine layering (in the x_1-direction) of two materials, each obtained by a finer layering process (in the x_2-direction) of two isotropic media.

Exercise 22.3 (Double layering) Let $n = 2$, $m = 1$, $p = 2$, and $\alpha_1, \alpha_2, \beta_1, \beta_2 > 0$. Let $a : \mathbf{R}^2 \to [0, +\infty)$ be the 1-periodic function satisfying

$$
a(x_1, x_2) = \begin{cases}
\alpha_1 & \text{if } 0 < x_1 \le \theta, \, 0 < x_2 \le t_1 \\
\beta_1 & \text{if } 0 < x_1 \le \theta, \, t_1 < x_2 \le 1 \\
\alpha_2 & \text{if } \theta < x_1 \le 1, \, 0 < x_2 \le t_2 \\
\alpha_1 & \text{if } \theta < x_1 \le 1, \, t_2 < x_2 \le 1,
\end{cases}
$$

where $0 < t_1, t_2, \theta < 1$. Compute the homogenized functional

$$
F(u, \Omega) = \Gamma\text{-}\lim_{\varepsilon \to 0^+} \int_\Omega a\Big(\frac{x_1}{\varepsilon}, \frac{x_2}{\varepsilon^2}\Big) |Du|^2 \, dx.
$$

Hint: using Exercise 14.3 we obtain

$$
P(x, A) = \begin{cases}
(t_1\alpha_1 + (1 - t_1)\beta_1)A_1^2 + \Big(\frac{t_1}{\alpha_1} + \frac{(1-t_1)}{\beta_1}\Big)^{-1} A_2^2 & \text{if } 0 < x_1 \le \theta \\[2mm]
(t_2\alpha_2 + (1 - t_2)\beta_2)A_1^2 + \Big(\frac{t_2}{\alpha_2} + \frac{(1-t_2)}{\beta_2}\Big)^{-1} A_2^2 & \text{if } \theta < x_1 \le 1.
\end{cases}
$$

The function φ in Theorem 22.1 is a quadratic form by Remark 7.12(ii). Moreover, by Remark 14.13 we get $\varphi(A) = \gamma A_1^2 + \delta A_2^2$. By formula (22.19) we have that the minimum problem defining $\gamma = \varphi(e_1)$ is solved by a function of the form $u(x_1, x_2) = v(x_1) + x_1$, where v solves the minimum problem

$$
\min\Big\{\alpha \int_0^\theta |v'|^2 \, dt + \beta \int_\theta^1 |v'|^2 \, dt : v(0) = 0, v(1) = 1\Big\},
$$

with $\alpha = (t_1\alpha_1 + (1 - t_1)\beta_1)$, $\beta = (t_2\alpha_2 + (1 - t_2)\beta_2)$. By Exercise 13.1 we then have

$$
\gamma = \Big(\frac{\theta}{t_1\alpha_1 + (1 - t_1)\beta_1} + \frac{1 - \theta}{t_2\alpha_2 + (1 - t_2)\beta_2}\Big)^{-1}
$$

The minimum problem defining $\delta = \varphi(e_2)$ is solved by $u(x_1, x_2) = x_2$ so that we obtain

$$
\delta = \theta\Big(\frac{t_1}{\alpha_1} + \frac{(1 - t_1)}{\beta_1}\Big)^{-1} + (1 - \theta)\Big(\frac{t_2}{\alpha_2} + \frac{(1 - t_2)}{\beta_2}\Big)^{-1}
$$

Exercise 22.4 Let $f : \mathbf{R}^n \times \mathbf{R}^N \times \mathbf{M}^{m \times n} \times \mathbf{M}^{m \times N} \to [0, +\infty)$ possess the following properties:

(i) $f(x, \cdot, A, B)$ is a measurable function for all $(x, A, B) \in \mathbf{R}^n \times \mathbf{M}^{m \times n} \times \mathbf{M}^{m \times N}$;

(ii) $f(x, y, \cdot, \cdot)$ is a strictly convex function for all $(x, y) \in \mathbf{R}^n \times \mathbf{R}^N$;

(iii) $|f(x, y, A, B) - f(x', y, A, B)| \le \omega(|x - x'|)\,(a(y) + f(x, y, A, B))$ for all $x, x' \in \mathbf{R}^n$, $y \in \mathbf{R}^N$, $A \in \mathbf{M}^{m \times n}$, $B \in \mathbf{M}^{m \times N}$, where ω is a continuous positive real function with $\omega(0) = 0$;

(iv) the function $f(\cdot, y, A, B)$ is X-periodic and $f(x, \cdot, A, B)$ is Y-periodic, where $X = \prod_{h=1}^{n}(0, x_h)$, $Y = \prod_{h=1}^{N}(0, y_h)$;

(v) $|A|^p + |B|^p \leq f(x, y, A, B) \leq C(1 + |A|^p + |B|^p)$ for all $x \in \mathbf{R}^n$, $y \in \mathbf{R}^N$, $A \in \mathbf{M}^{m \times n}$, $B \in \mathbf{M}^{m \times N}$.

For all bounded open subsets Ω of $\mathbf{R}^n \times \mathbf{R}^N$ and $u \in W^{1,p}(\Omega; \mathbf{R}^m)$ define

$$F_\varepsilon(u, \Omega) = \int_\Omega f\left(\frac{x}{\varepsilon}, \frac{x}{\varepsilon^2}, D_x u(x, y), D_y u(x, y)\right) dx\, dy.$$

Prove that the limit

$$\Gamma\text{-}\lim_{\varepsilon \to 0^+} F_\varepsilon(u, \Omega) = \int_\Omega \varphi(Du(x, y))\, dx\, dy$$

exists for all bounded open subsets Ω of $\mathbf{R}^n \times \mathbf{R}^N$, $u \in W^{1,p}(\Omega; \mathbf{R}^m)$, where φ satisfies

$$\varphi(A, B) = \min\left\{|X|^{-1} \int_X Pf(x, Du(x) + A, B)\, dx : u \in W_{\#}^{1,p}(X; \mathbf{R}^m)\right\}, \quad (22.23)$$

and Pf is a Carathéodory function given by

$$Pf(x, A, B) = \min\left\{|Y|^{-1} \int_Y f(x, y, A, Du(y) + B)\, dy : u \in W_{\#}^{1,p}(Y; \mathbf{R}^m)\right\}.$$
$$(22.24)$$

Hint: regard the function f as $(X \times Y)$-periodic, and apply Theorem 22.1 to obtain the Γ-convergence to an integral functional satisfying (22.1)–(22.4). With fixed x_0, A, B, consider the solution $u = u(x, y)$ of the minimum problem given by (22.20):

$$\min\left\{|Y|^{-1} \int_Y f(x_0, y, D_x u(x, y) + A, D_y u(x, y) + B)\, dy : u \in W_{\#}^{1,p}(X \times Y)\right\}.$$

If $x' \in \mathbf{R}^n$ and $u'(x, y) = u(x + x', y)$, then we have

$$\int_{X \times Y} f(x_0, y, D_x u'(x, y) + A, D_y u'(x, y) + B)\, dx\, dy$$
$$= \int_{X \times Y} f(x_0, y, D_x u(x + x', y) + A, D_y u(x + x', y) + B)\, dx\, dy$$
$$= \int_{(X - x') \times Y} f(x_0, y, D_x u(x, y) + A, D_y u(x, y) + B)\, dx\, dy$$
$$= \int_{X \times Y} f(x_0, y, D_x u(x, y) + A, D_y u(x, y) + B)\, dx\, dy.$$

By the strict convexity of f, we have $Du(x + x', y) = Du(x, y)$, hence the difference $u' - u$ must be constant a.e. By the periodicity of u we must have $u' = u$ a.e., and $u = u(y)$ by the arbitrariness of x'. We obtain (22.24), since

$$\int_{X\times Y} f(x_0, y, D_x u(x,y) + A, D_y u(x,y) + B)\, dx\, dy$$

$$= \int_X \int_Y f(x_0, y, A, D_y u(y) + B)\, dy\, dx = |X| \int_Y f(x_0, y, A, D_y u(y) + B)\, dy.$$

Deal with formula (22.23) in the same way.

Exercise 22.5 Compute the homogenized functional

$$F(u,\Omega) = \Gamma\text{-}\lim_{\varepsilon\to 0^+} \int_\Omega G\Big(\frac{x}{\varepsilon}\Big) G\Big(\frac{y}{\varepsilon^2}\Big)\Big(\Big(\frac{\partial u(x,y)}{\partial x}\Big)^2 + \Big(\frac{\partial u(x,y)}{\partial y}\Big)^2\Big)\, dx\, dy,$$

defined on $W^{1,2}(\Omega)$ ($\Omega \subset \mathbf{R}^2$), where G is given by Exercise 13.2.

    Hint: apply Exercise 22.4 with $n = N = m = 1$ and $p = 2$, modified by taking Remark 22.7 into account. If $f(x, y, A, B) = G(x)G(y)(A^2 + B^2)$, we have

$$Pf(x, A, B) = \min\Big\{ G(x) \int_0^1 G(y)((u'(y) + B)^2 + A^2)\, dy : u \in W^{1,2}_\#(0,1)\Big\}$$

$$= G(x)\Big(A^2 \int_0^1 G(y)\, dy + \min\Big\{\int_0^1 G(y)(u'(y) + B)^2\, dy : u \in W^{1,2}_\#(0,1)\Big\}\Big).$$

By Exercise 13.2

$$Pf(x, A, B) = G(x)\Big(\frac{\alpha + \beta}{2}A^2 + \frac{2\alpha\beta}{\alpha + \beta}B^2\Big)$$

and then

$$\varphi(A, B) = \min\Big\{\int_0^1 G(x)\Big(\frac{\alpha + \beta}{2}(u'(x) + A)^2 + \frac{2\alpha\beta}{\alpha + \beta}B^2\Big)\, dx : u \in W^{1,2}_\#(0,1)\Big\}$$

$$= \alpha\beta B^2 + \min\Big\{\frac{\alpha + \beta}{2}\int_0^1 G(x)(u'(x) + A)^2\, dx : u \in W^{1,2}_\#(0,1)\Big\}$$

$$= \alpha\beta(A^2 + B^2).$$

Exercise 22.6 Compute the homogenized functional

$$F(u,\Omega) = \Gamma\text{-}\lim_{\varepsilon\to 0^+} \int_\Omega \Big(G\Big(\frac{x}{\varepsilon}\Big) + G\Big(\frac{y}{\varepsilon^2}\Big)\Big)\Big(\Big(\frac{\partial u(x,y)}{\partial x}\Big)^2 + \Big(\frac{\partial u(x,y)}{\partial y}\Big)^2\Big)\, dx\, dy,$$

where G is given by Exercise 13.2.

    Hint: consider $f(x, y, A, B) = (G(x)+G(y))(A^2+B^2)$ and apply Exercise 22.4. A straightforward computation as in Exercises 22.2 and 22.5 shows that the integrand of F is

$$\varphi(A, B) = \frac{(3\alpha + \beta)(\alpha + 3\beta)}{4(\alpha + \beta)}A^2 + 2\frac{(\alpha + \beta)(\alpha^2 + \beta^2 + 6\alpha\beta)}{(3\alpha + \beta)(\alpha + 3\beta)}B^2.$$

23

CORRECTORS FOR THE HOMOGENIZATION

Let f be a function as in Chapter 14, satisfying a standard growth condition of order $p > 1$, and let

$$F_\varepsilon(u) = \int_\Omega f\left(\frac{x}{\varepsilon}, Du(x)\right) dx,$$

if $u \in W^{1,p}(\Omega)$. If in addition f is strictly convex and smooth, the unique solution u_ε to the Dirichlet boundary value problem

$$\begin{cases} -\mathrm{div}\left(D_\xi f\left(\dfrac{x}{\varepsilon}, Du_\varepsilon\right)\right) = g \quad \text{on } \Omega \\[2ex] u_\varepsilon \in W_0^{1,p}(\Omega) \end{cases} \tag{23.1}$$

is also the unique solution of the minimum problem involving F_ε whose Euler equation is (23.1). Here D_ξ denotes the derivation with respect to the gradient variable. By the homogenization results obtained in Chapter 14 for functionals and by Proposition 11.7 and Theorem 7.2, the solutions u_ε converge weakly in $W^{1,p}(\Omega)$ to the solution u of the homogenized problem

$$\begin{cases} -\mathrm{div}\left(D_\xi f_{\mathrm{hom}}(Du)\right) = g \quad \text{on } \Omega \\[2ex] u \in W_0^{1,p}(\Omega). \end{cases}$$

Clearly, we cannot expect Du_ε to converge strongly to Du. However, in this chapter we construct a family of 'correctors' depending only on the function $D_\xi f(x, \xi)$, which allow us to express Du_ε in terms of Du up to a remainder term which converges to 0 strongly in $L^p(\Omega; \mathbf{R}^n)$.

23.1 Convergence of momenta in homogenization

Let $f : \mathbf{R}^n \times \mathbf{R}^n \to [0, +\infty)$ be a function satisfying the following properties:

(i) for every $x \in \mathbf{R}^n$ the function $\xi \mapsto f(x, \xi)$ is convex and of class C^1 on $\mathbf{R}^n$; from now on we denote by f_ξ the vector $D_\xi f = (f_{\xi_1}, \ldots, f_{\xi_n})$ of the partial derivatives of f with respect to $\xi_1, \ldots, \xi_n$;

(ii) for every $\xi \in \mathbf{R}^n$ the function $f(\cdot, \xi)$ is measurable and 1-periodic on $\mathbf{R}^n$;

(iii) there exist $c_0, c_1 \in \mathbf{R}^n$ with $c_1 \geq c_0 > 0$ such that

$$c_0|\xi|^p \leq f(x, \xi) \leq c_1(1 + |\xi|^p) \tag{23.2}$$

for all $x \in \mathbf{R}^n$ and $\xi \in \mathbf{R}^n$;

(iv) there exist three constants $c_2, c_3, c_4 > 0$ and two constants α and β, with $0 \le \alpha \le 1 \wedge (p-1)$ and $p \vee 2 \le \beta < +\infty$, such that $f_\xi : \mathbf{R}^n \times \mathbf{R}^n \to \mathbf{R}^n$ fulfils the following boundedness, continuity and monotonicity assumptions:

$$|f_\xi(x,0)| \le c_2 \tag{23.3}$$

$$|f_\xi(x,\xi_1) - f_\xi(x,\xi_2)| \le c_3(1 + |\xi_1| + |\xi_2|)^{p-1-\alpha}|\xi_1 - \xi_2|^\alpha \tag{23.4}$$

$$\langle f_\xi(x,\xi_1) - f_\xi(x,\xi_2), \xi_1 - \xi_2 \rangle \ge c_4(1 + |\xi_1| + |\xi_2|)^{p-\beta}|\xi_1 - \xi_2|^\beta \tag{23.5}$$

for a.e. $x \in \mathbf{R}^n$ and for every $\xi \in \mathbf{R}^n$.

Remark 23.1 By (ii) it follows that $f_\xi(\cdot, \xi)$ is 1-periodic and Lebesgue measurable for every $\xi \in \mathbf{R}^n$.

We fix from now on a bounded open subset Ω of $\mathbf{R}^n$. By our assumptions on f for every $g \in \mathrm{L}^{p'}(\Omega)$ and for every $\varepsilon > 0$ there exists a unique solution $u_\varepsilon \in \mathrm{W}_0^{1,p}(\Omega)$ to the Dirichlet boundary value problem

$$\begin{cases} -\mathrm{div}\left(f_\xi\left(\dfrac{x}{\varepsilon}, Du_\varepsilon \right) \right) = g \quad \text{on } \Omega \\[2ex] u_\varepsilon \in \mathrm{W}_0^{1,p}(\Omega) . \end{cases} \tag{23.6}$$

The following homogenization result holds, which asserts the weak convergence of *momenta* $f_\xi(\frac{x}{\varepsilon}, Du_\varepsilon)$, besides the weak convergence of solutions.

Theorem 23.2 *Let $g \in \mathrm{L}^{p'}(\Omega)$ and let (ε_j) be a sequence of positive real numbers converging to 0. Let $a_j(x,\xi) = f_\xi(x/\varepsilon_j, \xi)$, and let $b = (f_{\mathrm{hom}})_\xi$, where f_{hom} is given by (14.16). Let u_j be the solution to the problem*

$$\begin{cases} -\mathrm{div}\left(a_j(x, Du_j) \right) = g \quad on\ \Omega \\[2ex] u_j \in \mathrm{W}_0^{1,p}(\Omega). \end{cases} \tag{23.7}$$

Then we have

$$u_j \rightharpoonup u \quad weakly\ in\ \mathrm{W}^{1,p}(\Omega) \tag{23.8}$$

$$a_j(x, Du_j) \rightharpoonup b(Du) \quad weakly\ in\ \mathrm{L}^{p'}(\Omega; \mathbf{R}^n), \tag{23.9}$$

where u is the solution to the homogenized problem

$$\begin{cases} -\mathrm{div}\left(b(Du) \right) = g \quad on\ \Omega \\[2ex] u \in \mathrm{W}_0^{1,p}(\Omega). \end{cases} \tag{23.10}$$

Remark 23.3 Using the Euler equation for the minimum problem defining f_{hom}, the function $b : \mathbf{R}^n \to \mathbf{R}^n$ can be expressed by

$$b(\xi) = \int_{(0,1)^n} f_\xi(x, \xi + Dv(x)) \, dx, \qquad (23.11)$$

where $v = v_\xi$ is the solution to the following cell problem on $(0,1)^n$:

$$\begin{cases} v \in \mathrm{W}^{1,p}_{\#}((0,1)^n) \\[2mm] \displaystyle\int_{(0,1)^n} \langle f_\xi(x, \xi + Dv(x)), Dw \rangle \, dx = 0 \quad \text{for every } w \in \mathrm{W}^{1,p}_{\#}((0,1)^n). \end{cases} \qquad (23.12)$$

In fact, under our assumptions on f it follows that if v is a solution to (23.12) then it coincides with the solution to the minimum problem (14.16) defining $f_{\text{hom}}(\xi)$. Moreover, since it follows from condition (iv) that $|f_\xi(x, \xi)| \le c(1 + |\xi|^{p-1})$, we have $v \in \mathrm{W}^{1,p}_{\text{loc}}(\mathbf{R}^n)$, Dv is 1-periodic, and $\operatorname{div}(f_\xi(x, \xi + Dv)) = 0$ on $\mathbf{R}^n$ in the sense of distributions (see Exercise 14.8).

While the convergence in (23.8) follows directly from the general homogenization theory, the convergence of the momenta in (23.9) needs a separate proof, which relies on the following compensated compactness argument.

Lemma 23.4 *Let (u_j) be a sequence converging weakly to u in $\mathrm{W}^{1,p}(\Omega)$, let (g_j) be a sequence in $\mathrm{L}^{p'}(\Omega; \mathbf{R}^n)$ converging weakly in $\mathrm{L}^{p'}(\Omega; \mathbf{R}^n)$, and let $(-\operatorname{div} g_j)$ converge strongly in $\mathrm{W}^{-1,p'}(\Omega)$ to $-\operatorname{div} g$. Then*

$$\int_\Omega g_j \, Du_j \varphi \, dx \to \int_\Omega g \, Du \varphi \, dx$$

for every $\varphi \in C_0^\infty(\Omega)$.

Proof It is enough to observe that

$$\int_\Omega g_j \, Du_j \varphi \, dx = \langle -\operatorname{div} g_j, u_j \varphi \rangle - \int_\Omega g_j u_j D\varphi \, dx$$

for every $\varphi \in C_0^\infty(\Omega)$. Here $\langle \cdot, \cdot \rangle$ denotes the duality pairing between $\mathrm{W}^{-1,p'}(\Omega)$ and $\mathrm{W}^{1,p}_0(\Omega)$. $\qquad\square$

Proof of Theorem 23.2 For every $\varepsilon > 0$ let $F_\varepsilon : \mathrm{L}^p(\Omega) \to [0, +\infty]$ be defined as in Chapter 14. By Theorem 14.5 and Theorem 14.7 (with $m = 1$) we have

$$\Gamma(\mathrm{L}^p)\text{-}\lim_{\varepsilon \to 0} F_\varepsilon(u) = \int_\Omega f_{\text{hom}}(Du) \, dx,$$

where $f_{\text{hom}} : \mathbf{R}^n \to [0, +\infty)$ is the convex function given by (14.16). By Remark 7.3(v) we have

$$\Gamma(L^p)\text{-}\lim_{\varepsilon\to 0}(F_\varepsilon(u)+G(u)) = \int_\Omega f_{\text{hom}}(Du)\,dx + G(u),$$

where $G : L^p(\Omega) \to \mathbf{R}$ is defined as $G(u) = -\int_\Omega gu\,dx$.

Let (ε_j) be any sequence of positive real numbers converging to 0. By Proposition 11.7 and Theorem 7.2 we see, on computing the corresponding Euler equations, that Γ-convergence implies that the solutions u_j to problem (23.7) weakly converge in $W^{1,p}(\Omega)$ to the solution of problem (23.10), which proves (23.8).

To prove (23.9), first note that, since $\|u_j\|_{W^{1,p}(\Omega)} \le c$, by (23.3) and (23.4) we get that

$$\|a_j(x, Du_j)\|_{L^{p'}(\Omega;\mathbf{R}^n)} = \|f_\xi(x/\varepsilon_j, Du_j)\|_{L^{p'}(\Omega;\mathbf{R}^n)} \le c',$$

where c and c' are suitable positive constants. Hence, up to a subsequence, we have that $(a_j(x, Du_j))_j$ converges weakly to a function $\psi \in L^{p'}(\Omega;\mathbf{R}^n)$, and $(-\text{div}\,(a_j(x, Du_j)))_j$ converges strongly in $W^{-1,p'}(\Omega)$ to $-\text{div}\,\psi$. We now prove that $\psi(x) = b(Du(x))$ a.e. for $x \in \Omega$. To this end we define suitable functions $v_j \in W^{1,p}(\Omega)$, $k_j \in L^{p'}(\Omega;\mathbf{R}^n)$, both ε_j-periodic, in the following way. Given $\xi \in \mathbf{R}^n$, let $v \in W^{1,p}_\#((0,1)^n)$ be a solution to problem (23.12) and set $k = f_\xi(\cdot, \xi + Dv(\cdot)) \in L^{p'}(\Omega;\mathbf{R}^n)$; we still denote by v and k the 1-periodic extensions of these functions to $\mathbf{R}^n$. By Remark 23.3 $v \in W^{1,p}_{\text{loc}}(\mathbf{R}^n)$, $k \in L^{p'}_{\text{loc}}(\mathbf{R}^n;\mathbf{R}^n)$ with

$$\int_{\mathbf{R}^n} \langle f_\xi(x, \xi + Dv), Dw(x)\rangle\,dx = 0$$

for every $w \in C_0^\infty(\mathbf{R}^n)$. Now we are in a position to define

$$v_j(x) = \varepsilon_j v\left(\frac{x}{\varepsilon_j}\right) + \langle \xi, x\rangle, \qquad k_j(x) = k\left(\frac{x}{\varepsilon_j}\right).$$

The periodicity properties of these functions yield (see Example 2.7) that

$$v_j \rightharpoonup \langle \xi, \cdot\rangle \qquad \text{weakly in } W^{1,p}(\Omega) \tag{23.13}$$

$$Dv_j \rightharpoonup \xi \qquad \text{weakly in } L^p(\Omega;\mathbf{R}^n) \tag{23.14}$$

$$k_j \rightharpoonup \int_{(0,1)^n} f_\xi(x, \xi + Dv(x))\,dx \quad \text{weakly in } L^{p'}(\Omega;\mathbf{R}^n) \tag{23.15}$$

$$k_j(x) = f_\xi\left(\frac{x}{\varepsilon_j}, Dv_j(x)\right) \qquad \text{for } x \in \Omega \text{ a.e.} \tag{23.16}$$

By the monotonicity of f_ξ in (23.5) we have

$$\int_\Omega \left\langle f_\xi\left(\frac{x}{\varepsilon_j}, Du_j(x)\right) - f_\xi\left(\frac{x}{\varepsilon_j}, Dv_j(x)\right), Du_j - Dv_j \right\rangle \varphi\,dx \ge 0$$

for every $\varphi \in C_0^\infty(\Omega)$, $\varphi \geq 0$. By passing to the limit as $j \to +\infty$, Lemma 23.4 implies that

$$\int_\Omega \left\langle \psi(x) - \int_{(0,1)^n} f_\xi(y, \xi + Dv(y))dy, Du(x) - \xi \right\rangle \varphi\, dx \geq 0$$

for every $\varphi \in C_0^\infty(\Omega)$, $\varphi \geq 0$; therefore, for every $\xi \in \mathbf{R}^n$ we have

$$\left\langle \psi(x) - \int_{(0,1)^n} f_\xi(y, \xi + Dv(y))dy, Du(x) - \xi \right\rangle \geq 0 \tag{23.17}$$

for a.e. $x \in \Omega$ By a density argument we easily get that

$$\left\langle \psi(x) - \int_{(0,1)^n} f_\xi(y, \xi + Dv(y))dy, Du(x) - \xi \right\rangle \geq 0 \tag{23.18}$$

for a.e. $x \in \mathbf{R}^n$ and for every $\xi \in \mathbf{R}^n$. By Exercise 5.5 (recall that $b(\xi) = (f_{\mathrm{hom}})_\xi$ and f_{hom} is convex and continuous), we conclude that $\psi(x) = b(Du(x))$ for a.e. $x \in \Omega$. $\qquad\square$

Remark 23.5 Some technicalities are required to prove that the function $b :$ $\mathbf{R}^n \to \mathbf{R}^n$ satisfies the inequalities

$$|b(\xi_1) - b(\xi_2)| \leq \tilde{c}_3(1 + |\xi_1| + |\xi_2|)^{p-1-\gamma}|\xi_1 - \xi_2|^\gamma, \tag{23.19}$$

$$\langle b(\xi_1) - b(\xi_2), \xi_1 - \xi_2 \rangle \geq \tilde{c}_4(1 + |\xi_1| + |\xi_2|)^{p-\beta}|\xi_1 - \xi_2|^\beta \tag{23.20}$$

for every $\xi_1, \xi_2 \in \mathbf{R}^n$, where $\gamma = \alpha/(\beta - \alpha)$, and $\tilde{c}_3$, $\tilde{c}_4$ are strictly positive constants depending on n, p, α, β, c_2, c_3, c_4. For the reader's convenience we will give their proof in Proposition 23.7. For simplicity, the letter c will denote various positive constants whose value can change from one line to the next.

By using (23.3)–(23.5) it follows that

$$|f_\xi(x, \xi)| \leq c(1 + |\xi|^{p-1}) \tag{23.21}$$

$$|\xi|^p \leq c(1 + \langle f_\xi(x, \xi), \xi \rangle) \tag{23.22}$$

hold for a.e. $x \in \Omega$ and for every $\xi \in \mathbf{R}^n$. The proof of (23.21) is trivial. As for the proof of (23.22), by Young's inequality with $\gamma > 0$ we have

$$|\xi|^p \leq \gamma^{(p-\beta)/\beta}\frac{p}{\beta}(1 + |\xi|)^{p-\beta}|\xi|^\beta + \gamma\frac{\beta - p}{\beta}(1 + |\xi|)^p;$$

this implies, for γ sufficiently small, together with (23.3) and (23.5), estimate (23.22). On the other hand, it is easy to show that the inequalities (23.21) and (23.22) are equivalent to

$$|f_\xi(x,\xi)|^{p'} \leq c(1 + \langle f_\xi(x,\xi), \xi \rangle) \tag{23.23}$$

$$|\xi|^p \leq c(1 + \langle f_\xi(x,\xi), \xi \rangle) \tag{23.24}$$

for a.e. $x \in \Omega$ and for every $\xi \in \mathbf{R}^n$. Moreover, the following inequalities can be deduced

$$|f_\xi(x,\xi_1) - f_\xi(x,\xi_2)| \leq c\big(1 + \langle f_\xi(x,\xi_1), \xi_1 \rangle + \langle f_\xi(x,\xi_2), \xi_2 \rangle\big)^{1-1/p-\alpha/\beta}$$
$$\times \langle f_\xi(x,\xi_1) - f_\xi(x,\xi_2), \xi_1 - \xi_2 \rangle^{\alpha/\beta} \tag{23.25}$$

and

$$\langle f_\xi(x,\xi_1) - f_\xi(x,\xi_2), \xi_1 - \xi_2 \rangle$$
$$\geq c\big(1 + \langle f_\xi(x,\xi_1), \xi_1 \rangle + \langle f_\xi(x,\xi_2), \xi_2 \rangle\big)^{(p-\beta)/p}|\xi_1 - \xi_2|^\beta \tag{23.26}$$

for a.e. $x \in \Omega$ and for every $\xi_1, \xi_2 \in \mathbf{R}^n$.

By using the following lemma and Theorem 23.2 we will show that these inequalities also hold with b in place of f_ξ. Estimates (23.19) and (23.20) for b are then easy consequences of them.

Lemma 23.6 *Let γ and δ be two non-negative constants with $\gamma+\delta \leq 1$. Let ψ, ζ, θ be functions in $\mathrm{L}^1(\Omega)$ and let (ψ_j), (ζ_j), (θ_j) be sequences in $\mathrm{L}^1(\Omega)$ converging to ψ, ζ, θ in the sense of distributions. Suppose that $\zeta_j \geq 0$, $\theta_j \geq 0$, and*

$$|\psi_j| \leq (\zeta_j)^\gamma(\theta_j)^\delta \qquad a.e. \ in \ \Omega. \tag{23.27}$$

Then

$$|\psi| \leq \zeta^\gamma\theta^\delta \qquad a.e. \ in \ \Omega. \tag{23.28}$$

Proof Let $\varepsilon = 1 - \gamma - \delta$. By (23.27), for every $\varphi \in C_0^\infty(\Omega)$, $\varphi \geq 0$, we have

$$\int_\Omega |\psi_j|\varphi \, dx \leq \left(\int_\Omega \zeta_j\varphi \, dx \right)^\gamma \left(\int_\Omega \theta_j\varphi \, dx \right)^\delta \left(\int_\Omega \varphi \, dx \right)^\varepsilon. \tag{23.29}$$

Since $(\psi_j\varphi)$ converges to $\psi\varphi$ in the sense of distributions we obtain

$$\int_\Omega |\psi|\varphi \, dx \leq \liminf_j \int_\Omega |\psi_j|\varphi \, dx.$$

Therefore, by taking the limit in (23.29) as $j \to +\infty$ we get

$$\int_\Omega |\psi|\varphi \, dx \leq \left(\int_\Omega \zeta\varphi \, dx \right)^\gamma \left(\int_\Omega \theta\varphi \, dx \right)^\delta \left(\int_\Omega \varphi \, dx \right)^\varepsilon \tag{23.30}$$

for every $\varphi \in C_0^\infty(\Omega)$, $\varphi \geq 0$. By a standard approximation argument we obtain (23.30) for every $\varphi \in L^\infty(\Omega)$, $\varphi \geq 0$. In particular, we have

$$\int_{B_\rho(x)} |\psi|\varphi\,dx \le \left(\int_{B_\rho(x)} \zeta\varphi\,dx\right)^\gamma \left(\int_{B_\rho(x)} \theta\varphi\,dx\right)^\delta \left(\int_{B_\rho(x)} \varphi\,dx\right)^\varepsilon$$

for every $x \in \Omega$ and $\rho \ge 0$ small enough and this implies (23.28) by the Lebesgue Derivation Theorem. $\qquad\square$

Proposition 23.7 *Inequalities* (23.23)–(23.26) *are satisfied with b in place of* f_ξ.

Proof We prove only (23.23) and (23.25), the proofs of (23.24) and (23.26) being completely analogous. Fix $\xi \in \mathbf{R}^n$. We denote by (v_j) the sequence defined as in the proof of Theorem 23.2 by

$$v_j(x) = \varepsilon_j v\left(\frac{x}{\varepsilon_j}\right) + \langle \xi, x\rangle \qquad \text{for a.e. } x \in \Omega,$$

where v is the solution to the cell problem (23.12) for ξ. By assumption we have

$$|a_j(x, Dv_j)|^{p'} \le c(1 + \langle a_j(x, Dv_j), Dv_j\rangle)$$

for a.e. $x \in \Omega$ Set

$$\begin{aligned}
&\psi_j = a_j(\cdot, Dv_j), && \psi = b(\xi),\\
&\zeta_j = c(1 + \langle a_j(\cdot, Dv_j), Dv_j\rangle), && \zeta = c(1 + \langle b(\xi), \xi\rangle),\\
&\theta_j = 1, && \theta = 1,
\end{aligned}$$

and $\gamma = 1/p$, $\delta = 0$. By Lemma 23.4 and the properties of (v_j), stated in the proof of Theorem 23.2, (ζ_j) converges to ζ and (θ_j) converges to θ in the sense of distributions. Therefore, $\zeta \ge 0$ a.e. in Ω and Lemma 23.6 yields

$$|\psi| \le \zeta^{1/p'} \qquad \text{a.e. in } \Omega,$$

proving (23.23).

The proof of (23.25) will follow the same argument. Fix $\xi_1, \xi_2 \in \mathbf{R}^n$. Denote by (v_j^1) and (v_j^2) two sequences defined, as in the proof of Theorem 23.2, by

$$v_j^1(x) = \varepsilon_j v^1\left(\frac{x}{\varepsilon_j}\right) + \langle \xi_1, x\rangle, \qquad v_j^2(x) = \varepsilon_j v^2\left(\frac{x}{\varepsilon_j}\right) + \langle \xi_2, x\rangle.$$

By assumption we have

$$\begin{aligned}
&|a_j(x, Dv_j^1) - a_j(x, Dv_j^2)|\\
&\le c\big(1 + \langle a_j(x, Dv_j^1), Dv_j^1\rangle + \langle a_j(x, Dv_j^2), Dv_j^2\rangle\big)^{1-1/p-\alpha/\beta}\\
&\qquad\times \langle a_j(x, Dv_j^1) - a_j(x, Dv_j^2), Dv_j^1 - Dv_j^2\rangle^{\alpha/\beta}
\end{aligned}$$

for a.e. $x \in \Omega$ Set

$$\psi_j = a_j(\cdot, Dv_j^1) - a_j(\cdot, Dv_j^2), \qquad \psi = b(\xi_1) - b(\xi_2),$$

$$\zeta_j = c(1 + \langle a_j(\cdot, Dv_j^1), Dv_j^1 \rangle + \langle a_j(\cdot, Dv_j^2), Dv_j^2 \rangle),$$
$$\zeta = c(1 + \langle b(\xi_1), \xi_1 \rangle + \langle b(\xi_2), \xi_2 \rangle),$$
$$\theta_j = \langle a_j(\cdot, Dv_j^1) - a_j(\cdot, Dv_j^2), Dv_j^1 - Dv_j^2 \rangle, \quad \theta = \langle b(\xi_1) - b(\xi_2), \xi_1 - \xi_2 \rangle,$$

and $\gamma = 1 - 1/p - \alpha/\beta$, $\delta = \alpha/\beta$. As above we get that (ζ_j) converges to ζ and (θ_j) converges to θ in the sense of distributions. Therefore, $\zeta \geq 0$ a.e. in Ω and Lemma 23.6 yields

$$|\psi| \leq \zeta^{1-1/p-\alpha/\beta} \theta^{\alpha/\beta} \quad \text{a.e. in } \Omega,$$

proving (23.25). $\square$

Finally, we state the following Meyers regularity result, which will be useful in the proof of the main theorem in the next section (see Appendix C).

Theorem 23.8 *Let* $w \in \mathrm{W}^{1,p}(\Omega)$ *be a weak solution of the equation*

$$-\operatorname{div}(f_\xi(x, Dw)) = 0 \quad \text{on } \Omega.$$

Then there exists $\eta > 0$ *such that for each open set* ω *with compact closure in* Ω *we have that* $w \in \mathrm{W}^{1,p+\eta}(\omega)$ *and*

$$\|w\|_{W^{1,p+\eta}(\omega)} \leq c\|w\|_{W^{1,p}(\Omega)}.$$

The constant η *depends only on* n, p, c_2, c_3, c_4, *while* $c > 0$ *depends, in addition, on* Ω *and* ω.

23.2 Definition and some properties of the correctors

We begin by defining the family (M_j) of approximations of the identity map on $\mathrm{L}^p(\Omega; \mathbf{R}^n)$. For simplicity, from now on we use the notation $Y = (0,1)^n$. Let (ε_j) be a sequence of positive real numbers converging to 0. We consider, for $\mathbf{z} \in \mathbf{Z}^n$ and $\varepsilon_j > 0$, the cube $Y_j^{\mathbf{z}} = \varepsilon_j(\mathbf{z} + Y)$, i.e. the translated image of $Y_j = \varepsilon_j Y$ by the vector $\varepsilon_j \mathbf{z}$. Given $\varphi \in \mathrm{L}^p(\Omega; \mathbf{R}^n)$, we define the function $M_j\varphi : \mathbf{R}^n \to \mathbf{R}^n$ by

$$(M_j\varphi)(x) = \sum_{\mathbf{z} \in I_j} \chi_{Y_j^{\mathbf{z}}}(x) \frac{1}{|Y_j^{\mathbf{z}}|} \int_{Y_j^{\mathbf{z}}} \varphi(y) \, dy, \tag{23.31}$$

where $I_j = \{\mathbf{z} \in \mathbf{Z}^n : Y_j^{\mathbf{z}} \subseteq \Omega\}$. Note that we have

$$M_j\varphi \to \varphi \quad \text{a.e. on } \Omega \text{ and strongly in } \mathrm{L}^p(\Omega; \mathbf{R}^n). \tag{23.32}$$

Moreover, by Jensen's inequality

$$\|M_j\varphi\|_{\mathrm{L}^p(\Omega;\mathbf{R}^n)} \leq \|\varphi\|_{\mathrm{L}^p(\Omega;\mathbf{R}^n)} \tag{23.33}$$

for every $\varphi \in \mathrm{L}^p(\Omega; \mathbf{R}^n)$.

Finally, we define the function $p : \mathbf{R}^n \times \mathbf{R}^n \to \mathbf{R}^n$ by

$$p(x, \xi) = \xi + Dv(x), \tag{23.34}$$

where $v = v_\xi$ is the unique solution to the cell problem (23.12). The function $p(\cdot, \xi)$ is 1-periodic, and the function $p_j : \mathbf{R}^n \times \mathbf{R}^n \to \mathbf{R}^n$ defined by

$$p_j(x, \xi) = p\left(\frac{x}{\varepsilon_j}, \xi\right) = \xi + Dv\left(\frac{x}{\varepsilon_j}\right) \tag{23.35}$$

is ε_j-periodic in x. These facts yield

$$\int_Y p(x, \xi)\, dx = \xi$$

and

$$p_j(\cdot, \xi) \rightharpoonup \xi \quad \text{weakly in } \mathrm{L}^p(\Omega; \mathbf{R}^n). \tag{23.36}$$

The functions $p_j(\cdot, M_j Du(\cdot))$ will be called the *correctors* of Du and will give a strong approximation of Du_j. The meaning of this term will be clear after Theorem 23.14.

Using the notation introduced above, the definition of b can be written in the form

$$b(\xi) = \int_Y f_\xi(x, p(x, \xi))\, dx. \tag{23.37}$$

Moreover, with equation (23.12) satisfied by v, we have

$$\int_Y \langle f_\xi(x, p(x, \xi)), p(x, \xi)\rangle\, dx = \int_Y \langle f_\xi(x, p(x, \xi)), \xi\rangle\, dx = \langle b(\xi), \xi\rangle \tag{23.38}$$

and

$$-\operatorname{div}\left(f_\xi(x, p(x, \xi))\right) = 0 \tag{23.39}$$

in the sense of distributions for every $\xi \in \mathbf{R}^n$.

In the remaining part of this section we prove some technical estimates of the functions p_j defined by (23.35). The letter c will denote various positive constants whose value can change from one line to the next.

Lemma 23.9 *Let φ_1 and φ_2 be two functions belonging to $\mathrm{L}^p(\Omega; \mathbf{R}^n)$. Then*

$$\|\varphi_1 - \varphi_2\|_{\mathrm{L}^p(\Omega;\mathbf{R}^n)} \leq c\left(\int_\Omega |\varphi_1 - \varphi_2|^\beta (1 + |\varphi_1| + |\varphi_2|)^{p-\beta}\, dx\right)^{1/\beta}$$

$$\times \left(|\Omega|^{1/p} + \|\varphi_1\|_{\mathrm{L}^p(\Omega;\mathbf{R}^n)} + \|\varphi_2\|_{\mathrm{L}^p(\Omega;\mathbf{R}^n)}\right)^{(\beta-p)/\beta} \tag{23.40}$$

with $c > 0$ depending only on n and p.

Proof By using Hölder's inequality with $r = \beta/p$ and $s = \beta/(\beta - p)$, it follows that

$$\|\varphi_1 - \varphi_2\|_{L^p(\Omega;\mathbf{R}^n)} = \int_\Omega \frac{|\varphi_1 - \varphi_2|^p}{(1 + |\varphi_1| + |\varphi_2|)^\lambda}(1 + |\varphi_1| + |\varphi_2|)^\lambda \, dx$$

$$\leq \left(\int_\Omega \frac{|\varphi_1 - \varphi_2|^\beta}{(1 + |\varphi_1| + |\varphi_2|)^{\lambda\beta/p}} \, dx\right)^{p/\beta} \left(\int_\Omega (1 + |\varphi_1| + |\varphi_2|)^{\lambda\beta/(\beta-p)} \, dx\right)^{(\beta-p)/\beta}$$

$$(23.41)$$

for every $\lambda \geq 0$. In particular, for $\lambda = p(\beta - p)/\beta$ we easily obtain (23.40). $\quad\square$

Lemma 23.10 *For every $\xi \in \mathbf{R}^n$ we have*

$$\|p_j(\cdot, \xi)\|_{L^p(Y_j;\mathbf{R}^n)}^p \leq c(1 + |\xi|^p)|Y_j|, \tag{23.42}$$

where $c > 0$ depends only on n, p, β, c_2, c_3, c_4.

Proof Let $\xi \in \mathbf{R}^n$. By the definition of p_j and by a simple change of variables, it is enough to prove that

$$\|p(\cdot, \xi)\|_{L^p(Y;\mathbf{R}^n)}^p \leq c(1 + |\xi|^p). \tag{23.43}$$

By Lemma 23.9 we have

$$\|p(\cdot, \xi)\|_{L^p(Y;\mathbf{R}^n)}^p$$

$$\leq c\left(\int_Y |p(x, \xi)|^\beta (1 + |p(x, \xi)|)^{p-\beta} \, dx\right)^{p/\beta} \left(1 + \|p(\cdot, \xi)\|_{L^p(Y;\mathbf{R}^n)}^p\right)^{(\beta-p)/\beta}.$$

Moreover, by using Young's inequality with $\eta > 0$, we get

$$\|p(\cdot, \xi)\|_{L^p(Y;\mathbf{R}^n)}^p$$

$$\leq c\eta^{(p-\beta)/p}\left(\int_Y |p(x, \xi)|^\beta (1 + |p(x, \xi)|)^{p-\beta} \, dx\right) + c\eta\left(1 + \|p(\cdot, \xi)\|_{L^p(Y;\mathbf{R}^n)}^p\right).$$

Now, assumption (23.5) and a suitable choice of η ensure the estimate

$$\|p(\cdot, \xi)\|_{L^p(Y;\mathbf{R}^n)}^p \leq c\left(1 + \int_Y \langle f_\xi(x, p(x, \xi)) - f_\xi(x, 0), p(x, \xi)\rangle \, dx\right).$$

By using (23.38) and (23.4) we obtain

$$\|p(\cdot, \xi)\|_{L^p(Y;\mathbf{R}^n)}^p$$

$$\leq c\left(1 + \int_Y (1 + |p(x, \xi)|)^{p-1}|\xi| \, dx + \int_Y |f_\xi(x, 0)|(|\xi| + |p(x, \xi)|) \, dx\right).$$

Thus, applying Hölder's inequality first and, successively, Young's inequality with $\gamma > 0$, and taking (23.3) into account, we may conclude that

$$\|p(\cdot,\xi)\|^p_{L^p(Y;\mathbf{R}^n)} \le c\gamma\|p(\cdot,\xi)\|^p_{L^p(Y;\mathbf{R}^n)} + c\gamma^{-1/(p-1)}(1+|\xi|^p). \qquad (23.44)$$

Finally, by taking γ sufficiently small, (23.44) immediately implies (23.43). $\qquad\square$

Corollary 23.11 *There exist $\eta > 0$ (depending only on n, p, c_2, c_3, c_4) and $c > 0$ (depending also on β) such that*

$$\|p_j(\cdot,\xi)\|^{p+\eta}_{L^{p+\eta}(Y_j;\mathbf{R}^n)} \le c(1+|\xi|^p)^{(p+\eta)/p}|Y_j| \qquad (23.45)$$

for every $\xi \in \mathbf{R}^n$.

Proof Let $\xi \in \mathbf{R}^n$. By the definition of $p_j(\cdot,\xi)$ and by a simple change of variables, it is enough to prove that

$$\|p(\cdot,\xi)\|^{p+\eta}_{L^{p+\eta}(Y;\mathbf{R}^n)} \le c(1+|\xi|^p)^{(p+\eta)/p}. \qquad (23.46)$$

If we apply Theorem 23.8 with $\omega = Y$ and Ω equal to the cube with the same centre as Y and sides with triple length, by taking the periodicity of $p(\cdot,\xi)$ into account, we obtain

$$\|p(\cdot,\xi)\|_{L^{p+\eta}(Y;\mathbf{R}^n)} \le c\|p(\cdot,\xi)\|_{L^p(Y;\mathbf{R}^n)}, \qquad (23.47)$$

where η and c are positive constants depending only on n, p, c_2, c_3, c_4. Therefore, (23.46) follows from (23.47) as an easy application of (23.43). $\qquad\square$

Lemma 23.12 *For every $\xi_1,\xi_2 \in \mathbf{R}^n$ we have*

$$\|p_j(\cdot,\xi_1) - p_j(\cdot,\xi_2)\|^p_{L^p(Y_j;\mathbf{R}^n)}$$
$$\le c(1+|\xi_1|^p + |\xi_2|^p)^{(\beta-\alpha-1)/(\beta-\alpha)}|\xi_1 - \xi_2|^{p/(\beta-\alpha)}|Y_j|, \qquad (23.48)$$

where $c > 0$ depends on n, p, α, β, c_2, c_3, c_4.

Proof Let $\xi_1,\xi_2 \in \mathbf{R}^n$. As in the proof of the previous results, it is enough to show that

$$\|p(\cdot,\xi_1) - p(\cdot,\xi_2)\|^p_{L^p(Y;\mathbf{R}^n)}$$
$$\le c(1+|\xi_1|^p + |\xi_2|^p)^{(\beta-\alpha-1)/(\beta-\alpha)}|\xi_1 - \xi_2|^{p/(\beta-\alpha)}. \qquad (23.49)$$

From Lemmas 23.9 and 23.10 and from the strict monotonicity assumption (23.5), we get the following estimate:

$$\|p(\cdot,\xi_1) - p(\cdot,\xi_2)\|^p_{L^p(Y;\mathbf{R}^n)}$$
$$\le c\left(\int_Y |p(x,\xi_1) - p(x,\xi_2)|^\beta (1 + |p(x,\xi_1)| + |p(x,\xi_2)|)^{p-\beta}\,dx\right)^{p/\beta}$$
$$\times (1+|\xi_1|^p + |\xi_2|^p)^{(\beta-p)/\beta}$$

$$\leq c\Big(\int_Y \langle f_\xi(x, p(x,\xi_1)) - f_\xi(x, p(x,\xi_2)), p(x,\xi_1) - p(x,\xi_2)\rangle\, dx\Big)^{p/\beta}$$
$$\times (1 + |\xi_1|^p + |\xi_2|^p)^{(\beta-p)/\beta}.$$

By (23.38) and (23.12), the last inequality implies that

$$\|p(\cdot,\xi_1) - p(\cdot,\xi_2)\|^p_{L^p(Y;\mathbf{R}^n)}$$
$$\leq c\Big(\int_Y \langle f_\xi(x, p(x,\xi_1)) - f_\xi(x, p(x,\xi_2)), \xi_1 - \xi_2\rangle\, dx\Big)^{p/\beta}$$
$$\times (1 + |\xi_1|^p + |\xi_2|^p)^{(\beta-p)/\beta}. \tag{23.50}$$

Now by first applying the continuity assumption (23.4) and Hölder's inequality successively (with $r_1 = p/(p-1-\alpha)$, $r_2 = p/\alpha$ and $r_3 = p$), we obtain

$$\Big(\int_Y \langle f_\xi(x, p(x,\xi_1)) - f_\xi(x, p(x,\xi_2)), \xi_1 - \xi_2\rangle\, dx\Big)^{p/\beta}$$
$$\leq c\Big(\int_Y (1 + |p(x,\xi_1)|^p + |p(x,\xi_2)|^p)\, dx\Big)^{(p-1-\alpha)/\beta}$$
$$\times \Big(\int_Y |p(x,\xi_1) - p(x,\xi_2)|^p\, dx\Big)^{\alpha/\beta} \Big(\int_Y |\xi_1 - \xi_2|^p\, dx\Big)^{1/\beta}$$
$$\leq c(1 + |\xi_1|^p + |\xi_2|^p)^{(p-1-\alpha)/\beta}\|p(\cdot,\xi_1) - p(\cdot,\xi_2)\|^{p\alpha/\beta}_{L^p(Y;\mathbf{R}^n)}|\xi_1 - \xi_2|^{p/\beta},$$

where in the last inequality we also use Lemma 23.10. Thus, by (23.50) we get

$$\|p(\cdot,\xi_1) - p(\cdot,\xi_2)\|^p_{L^p(Y;\mathbf{R}^n)}$$
$$\leq c(1 + |\xi_1|^p + |\xi_2|^p)^{(\beta-1-\alpha)/\beta}\|p(\cdot,\xi_1) - p(\cdot,\xi_2)\|^{p\alpha/\beta}_{L^p(Y;\mathbf{R}^n)}|\xi_1 - \xi_2|^{p/\beta},$$

which immediately implies (23.49). $\square$

Lemma 23.13 *Let* $\varphi \in L^p(\Omega; \mathbf{R}^n)$ *and let* Ψ *be a piecewise constant measurable function of the form*

$$\Psi = \sum_{h=1}^m \eta_h \chi_{\Omega_h}$$

with $\eta_h \in \mathbf{R}^n \setminus \{0\}$, $\Omega_h \subset\subset \Omega$, $|\partial\Omega_h| = 0$, $\Omega_h \cap \Omega_k = \emptyset$ *for* $h \neq k$. *Then*

$$\limsup_j \|p_j(\cdot, M_j\varphi) - p_j(\cdot, \Psi)\|_{L^p(\Omega;\mathbf{R}^n)}$$
$$\leq c(|\Omega| + \|\varphi\|_{L^p(\Omega;\mathbf{R}^n)} + \|\Psi\|_{L^p(\Omega;\mathbf{R}^n)})^{(\beta-\alpha-1)/(\beta-\alpha)}\|\varphi - \Psi\|^{1/(\beta-\alpha)}_{L^p(\Omega;\mathbf{R}^n)}, \tag{23.51}$$

where the constant c *depends only on* n, p, α, β, c_2, c_3, c_4.

Proof Let $\Omega_0 = \Omega \setminus \bigcup_{h=1}^m \Omega_h$ and $\eta_0 = 0$, so that

$$\Psi = \sum_{h=0}^m \eta_h \chi_{\Omega_h}.$$

Moreover, for every j, we denote by Ω_j the union of all closed cubes $\overline{Y}_j^{\mathbf{z}}$, such that $Y_j^{\mathbf{z}} \subset \Omega$, and for every $h = 0, 1, \ldots, m$, we set

$$I_j^h = \{\mathbf{z} \in I_j : Y_j^{\mathbf{z}} \subseteq \Omega_h\}, \qquad J_j^h = \{\mathbf{z} \in I_j : Y_j^{\mathbf{z}} \cap \Omega_h \neq \emptyset, Y_j^{\mathbf{z}} \setminus \Omega_h \neq \emptyset\},$$

where I_j is defined after (23.31). Furthermore, by E_j^h (F_j^h, respectively) we denote the union of all closed cubes $\overline{Y}_j^{\mathbf{z}}$ with $\mathbf{z} \in I_j^h$ ($\mathbf{z} \in J_j^h$, respectively). Since, for j sufficiently large, we have that $\Omega_h \subseteq \Omega_j$ for $h \neq 0$, by (23.31) and by the definition of Ψ it follows that

$$\|p_j(\cdot, M_j\varphi) - p_j(\cdot, \Psi)\|_{L^p(\Omega;\mathbf{R}^n)}^p = \int_{\Omega_j} |p_j(x, M_j\varphi) - p_j(x, \Psi)|^p \, dx$$

$$\leq \sum_{h=0}^m \int_{E_j^h} |p_j(x, M_j\varphi) - p_j(x, \eta_h)|^p \, dx + \sum_{h=0}^m \int_{F_j^h} |p_j(x, M_j\varphi) - p_j(x, \eta_h)|^p \, dx.$$

If we set

$$\xi_j^{\mathbf{z}} = \frac{1}{|Y_j^{\mathbf{z}}|} \int_{Y_j^{\mathbf{z}}} \varphi(x) \, dx,$$

by repeatedly applying Hölder's and Jensen's inequalities we obtain from Lemma 23.12 that

$$\|p_j(\cdot, M_j\varphi) - p_j(\cdot, \Psi)\|_{L^p(\Omega;\mathbf{R}^n)}^p$$

$$\leq c \sum_{h=0}^m \Big(\sum_{\mathbf{z} \in I_j^h} (1 + |\xi_j^{\mathbf{z}}|^p + |\eta_h|^p)^{(\beta-\alpha-1)/(\beta-\alpha)} |\xi_j^{\mathbf{z}} - \eta_h|^{p/(\beta-\alpha)} |Y_j^{\mathbf{z}}| \Big)$$

$$+ c \sum_{h=0}^m \Big(\sum_{\mathbf{z} \in J_j^h} (1 + |\xi_j^{\mathbf{z}}|^p + |\eta_h|^p)^{(\beta-\alpha-1)/(\beta-\alpha)} |\xi_j^{\mathbf{z}} - \eta_h|^{p/(\beta-\alpha)} |Y_j^{\mathbf{z}}| \Big)$$

$$\leq c \Big(\sum_{h=0}^m \Big(|E_j^h| + \int_{E_j^h} |\varphi|^p \, dx + |\eta_h|^p |E_j^h| \Big) \Big)^{(\beta-\alpha-1)/(\beta-\alpha)} \|\varphi - \Psi\|_{L^p(\Omega;\mathbf{R}^n)}^{p/(\beta-\alpha)}$$

$$+ c \Big(\sum_{h=0}^m \Big(|F_j^h| + \int_{F_j^h} |\varphi|^p \, dx + |\eta_h|^p |F_j^h| \Big) \Big)^{(\beta-\alpha-1)/(\beta-\alpha)} \|\varphi - \eta_h\|_{L^p(\Omega;\mathbf{R}^n)}^{p/(\beta-\alpha)}.$$

Therefore, we get

$$\|p_j(\cdot, M_j\varphi) - p_j(\cdot, \Psi)\|_{L^p(\Omega;\mathbf{R}^n)}^p$$

$$\leq c(|\Omega| + \|\varphi\|^p_{L^p(\Omega;\mathbf{R}^n)} + \|\Psi\|^p_{L^p(\Omega;\mathbf{R}^n)})^{(\beta-\alpha-1)/(\beta-\alpha)} \|\varphi - \Psi\|^{p/(\beta-\alpha)}_{L^p(\Omega;\mathbf{R}^n)}$$

$$+ c \sum_{h=0}^{m} \left(|F_j^h| + \int_{F_j^h} |\varphi|^p \, dx + |\eta_h|^p |F_j^h| \right)^{(\beta-\alpha-1)/(\beta-\alpha)} \|\varphi - \eta_h\|^{p/(\beta-\alpha)}_{L^p(\Omega;\mathbf{R}^n)}.$$

$$(23.52)$$

Since $|\partial\Omega_h| = 0$ for $h \neq 0$, we have that $|F_j^h|$ tends to 0 as $j \to +\infty$ for every $h = 0, 1, \ldots, m$; therefore (23.51) is an easy consequence of (23.52). $\qquad\square$

23.3   Statement and proof of the correctors result

Now we are in a position to state the main result of the chapter.

Theorem 23.14 *Let $g \in L^{p'}(\Omega)$, let u_j be the solutions to problem (23.7), and let u be the solution to problem (23.10). Then*

$$Du_j = p_j(\cdot, M_j Du) + r_j, \qquad (23.53)$$

where p_j is defined by (23.35) and (r_j) converges strongly to 0 in $L^p(\Omega; \mathbf{R}^n)$.

Note that the previous theorem includes the homogenization result stated in Theorem 23.2 as the following proposition shows.

Proposition 23.15 *Let $p_j : \mathbf{R}^n \times \mathbf{R}^n \to \mathbf{R}^n$ be the function defined by (23.35). Then $p_j(\cdot, M_j\varphi) \rightharpoonup \varphi$ weakly in $L^p(\Omega; \mathbf{R}^n)$ for every $\varphi \in L^p(\Omega; \mathbf{R}^n)$.*

Proof of Proposition 23.15 By (23.36) we have that

$$p_j(\cdot, \Psi) \rightharpoonup \Psi \quad \text{weakly in } L^p(\Omega; \mathbf{R}^n) \qquad (23.54)$$

for every step function Ψ on Ω. Given $\varphi \in L^p(\Omega; \mathbf{R}^n)$, for every $\delta > 0$ there exists a piecewise constant measurable function Ψ satisfying the assumptions of Lemma 23.13 and such that

$$\|\varphi - \Psi\|_{L^p(\Omega;\mathbf{R}^n)} \leq \delta.$$

Now, since $p_j(\cdot, M_j\varphi) - \varphi = (p_j(\cdot, M_j\varphi) - p_j(\cdot, \Psi)) + (p_j(\cdot, \Psi) - \Psi) + (\Psi - \varphi)$, the conclusion follows immediately from Lemma 23.13, from (23.54) and from the arbitrariness of δ. $\qquad\square$

Proof of Theorem 23.14 By Lemma 23.9 and (23.5), we have that

$$\|p_j(\cdot, M_j Du) - Du_j\|_{L^p(\Omega;\mathbf{R}^n)}$$

$$\leq c\left(\int_\Omega \langle a_j(x, p_j(x, M_j Du)) - a_j(x, Du_j), p_j(x, M_j Du) - Du_j \rangle \, dx \right)^{1/\beta}$$

$$\times (|\Omega|^{1/p} + \|p_j(\cdot, M_j Du)\|_{L^p(\Omega;\mathbf{R}^n)} + \|Du_j\|_{L^p(\Omega;\mathbf{R}^n)})^{(\beta-p)/\beta}, \quad (23.55)$$

where the constant c does not depend on j. To estimate the right-hand side of (23.55) we use the following notation:

$$\xi_j^{\mathbf{z}} = \frac{1}{|Y_j^{\mathbf{z}}|} \int_{Y_j^{\mathbf{z}}} Du\, dx; \qquad J_j = \{\mathbf{z} \in \mathbf{Z}^n : Y_j^{\mathbf{z}} \cap \Omega \neq \emptyset, Y_j^{\mathbf{z}} \setminus \Omega \neq \emptyset\}.$$

By applying Lemma 23.10, Corollary 23.11 and the inequality (23.33), we obtain

$$\|p_j(\cdot, M_j Du)\|_{L^p(\Omega;\mathbf{R}^n)}^p$$

$$= \sum_{\mathbf{z} \in I_j} \int_{Y_j^{\mathbf{z}}} |p_j(x, \xi_j^{\mathbf{z}})|^p\, dx + \int_{\Omega \setminus \Omega_j} |p_j(x, 0)|^p\, dx$$

$$\leq \sum_{\mathbf{z} \in I_j} c(1 + |\xi_j^{\mathbf{z}}|^p)|Y_j^{\mathbf{z}}| + \Big(\sum_{\mathbf{z} \in J_j} \|p_j(\cdot, 0)\|_{L^{p+\eta}(Y_j^{\mathbf{z}};\mathbf{R}^n)}^{p+\eta}\Big)^{p/(p+\eta)} |\Omega \setminus \Omega_j|^{\eta/(p+\eta)}$$

$$\leq c|\Omega| + c\|M_j Du\|_{L^p(\Omega;\mathbf{R}^n)}^p + c\Big(\sum_{\mathbf{z} \in J_j} |Y_j^{\mathbf{z}}|\Big)^{p/(p+\eta)} |\Omega \setminus \Omega_j|^{\eta/(p+\eta)}$$

$$\leq c|\Omega| + c\|Du\|_{L^p(\Omega;\mathbf{R}^n)}^p + c\Big(\sum_{\mathbf{z} \in J_j} |Y_j^{\mathbf{z}}|\Big)^{p/(p+\eta)} |\Omega \setminus \Omega_j|^{\eta/(p+\eta)}, \qquad (23.56)$$

where Ω_j denotes the union of closed cubes $\overline{Y}_j^{\mathbf{z}}$ with $Y_j^{\mathbf{z}} \subseteq \Omega$. Since $\sum_{\mathbf{z} \in J_j} |Y_j^{\mathbf{z}}|$ tends to $|\partial\Omega|$ and $|\Omega \setminus \Omega_j|$ tends to 0, by (23.56) it follows that $(p_j(\cdot, M_j Du))$ is bounded in $L^p(\Omega;\mathbf{R}^n)$. Since by Theorem 23.2 the sequence (u_j) converges weakly to u in $W^{1,p}(\Omega)$, the sequence (Du_j) is bounded in $L^p(\Omega;\mathbf{R}^n)$. Therefore, by (23.55), Theorem 23.14 is completely proved if we show that

$$\lim_j \int_\Omega \langle a_j(x, p_j(x, M_j Du)) - a_j(x, Du_j), p_j(x, M_j Du) - Du_j \rangle\, dx = 0. \quad (23.57)$$

This will be done in four steps.

Step 1: $\lim_j \int_\Omega \langle a_j(x, p_j(x, M_j Du)), p_j(x, M_j Du) \rangle\, dx = \int_\Omega \langle b(Du), Du \rangle\, dx.$
We write

$$\int_\Omega \langle a_j(x, p_j(x, M_j Du)), p_j(x, M_j Du) \rangle\, dx$$

$$= \sum_{\mathbf{z} \in I_j} \int_{Y_j^{\mathbf{z}}} \langle f_\xi\Big(\frac{x}{\varepsilon_j}, p\Big(\frac{x}{\varepsilon_j}, \xi_j^{\mathbf{z}}\Big)\Big), p\Big(\frac{x}{\varepsilon_j}, \xi_j^{\mathbf{z}}\Big) \rangle\, dx + \int_{\Omega \setminus \Omega_j} \langle a_j(x, p_j(x, 0)), p_j(x, 0) \rangle\, dx$$

$$= \varepsilon_j^n \sum_{\mathbf{z} \in I_j} \int_Y \langle f_\xi(x, p(x, \xi_j^{\mathbf{z}})), p(x, \xi_j^{\mathbf{z}}) \rangle\, dx + \int_{\Omega \setminus \Omega_j} \langle a_j(x, p_j(x, 0)), p_j(x, 0) \rangle\, dx$$

$$= \sum_{\mathbf{z} \in I_j} \int_\Omega \chi_{Y_j^{\mathbf{z}}}(x) \langle b(\xi_j^{\mathbf{z}}), \xi_j^{\mathbf{z}} \rangle\, dx + \int_{\Omega \setminus \Omega_j} \langle a_j(x, p_j(x, 0)), p_j(x, 0) \rangle\, dx, \qquad (23.58)$$

where in the last equality we use (23.38). Since, by (23.19), the function b is continuous and satisfies suitable upper bounds, it turns out that the map $\varphi \mapsto$

$b(\varphi)$ is continuous from $L^p(\Omega; \mathbf{R}^n)$ into $L^{p'}(\Omega; \mathbf{R}^n)$. By using (23.32), this yields that

$$b(M_j Du) \to b(Du) \tag{23.59}$$

strongly in $L^{p'}(\Omega; \mathbf{R}^n)$ and

$$\lim_j \sum_{\mathbf{z} \in I_j} \int_\Omega \chi_{Y_j^{\mathbf{z}}}(x) \langle b(\xi_j^{\mathbf{z}}), \xi_j^{\mathbf{z}} \rangle \, dx = \lim_j \int_\Omega \langle b(M_j Du), M_j Du \rangle \, dx$$

$$= \int_\Omega \langle b(Du), Du \rangle \, dx. \tag{23.60}$$

On the other hand, by the continuity assumption (23.4), we have

$$\left| \int_{\Omega \setminus \Omega_j} \langle a_j(x, p_j(x, 0)), p_j(x, 0) \rangle \, dx \right|$$

$$\leq c_3 \int_{\Omega \setminus \Omega_j} (1 + |p_j(x, 0)|)^p \, dx + \left| \int_{\Omega \setminus \Omega_j} \langle a_j(x, 0), p_j(x, 0) \rangle \, dx \right|$$

$$\leq c |\Omega \setminus \Omega_j| + c \int_{\Omega \setminus \Omega_j} |p_j(x, 0)|^p \, dx + c \Big(\sum_{\mathbf{z} \in J_j} |Y_j^{\mathbf{z}}| \Big)^{1/p'} \Big(\int_{\Omega \setminus \Omega_j} |p_j(x, 0)|^p \, dx \Big)^{1/p}$$

By arguing as in (23.56) it follows that

$$\lim_j \int_{\Omega \setminus \Omega_j} \langle a_j(x, p_j(x, 0)), p_j(x, 0) \rangle \, dx = 0,$$

which together with (23.60) and (23.58) proves Step 1.

Step 2: $\lim_j \int_\Omega \langle a_j(x, p_j(x, M_j Du)), Du_j \rangle \, dx = \int_\Omega \langle b(Du), Du \rangle \, dx.$

Let $\delta > 0$. Since $Du \in L^p(\Omega; \mathbf{R}^n)$, there exists a piecewise constant measurable function $\Psi = \sum_{h=1}^m \eta_h \chi_{\Omega_h}$ satisfying the assumptions of Lemma 23.13 such that

$$\|Du - \Psi\|_{L^p(\Omega; \mathbf{R}^n)} \leq \delta. \tag{23.61}$$

We write

$$\int_\Omega \langle a_j(x, p_j(x, M_j Du)), Du_j \rangle \, dx = \int_\Omega \langle a_j(x, p_j(x, \Psi)), Du_j \rangle \, dx$$

$$+ \int_\Omega \langle a_j(x, p_j(x, M_j Du)) - a_j(x, p_j(x, \Psi)), Du_j \rangle \, dx. \tag{23.62}$$

We have

$$\int_\Omega \langle a_j(x, p_j(x, \Psi)), Du_j \rangle \, dx = \sum_{h=0}^m \int_{\Omega_h} \langle a_j(x, p_j(x, \eta_h)), Du_j \rangle \, dx, \tag{23.63}$$

where η_0 and $\Omega_0 = \Omega \setminus \bigcup_{h=1}^m \Omega_h$. Since by Corollary 23.11 the sequence $(p_j(\cdot, \eta_h))$ is bounded in $L^{p+\eta}(\Omega; \mathbf{R}^n)$, by (23.3) and (23.4) it follows that the sequence $(a_j(\cdot, p_j(\cdot, \eta_h)))$ is bounded in $L^s(\Omega; \mathbf{R}^n)$ for some $s > p'$. As we already observed, the sequence (Du_j) is bounded in $L^p(\Omega; \mathbf{R}^n)$ by Theorem 23.2. This implies that there exists $\sigma > 1$ such that

$$\|\langle a_j(\cdot, p_j(\cdot, \eta_h)), Du_j \rangle\|_{L^\sigma(\Omega)} \leq c,$$

for all j. Then, up to a subsequence, the scalar product $\langle a_j(\cdot, p_j(\cdot, \eta_h)), Du_j(\cdot) \rangle$ converges weakly to a function $g_h \in L^\sigma(\Omega)$. Since $(a_j(\cdot, p_j(\cdot, \eta_h)))$ converges to $b(\eta_h)$ weakly in $L^{p'}(\Omega; \mathbf{R}^n)$ by (23.37) and $\mathrm{div}\,(a_j(x, p_j(x, \eta_h))) = 0$ by (23.39), the compensated compactness Lemma 23.4 implies that $\langle a_j(\cdot, p_j(\cdot, \eta_h)), Du_j(\cdot) \rangle$ converges to $\langle b(\eta_h), Du \rangle$ in the sense of distributions. Therefore, we may conclude that $g_h = \langle b(\eta_h), Du \rangle$, and hence

$$\lim_j \sum_{h=0}^m \int_{\Omega_h} \langle a_j(x, p_j(x, \eta_h)), Du_j \rangle \, dx = \sum_{h=0}^m \int_{\Omega_h} \langle b(\eta_h), Du \rangle \, dx.$$

Thus, by (23.63) we get

$$\lim_j \int_\Omega \langle a_j(x, p_j(x, \Psi)), Du_j \rangle \, dx = \int_\Omega \langle b(\Psi), Du \rangle \, dx. \tag{23.64}$$

We now estimate the second integral on the right-hand side of (23.62). By (23.4) and by Hölder's inequality (applied with $r_1 = p/(p-1-\alpha)$, $r_2 = p/\alpha$, and $r_3 = p$) we obtain

$$\left| \int_\Omega \langle a_j(x, p_j(x, M_j Du)) - a_j(x, p_j(x, \Psi)), Du_j \rangle \, dx \right|$$

$$\leq c \int_\Omega (1 + |p_j(x, M_j Du)|^p + |p_j(x, \Psi)|^p)^{(p-1-\alpha)/p}$$

$$\times |p_j(x, M_j Du) - p_j(x, \Psi)|^\alpha |Du_j| \, dx$$

$$\leq c \left(\int_\Omega (1 + |p_j(x, M_j Du)|^p + |p_j(x, \Psi)|^p) \, dx \right)^{(p-1-\alpha)/p}$$

$$\times \left(\int_\Omega |Du_j|^p \, dx \right)^{1/p} \left(\int_\Omega |p_j(x, M_j Du) - p_j(x, \Psi)|^p \, dx \right)^{\alpha/p} \tag{23.65}$$

Since the sequences $(p_j(\cdot, M_j Du))$ and $(p_j(\cdot, \Psi))$ are bounded in $L^p(\Omega; \mathbf{R}^n)$ (see (23.56) and Lemma 23.10), and the same holds for (Du_j) (by Theorem 23.2), taking (23.61) into account, the last inequality and Lemma 23.13 imply that

$$\limsup_j \left| \int_\Omega \langle a_j(x, p_j(x, M_j Du)) - a_j(x, p_j(x, \Psi)), Du_j \rangle \, dx \right| \leq c \delta^{\alpha/\beta - \alpha}, \tag{23.66}$$

where c does not depend on δ. Finally, by using (23.19) we obtain

$$\left| \int_\Omega \langle b(Du) - b(\Psi), Du \rangle \, dx \right| \le c \int_\Omega (1 + |Du|^p + |\Psi|^p)^{(p-1-\gamma)/p} |Du - \Psi|^\gamma |Du| \, dx$$

with $\gamma = \alpha/(\beta - \alpha)$. By applying Hölder's inequality (with $r_1 = p/(p-1-\gamma)$, $r_2 = p/\gamma$ and $r_3 = p$) and (23.61), we get

$$\left| \int_\Omega \langle b(Du) - b(\Psi), Du \rangle \, dx \right|$$
$$\le c \left(\int_\Omega (1 + |u|^p + |\Psi|^p) \, dx \right)^{(p-1-\gamma)/p} \left(\int_\Omega |Du|^p \, dx \right)^{1/p} \delta^\gamma \le c\delta^\gamma, \quad (23.67)$$

where c does not depend on δ. Now, since δ is arbitrarily small, by taking (23.62), (23.64), (23.66) and (23.67) into account we conclude the proof of Step 2.

Step 3: $\lim_j \int_\Omega \langle a_j(x, Du_j), p_j(x, M_j Du) \rangle \, dx = \int_\Omega \langle b(Du), Du \rangle \, dx$.

As in the proof of Step 2, fix $\delta > 0$ and consider a piecewise constant measurable function $\Psi = \sum_{h=1}^m \eta_h \chi_{\Omega_h}$ satisfying (23.61) and the assumptions of Lemma 23.13. We write

$$\int_\Omega \langle a_j(x, Du_j), p_j(x, M_j Du) \rangle \, dx = \sum_{h=0}^m \int_{\Omega_h} \langle a_j(x, Du_j), p_j(x, \eta_h) \rangle \, dx$$
$$+ \int_\Omega \langle a_j(x, Du_j), p_j(x, M_j Du) - p_j(x, \Psi) \rangle \, dx, \quad (23.68)$$

where $\eta_0 = 0$ and $\Omega_0 = \Omega \setminus \bigcup_{h=1}^m \Omega_h$. Since $(a_j(\cdot, Du_j))$ converges weakly to $b(Du)$ in $L^{p'}(\Omega; \mathbf{R}^n)$ (by Theorem 23.2) and $(p_j(\cdot, \eta_h))$ converges weakly to η_h in $L^p(\Omega; \mathbf{R}^n)$ (by (23.36)), by using (23.7) and the boundedness of $(p_j(\cdot, \eta_h))$ in $L^{p+\eta}(\Omega; \mathbf{R}^n)$ (see Corollary 23.11) we get as in Step 2

$$\lim_j \sum_{h=0}^m \int_{\Omega_h} \langle a_j(x, Du_j), p_j(x, \eta_h) \rangle \, dx = \sum_{h=0}^m \int_{\Omega_h} \langle b(Du), \eta_h \rangle \, dx$$
$$= \int_\Omega \langle b(Du), \Psi \rangle \, dx. \quad (23.69)$$

Moreover,

$$\left| \int_\Omega \langle a_j(x, Du_j), p_j(x, M_j Du) - p_j(x, \Psi) \rangle \, dx \right|$$
$$\le \left(\int_\Omega |a_j(x, Du_j)|^{p'} \, dx \right)^{1/p'} \left(\int_\Omega |p_j(x, M_j Du) - p_j(x, \Psi)|^p \, dx \right)^{1/p}$$

Thus, Lemma 23.13 yields

$$\limsup_j \left| \int_\Omega \langle a_j(x, Du_j), p_j(x, M_j Du) - p_j(x, \Psi) \rangle \, dx \right| \le c\delta^{1/\beta - \alpha}. \quad (23.70)$$

Hence, proceeding as in Step 2, from (23.69) and (23.61), we obtain that

$$\limsup_j \left| \int_\Omega \langle a_j(x, Du_j), p_j(x, M_j Du) \rangle \, dx - \int_\Omega \langle b(Du), Du \rangle \, dx \right|$$

$$\leq c\delta^{1/\beta - \alpha} + \lim_j \left| \int_\Omega \langle a_j(x, Du_j), p_j(x, \Psi) \rangle \, dx - \int_\Omega \langle b(Du), \Psi \rangle \, dx \right|$$

$$+ c\delta \|b(Du)\|_{L^{p'}(\Omega; \mathbf{R}^n)}, \tag{23.71}$$

where c does not depend on δ. Now, since δ is arbitrarily small and (23.69) holds, the proof of Step 3 is complete.

Step 4: $\lim_j \int_\Omega \langle a_j(x, Du_j), Du_j \rangle \, dx = \int_\Omega \langle b(Du), Du \rangle \, dx.$
Since

$$\int_\Omega \langle a_j(x, Du_j), Du_j \rangle \, dx = \langle -\mathrm{div}\,(a_j(x, Du_j)), u_j \rangle = \langle g, u_j \rangle$$

$$\int_\Omega \langle b(Du), Du \rangle \, dx = \langle -\mathrm{div}\,(b(Du)), u \rangle = \langle g, u \rangle$$

and (u_j) converges weakly to u in $W^{1,p}(\Omega)$ (by Theorem 23.2), we immediately obtain Step 4. In the previous equation, by a slight abuse of notation, we denote by $\langle \cdot, \cdot \rangle$ the scalar product on $\mathbf{R}^n$ as well as the duality pairing between $W^{-1,p'}(\Omega)$ and $W_0^{1,p}(\Omega)$.

Eventually, by Steps 1–4, we obtain (23.57) and conclude the proof of Theorem 23.14. $\qquad\square$

Note that Theorem 23.2 is still true in the case of non-homogeneous Dirichlet boundary conditions, with the same proof. In this case, the correctors result holds as follows.

Theorem 23.16 *Let* $g \in L^{p'}(\Omega; \mathbf{R}^n)$, *and let* $\varphi \in W^{1,p}(\Omega)$. *Let* ε_j, f, a_j *and* b *be as in Theorem 23.2. If* u_j *is the solution to the Dirichlet boundary value problem*

$$\begin{cases} -\mathrm{div}\,(a_j(x, Du_j)) = g & on\ \Omega \\ \\ u_j - \varphi \in W_0^{1,p}(\Omega), \end{cases} \tag{23.72}$$

and u *is the solution to the homogenized problem*

$$\begin{cases} -\mathrm{div}\,(b(Du)) = g & on\ \Omega \\ \\ u - \varphi \in W_0^{1,p}(\Omega), \end{cases} \tag{23.73}$$

then (23.53) *still holds.*

Proof As in the proof of Theorem 23.14 we will show that

$$\lim_j \int_\Omega \langle a_j(x, p_j(x, M_j Du)) - a_j(x, Du_j), p_j(x, M_j Du) - Du_j \rangle \, dx = 0. \quad (23.74)$$

We immediately obtain that Steps 1, 2 and 3 in Theorem 23.14 still hold. Therefore, to conclude the proof of (23.74) it remains to verify that

$$\lim_j \int_\Omega \langle a_j(x, Du_j), Du_j \rangle \, dx = \int_\Omega \langle b(Du), Du \rangle \, dx. \quad (23.75)$$

Since

$$\int_\Omega \langle a_j(x, Du_j), Du_j \rangle \, dx = \int_\Omega \langle a_j(x, Du_j), D(u_j - \varphi) \rangle \, dx + \int_\Omega \langle a_j(x, Du_j), D\varphi \rangle \, dx,$$

where $u_j - \varphi \in W_0^{1,p}(\Omega)$, (23.75) can be obtained arguing as in Step 4 of the proof of Theorem 23.14. $\qquad\square$

23.4 Correctors in the quasiperiodic case

A correctors result analogous to Theorem 23.14 can be proven in the general setting of functions $f(x, \xi)$ almost periodic in x, uniformly with respect to ξ. In this section, we state the correctors result in the quasiperiodic case. The proof follows the line of the previous sections, and is omitted. We refer to Braides (1991) for more details.

In the rest of the section we assume that f satisfies (i), (ii) and (iv) as in Section 23.1 and that for every $\xi \in \mathbf{R}^n$ the function $f(\cdot, \xi)$ is quasiperiodic on $\mathbf{R}^n$, uniformly with respect to ξ; that is, $f(x, \xi) = F(x, \ldots, x, \xi)$, where $F : \mathbf{R}^{nN} \times \mathbf{R}^n \to [0, +\infty)$ is a continuous function, Y-periodic in the first nN variables, with the periodicity cell Y independent of ξ. Clearly, $f_\xi(\cdot, \xi)$ is also quasiperiodic, uniformly with respect to ξ.

In trying to follow the periodic case, one attempts to build up the correctors by a patchwork procedure. Consider the cubes $(0, t)^n$ and the corresponding solutions $v_t = v_t^\xi$ to the minimum problem in (15.4), or equivalently to

$$\begin{cases} -\mathrm{div}\, (f_\xi(x, \xi + Dv_t(x))) = 0 & \text{on } (0, t)^n \\[2mm] v_t \in W_0^{1,p}((0, t)^n), \end{cases} \quad (23.76)$$

and define $p^t : (0, t)^n \times \mathbf{R}^n \to \mathbf{R}^n$ as

$$p^t(x, \xi) = \xi + Dv_t(x). \quad (23.77)$$

To extend p^t to the whole $\mathbf{R}^n$, we use the fact that f_ξ behaves 'almost' like a periodic function when restricted to a cylinder $\{|\xi| \leq R\}$. In fact, using Kronecker's lemma (see Appendix A) it can be easily seen that for all $R > 0$ and for all $\eta > 0$ there exists a relatively dense set of η-almost periods τ such that

$$|f_\xi(x+\tau,\xi) - f_\xi(x,\xi)| < \eta \tag{23.78}$$

for every $x \in \mathbf{R}^n$, and every $\xi \in \mathbf{R}^n$ with $|\xi| \leq R$. Then, we can fill the space with cubes of side-length t, which are translations of $(0,t)^n$ by an η-almost period of the restriction of f_ξ to $\{|\xi| \leq R\}$. Fix R, η and t. For each $\mathbf{z} \in \mathbf{Z}^n$ choose

$$\begin{cases} \tau_\mathbf{z} \in (t + L_\eta^R)\mathbf{z} + [0, L_\eta^R)^n \\[2mm] \tau_\mathbf{z} \quad \eta\text{-almost period of } f_\xi \text{ in } \{|\xi| \leq R\}, \end{cases} \tag{23.79}$$

and define

$$Q_\mathbf{z}^{R,\eta,t} = \tau_\mathbf{z} + (0,t)^n.$$

The extension of p^t is the function $p^{R,\eta,t} : \mathbf{R}^n \times \mathbf{R}^n \to \mathbf{R}^n$ defined by

$$p^{R,\eta,t}(x,\xi) = \begin{cases} \xi + Dv_t(x - \tau_\mathbf{z}) & \text{if } x \in Q_\mathbf{z}^{R,\eta,t} \\[2mm] \xi & \text{otherwise,} \end{cases} \tag{23.80}$$

where v_t is the solution to (23.76). Since the region outside the cubes $Q_\mathbf{z}^{R,\eta,t}$ has relative measure proportional to L_η^R/t, it makes no contribution as $t \to +\infty$. For every $\varepsilon_j > 0$, set

$$p_j^{R,\eta,t}(x,\xi) = p^{R,\eta,t}\left(\frac{x}{\varepsilon_j}, \xi\right) \tag{23.81}$$

Note that

$$\fint p^{R,\eta,t}(x,\xi)\, dx = \xi,$$

$$p_j^{R,\eta,t}(\cdot,\xi) \rightharpoonup \xi \quad \text{weakly in } \mathrm{L}_{\mathrm{loc}}^p(\mathbf{R}^n) \text{ as } j \to +\infty.$$

We can construct, using the same method as above, an approximation of the identity in $\mathrm{L}^p(\Omega; \mathbf{R}^n)$. For $\varepsilon_j > 0$, set

$$I_j^{R,\eta,t} = \{\mathbf{z} \in \mathbf{Z}^n : \varepsilon_j Q_\mathbf{z}^{R,\eta,t} \subset \Omega\}$$

$$J_j^{R,\eta,t} = \{\mathbf{z} \in \mathbf{Z}^n : \varepsilon_j Q_\mathbf{z}^{R,\eta,t} \cap \Omega \neq \emptyset, \varepsilon_j Q_\mathbf{z}^{R,\eta,t} \setminus \Omega \neq \emptyset\}$$

$$\Omega_j^{R,\eta,t} = \bigcup\{\overline{\varepsilon_j Q_\mathbf{z}^{R,\eta,t}} : \mathbf{z} \in I_j^{R,\eta,t}\}$$

$$F_j^{R,\eta,t} = \bigcup\{\varepsilon_j Q_\mathbf{z}^{R,\eta,t} \cap \Omega : \mathbf{z} \in J_j^{R,\eta,t}\}.$$

Note that $\lim_{t\to+\infty} \lim_j |\Omega \setminus \Omega_j^{R,\eta,t}| = 0$.

If $\varphi \in \mathrm{L}^p(\Omega; \mathbf{R}^n)$, we define $M_j^{R,\eta,t}\varphi : \Omega \to \mathbf{R}^n$ as follows:

$$(M_j^{R,\eta,t}\varphi)(x) = \sum_{\mathbf{z} \in I_j^{R,\eta,t}} \chi_{\varepsilon_j Q_\mathbf{z}^{R,\eta,t}}(x) \frac{1}{\varepsilon_j^n t^n} \int_{\varepsilon_j Q_\mathbf{z}^{R,\eta,t}} \varphi(y)\, dy.$$

It turns out that

$$(M_j^{R,\eta,t}\varphi)(x) = 0 \quad \text{if } \Omega \setminus \Omega_j^{R,\eta,t};$$

moreover, by Jensen's inequality

$$\|M_j^{R,\eta,t}\varphi\|_{L^p(\Omega;\mathbf{R}^n)} \le \|\varphi\|_{L^p(\Omega;\mathbf{R}^n)}, \tag{23.82}$$

and

$$\lim_{t\to+\infty} \limsup_j \|M_j^{R,\eta,t}\varphi - \varphi\|_{L^p(\Omega;\mathbf{R}^n)} = 0.$$

The last inequality is immediate when φ is continuous; in the general case we can proceed by approximation, using (23.82) and the linearity of $M_j^{R,\eta,t}$.

Finally, we can state the correctors result under quasiperiodicity assumptions on f.

Theorem 23.17 *Let f be as above and let f_{hom} be as in Theorem 14.5. Let (ε_j) be a sequence of positive real numbers converging to 0, and let $a_j(x,\xi) = f_\xi(x/\varepsilon_j,\xi)$ and $b(\xi) = (f_{\text{hom}})_\xi$. Fix $g \in L^{p'}(\Omega)$, and let u_j, $u \in W_0^{1,p}(\Omega)$ be the solutions to (23.7) and (23.10), respectively. Then $u_j \rightharpoonup u$ weakly in $W^{1,p}(\Omega)$, $a_j(x, Du_j) \rightharpoonup b(Du)$ weakly in $L^{p'}(\Omega;\mathbf{R}^n)$, and*

$$\lim_{R\to+\infty} \limsup_{\eta\to 0} \limsup_{t\to+\infty} \limsup_{j\to+\infty} \|p_j^{R,\eta,t}(\cdot, M_j^{R,\eta,t}Du) - Du_j\|_{L^p(\Omega;\mathbf{R}^n)} = 0.$$

Remark 23.18 (a) The weak convergence $u_j \rightharpoonup u$ follows from Theorem 17.10, Proposition 11.7 and Theorem 7.2.

(b) Using the asymptotic homogenization formula for f_{hom}, the function b can be expressed as

$$b(\xi) = \lim_{t\to+\infty} \frac{1}{t^n} \int_{(0,t)^n} f_\xi(x, \xi + Dv_t(x))\, dx, \tag{23.83}$$

where $v_t = v_t^\xi$ is the solution to (23.76).

23.5 Exercises

Exercise 23.1 Let p and p_j be the functions defined in (23.34) and (23.35), and let u_j be the solution to problem (23.7). Assume that

$$|p(x,\xi) - p(x,\xi')| \le \lambda(x)|\xi - \xi'|$$

for a.e. $x \in (0,1)^n$ and for every ξ, $\xi' \in \mathbf{R}^n$, where $\lambda \in L_{\text{loc}}^r(\mathbf{R}^n)$ is 1-periodic and $p' \le r \le +\infty$. Prove that $Du_j = p_j(x, Du) + r_j$, with $r_j \to 0$ strongly in $L^s(\Omega;\mathbf{R}^n)$, where $\frac{1}{s} = \frac{1}{p} + \frac{1}{r}$.

Hint: use Hölder's inequality to show that

$$\|p_j(\cdot, M_j\varphi) - p_j(\cdot, \varphi)\|_{L^s(\Omega;\mathbf{R}^n)} \le c\|\lambda\|_{L^r((0,1)^n)}\|M_j\varphi - \varphi\|_{L^p(\Omega;\mathbf{R}^n)}$$

for all $\varphi \in L^p(\Omega;\mathbf{R}^m)$ and $c = c(\Omega)$ is a constant depending on the geometry of Ω. Use the convergence properties of (M_j) and Theorem 23.14 to conclude.

24

HOMOGENIZATION OF MULTI-DIMENSIONAL STRUCTURES

In this chapter we describe some homogenization processes that may model the macroscopic properties of media whose microscopic behaviour takes lower-dimensional or multi-dimensional structures into account. We will show that the method for homogenization described in Chapter 14 can be adapted to the asymptotic analysis of these functionals. In addition to this study, we also have to face the problem of the choice of a proper description of the small-scale functionals themselves, whose natural domain is not the usual Sobolev-space setting. In Sections 24.1 and 24.2 we deal with energies defined on lower-dimensional structures whose homogenization gives rise to integral functionals defined on full-dimensional domains. We are going to describe two different and, to some extent, complementary approaches to this problem. We remark that while the limit analysis will be performed in detail, the description of the energies at length scale ε is only outlined, leaving some details to the reader. In Section 24.3 we consider energies which describe structures as rods, plates or thin films, where one or more dimensions are negligible with respect to the others. These lower-dimensional energies are obtained by a Γ-limit procedure from (normalized) energies defined on full-dimensional domains contracting to lower-dimensional ones. We treat the case when the full dimensional structures possess in addition a (periodic) cellular structure. For the sake of illustration, we only deal with the simplified case when the period and the 'thickness' of the structure are on a comparable scale.

24.1 A smooth approach

We start with an example which illustrates the types of functionals under consideration.

Example 24.1 (An elastic network) We consider an elastic 1-dimensional network in $\mathbf{R}^n$ that we can parameterize by the set

$$E = \{y \in \mathbf{R}^n : \exists i \in \{1, \ldots, n\} \text{ such that } \widehat{y}_i \in \mathbf{Z}^{n-1}\}, \qquad (24.1)$$

where $\widehat{x}_i = (x_1, \ldots, x_{i-1}, x_{i+1}, \ldots, x_n)$. If $u : E \cap \Omega \to \mathbf{R}^n$ represents the displacement of the portion of the network contained in the bounded open set Ω, which we suppose sufficiently smooth, then we can describe its elastic energy as depending on the displacement gradient. In this case, we locally identify E (except at 'knots', i.e. $\mathbf{Z}^n$) with an interval of $\mathbf{R}$, so that the displacement can be seen locally as a function, for example, in $C^1(\mathbf{R}; \mathbf{R}^n)$. Using this representation, after choosing a tangent $\tau(x)$ at every point of $E \setminus \mathbf{Z}^n$, we can define everywhere except on $\mathbf{Z}^n$ the tangential gradient of u along E, which we denote by $D_\tau^E u$. We can then write the elastic energy of the network's deformation as

$$F(u) = \int_{E \cap \Omega} f(x, D_\tau^E u(x)) \, d\mathcal{H}^1, \tag{24.2}$$

where $\mathcal{H}^1$ represents the 1-dimensional line measure and f is an appropriate elastic energy density.

We can slightly modify the setting of the problem and consider a different functional of the form

$$\widetilde{F}(u) = \int_\Omega \widetilde{f}(x, Du(x)) \, d\mu, \qquad u \in C^1(\Omega; \mathbf{R}^n), \tag{24.3}$$

where $\mu = \mathcal{H}^1 \, \llcorner \, E$ is the restriction of $\mathcal{H}^1$ to E, and the function $\widetilde{f}$ is such that

$$f(x, a) = \inf\{\widetilde{f}(x, a \otimes \tau(x) + B) : \ B\tau(x) = 0\}.$$

For a wide class of integrands $\widetilde{f}$, Bouchitté *et al.* (1997) have shown that the two formulations above are equivalent in the sense that minima for analogous problems for the two functionals are equal. In this case the equivalence is easily explained as minimum energy configurations in (24.3) will have zero normal derivative.

The example above justifies the choice of functionals of the form (24.3) as modelling some types of problems involving lower-dimensional structures. We will treat the problem of the homogenization of such functionals.

In the rest of the chapter μ stands for a non-zero σ-finite positive Borel measure on $\mathbf{R}^n$ which is 1-periodic, i.e.

$$\mu(B + e_i) = \mu(B)$$

for all Borel subsets B of $\mathbf{R}^n$ and for all $i = 1, \ldots, n$. We will assume the normalization

$$\mu([0, 1)^n) = 1. \tag{24.4}$$

For all $\varepsilon > 0$ we define the ε-periodic σ-finite positive Borel measure μ_ε by

$$\mu_\varepsilon(B) = \varepsilon^n \mu\left(\frac{1}{\varepsilon} B\right) \tag{24.5}$$

for all Borel sets B. Note that by (24.4) the family (μ_ε) converges locally weakly* in the sense of measures to the Lebesgue measure as $\varepsilon \to 0$.

In the sequel $f : \mathbf{R}^n \times \mathbf{M}^{m \times n} \to [0, +\infty)$ will be a fixed Borel function satisfying the growth condition of order $p \geq 1$ in (14.2), and 1-periodic in the first variable. For every bounded open set Ω, we define the functionals at length scale $\varepsilon > 0$ as

$$F_\varepsilon(u, \Omega) = \begin{cases} \int_\Omega f\left(\dfrac{x}{\varepsilon}, Du\right) d\mu_\varepsilon & \text{if } u \in C^\infty(\Omega; \mathbf{R}^m) \\ +\infty & \text{otherwise.} \end{cases} \qquad (24.6)$$

Example 24.2 As an example, we can consider (up to a normalization factor) the measure $\mu = \mathcal{H}^1 \llcorner E$ of Example 24.1. In this case we have

$$\mu_\varepsilon = \varepsilon^{n-1} \mathcal{H}^1 \llcorner (\varepsilon E);$$

note the scaling factor ε^{n-1}. On smooth functions the functional F_ε can be rewritten as

$$F_\varepsilon(u, \Omega) = \varepsilon^{n-1} \int_{\Omega \cap \varepsilon E} f\left(\frac{x}{\varepsilon}, Du\right) d\mathcal{H}^1.$$

Note that we cannot use L^p-topologies with p finite to describe the limit as $\varepsilon \to 0$ of the family of functionals F_ε, since in our general assumptions all the measures μ_ε might be singular with respect to the Lebesgue measure. We will then use the L^∞-convergence, noting that this is indeed essentially used on sequences of smooth functions, so that it is equivalent to the C^0-convergence.

Theorem 24.3 *Let $p > n$. For every $\varepsilon > 0$ let F_ε be defined on $\mathrm{L}^\infty(\Omega; \mathbf{R}^m)$ by (24.6). Then for every bounded open set Ω with Lipschitz boundary the Γ-limit*

$$F_{\mathrm{hom}}(u, \Omega) = \Gamma(\mathrm{L}^\infty)\text{-}\lim_{\varepsilon \to 0} F_\varepsilon(u, \Omega) \qquad (24.7)$$

exists for all $u \in \mathrm{W}^{1,p}(\Omega; \mathbf{R}^m)$, and it can be represented as

$$F_{\mathrm{hom}}(u, \Omega) = \int_\Omega f_{\mathrm{hom}}(Du)\, dx, \qquad (24.8)$$

where the homogenized integrand satisfies the asymptotic formula

$$f_{\mathrm{hom}}(A) = \lim_{T \to +\infty} \inf\left\{ \frac{1}{T^n} \int_{(0,T)^n} f(x, Du + A)\, d\mu : \; u \in C_0^\infty((0,T)^n; \mathbf{R}^m) \right\}. \qquad (24.9)$$

Furthermore, if f is convex then the cell-problem formula holds

$$f_{\mathrm{hom}}(A) = \inf\left\{ \int_{[0,1)^n} f(x, Du + A)\, d\mu : \; u \in C_\#^\infty((0,1)^n; \mathbf{R}^m) \right\} \qquad (24.10)$$

for all $A \in \mathbf{M}^{m \times n}$.

Proof The L^∞-fundamental estimate, as defined in Remark 11.3, can be proven as in Proposition 12.2 (see also Exercise 24.1), and the existence of the Γ-limit, up to subsequences, follows as in Proposition 12.3, as well as the measure properties for the limit. The integral representation is obtained directly from Proposition 9.2, once we note that the limit functional satisfies a growth condition from above (see Remark 24.4(a)) and is weakly lower semicontinuous on $W^{1,p}(\Omega; \mathbf{R}^m)$ by Rellich's Theorem. Finally, the proofs of the asymptotic homogenization formula and of the cell-problem formula are the same as in Proposition 14.4 and Theorem 14.7. $\qquad\square$

Remark 24.4 (a) From the homogenization formula (24.9) we easily deduce that

$$f_{\text{hom}}(A) \le \beta(1 + |A|^p). \tag{24.11}$$

In fact, it suffices to take $u = 0$ to get

$$f_{\text{hom}}(A) \le \int f(x, A)\, d\mu \le \beta(1 + |A|^p). \tag{24.12}$$

(b) In general f_{hom} is not coercive. As an example, take

$$\mu(U) = \sum_{i \in \mathbf{Z}^n} \lambda(i + U), \tag{24.13}$$

where λ is any probability measure with spt λ contained in $(0,1)^n$. In this case, consider any $v \in C_0^\infty((0,1)^n; \mathbf{R}^m)$ with $v = -Ax$ on spt λ, extended by periodicity to the whole $\mathbf{R}^n$. We can use as a test function in formula (24.9) the function $u \in C_0^\infty(\Omega; \mathbf{R}^m)$ defined by

$$u(x) = \begin{cases} v(x) & \text{if } x \in (0, [T])^n \\ \\ 0 & \text{otherwise.} \end{cases}$$

We then get

$$\begin{aligned} f_{\text{hom}}(A) &\le \int f(x, Du + A)\, d\mu \\ &= \int_{\text{spt }\lambda} f(x, Du + A)\, d\lambda = f(x, 0) \le c \end{aligned} \tag{24.14}$$

for all $A \in \mathbf{M}^{m \times n}$, so that actually f_{hom} is a constant.

(c) A special case is $\mu(B) = |B \cap E|$, where E is a periodic set. If E is connected and has Lipschitz boundary, the results of Chapter 19 show that f_{hom} is coercive, so that Theorem 24.3 completely describes the homogenization process. Further results can be found in Zhikov (1995) when E is *p-connected*, that is, u is constant on E if $u \in W^{1,p}_{\text{loc}}(E)$ and $Du = 0$ on E, in which case Theorem 24.3 also holds if $p \le n$. Note that in this case the extension results in Chapter 19 are in general false.

Example 24.5 We choose $n = 2$, $m = 1$ and $\mu = \frac{1}{2}\mathcal{H}^1 \llcorner E$, with E as in Example 24.1. If $f(x, A) = |A|^p$ then formula (24.10) gives

$$f_{\text{hom}}(A) = \frac{1}{2}\inf\left\{\int_{[0,1)^2 \cap E} |Du + A|^p\, d\mathcal{H}^1 : u \in C^{\infty}_{\#}((0,1)^2)\right\}$$

$$= \frac{1}{2}\sum_{i=1}^{2}\inf\left\{\int_0^1 |v'|^p\, dt : v(0) = 0,\ v(1) = a_i\right\}$$

$$= \frac{1}{2}(|a_1|^p + |a_2|^p), \tag{24.15}$$

for $A = (a_1, a_2)$.

24.2   A measure Sobolev-space approach

We begin with an example to pinpoint the differences with the previous approach. In the rest of the chapter, with a slight abuse of notation, Du stands for the *distributional derivative* of u. For all functions considered in this section, Du is a *vector measure*, i.e. there exist measures μ_{ij} such that

$$\int_\Omega u^j \frac{\partial\phi}{\partial x_i}\, dx = -\int_\Omega \phi\, d\mu_{ij} \quad \text{for all } \phi \in C_0^\infty(\Omega),$$

which define the entries of the (vector or matrix-valued) measure Du. For Du we use the measure-theoretic notation; in particular $|Du|$ denotes the *total variation* of Du, $Du \ll \lambda$ means that Du is *absolutely continuous* with respect to λ, i.e. $Du(B) = 0$ if $\lambda(B) = 0$, and in this case $dDu/d\lambda$ denotes the *Radon Nikodym derivative* of Du with respect to λ, characterized by the equality

$$Du(B) = \int_B \frac{dDu}{d\lambda}d\lambda \quad \text{for all Borel sets } B.$$

If Du is absolutely continuous with respect to the Lebesgue measure (i.e. u is $W^{1,1}$), then we may use the same symbol to denote the weak gradient as well, since no confusion will arise.

Example 24.6 (Perfectly rigid bars connected by elastic springs) Let the interval (a, b) parameterize an array of N perfectly rigid bars, the i-th of which corresponds to the interval (x_{i-1}, x_i) ($a = x_0 < x_1 < \cdots < x_N = b$). If we suppose that each bar is connected by an elastic spring to the subsequent one, then the elastic energy of the system is given by

$$\sum_{i=1}^{N-1} \Psi_i(u_{i+1} - u_i), \tag{24.16}$$

where u_i is the displacement of the i-th bar, and Ψ_i represents the elastic energy due to the i-th spring.

We can regard the displacement as a piecewise constant function $u : (a,b) \to \mathbf{R}^3$, with constant value $u(t) = u_i$ on (x_{i-1}, x_i). The distributional derivative Du is then the measure

$$Du = \sum_{i=1}^{N-1} (u_{i+1} - u_i)\delta_{x_i}, \tag{24.17}$$

where δ_x is the *Dirac mass* at x. We can interpret the measure

$$\mu = \sum_{i=1}^{N-1} \delta_{x_i} \tag{24.18}$$

as describing the properties of the system of bars, since it highlights the location of the endpoints of each bar, while the Radon Nikodym derivative of Du with respect to μ is given by

$$\frac{dDu}{d\mu}(x_i) = u_{i+1} - u_i, \tag{24.19}$$

$i = 1, \dots, N-1$, and gives the elongation of each spring. Hence, we can rewrite the energy in (24.16) as the integral with respect to μ

$$\int_{(a,b)} f\left(x, \frac{dDu}{d\mu}\right) d\mu, \tag{24.20}$$

where $f(x,z) = \Psi_i(z)$ if $x = x_i$. Furthermore, if we also include in this formulation the requirement that

$$Du \ll \mu, \tag{24.21}$$

we obtain precisely that $Du = 0$, and hence that u is constant, on each (x_{i-1}, x_i), i.e. u is an admissible displacement for the system of bars. In conclusion, (24.20) and (24.21) provide a complete description of our model that can be framed in an abstract setting.

The following notion of Sobolev space with respect to a measure was introduced by Ambrosio *et al.* (1996).

Definition 24.7 *Let λ be a finite Borel measure on the open set $\Omega \subset \mathbf{R}^n$, and let $1 \leq p \leq +\infty$. The Sobolev space with respect to λ, $\mathrm{W}_\lambda^{1,p}(\Omega; \mathbf{R}^m)$, is defined as*

$$\mathrm{W}_\lambda^{1,p}(\Omega; \mathbf{R}^m) = \left\{ u \in \mathrm{L}^p(\Omega; \mathbf{R}^m) : \ Du \ll \lambda, \ \frac{dDu}{d\lambda} \in \mathrm{L}_\lambda^p(\Omega; \mathbf{M}^{m \times n}) \right\}, \tag{24.22}$$

where $\mathrm{L}_\lambda^p(\Omega; \mathbf{R}^N)$ stands for the usual Lebesgue space of p-summable $\mathbf{R}^N$-valued functions with respect to λ.

Remark 24.8 By definition, functions in $\mathrm{W}_\lambda^{1,p}(\Omega; \mathbf{R}^m)$ are functions of bounded variation, i.e. their distributional derivative is a measure. We will not treat in depth the properties of the space $BV(\Omega; \mathbf{R}^m)$ of $\mathbf{R}^m$-valued functions of bounded variation, for which we refer to Evans and Gariepy (1992). From these properties the following two facts can be easily deduced, and are used in the sequel.

(a) $W^{1,p}_\lambda(\Omega; \mathbf{R}^m)$ is imbedded in $L^{n/(n-1)}(\Omega; \mathbf{R}^m)$.

(b) Let $\tilde{u}$ be the *precise representative* of u, whose components are defined by

$$\tilde{u}_i(x) = \limsup_{\rho \to 0^+} \fint_{B(x,\rho)} u_i(y)\, dy. \tag{24.23}$$

If $u \in W^{1,p}_\lambda(\Omega; \mathbf{R}^m)$ and $v \in W^{1,\infty}_\lambda(\Omega)$, then $uv \in W^{1,p}_\lambda(\Omega; \mathbf{R}^m)$, and

$$\frac{dD(uv)}{d\lambda} = \tilde{v}\frac{dDu}{d\lambda} + \tilde{u} \otimes \frac{dDv}{d\lambda}. \tag{24.24}$$

Note that in (24.24) it is necessary to consider the precise representatives, since the measure λ may also take into account sets of zero Lebesgue measure.

From the properties of functions of bounded variation, it is easy to deduce that if $u \in W^{1,p}_\lambda(\Omega; \mathbf{R}^m)$, then $Du(B) = 0$ when B is a set of dimension lower than $n - 1$. Hence, $W^{1,p}_\lambda(\Omega; \mathbf{R}^m) = W^{1,p}_{\lambda'}(\Omega; \mathbf{R}^m)$ if $\lambda - \lambda'$ is concentrated on a set of dimension lower than $n - 1$, e.g. points in $\mathbf{R}^3$.

Properties of lower semicontinuity and relaxation for functionals defined on Sobolev spaces with respect to a measure have been studied by Ambrosio *et al.* (1996). We will deal with their homogenization.

Let μ, μ_ε and f be as in Section 24.1. We define the functionals at scale $\varepsilon > 0$ as

$$F_\varepsilon(u, \Omega) = \begin{cases} \displaystyle\int_\Omega f\left(\frac{x}{\varepsilon}, \frac{dDu}{d\mu_\varepsilon}\right) d\mu_\varepsilon & \text{if } u \in W^{1,p}_{\mu_\varepsilon}(\Omega; \mathbf{R}^m) \\ +\infty & \text{otherwise.} \end{cases} \tag{24.25}$$

Example 24.9 (a) (Perfectly rigid bodies connected by springs) We take

$$E = \{y \in \mathbf{R}^n : \exists i \in \{1, \ldots, n\} \text{ such that } y_i \in \mathbf{Z}\},$$

that is, the union of all the boundaries of cubes $Q_i = i + (0,1)^n$ with $i \in \mathbf{Z}^n$. E is an $(n-1)$-dimensional set in $\mathbf{R}^n$. We take

$$\mu = \frac{1}{n}\mathcal{H}^{n-1} \mathbin{\llcorner} E,$$

where $\mathcal{H}^{n-1}$ stands for the $(n-1)$-dimensional surface measure. For every $\varepsilon > 0$ we have

$$\mu_\varepsilon = \frac{1}{n}\varepsilon\mathcal{H}^{n-1} \mathbin{\llcorner} \varepsilon E.$$

In this case $W^{1,p}_{\mu_\varepsilon}$ consists of all functions which are constant on every connected component of each $\varepsilon Q_i \cap \Omega$, since we must have $Du = 0$ on every εQ_i. In the case where u is constant on each εQ_i, we have

$$\frac{dDu}{d\mu_\varepsilon} = \frac{n}{\varepsilon}\frac{dDu}{d\mathcal{H}^{n-1}} = \frac{n}{\varepsilon}(u_i - u_j) \otimes (i - j) \quad \text{on } \partial(\varepsilon Q_i) \cap \partial(\varepsilon Q_j) \cap \Omega,$$

where u_i is the value of u on εQ_i. In this case the functionals F_ε take the form

$$\varepsilon \int_{\Omega \cap \varepsilon E} g\left(\frac{x}{\varepsilon}, \frac{1}{\varepsilon}\frac{dDu}{d\mathcal{H}^{n-1}}\right) d\mathcal{H}^{n-1},$$

up to a normalization factor. Note that if Ω is bounded then $W^{1,p}_{\mu_\varepsilon}(\Omega; \mathbf{R}^m) = W^{1,\infty}_{\mu_\varepsilon}(\Omega; \mathbf{R}^m)$ for all p if the number of connected components of each $\Omega \cap \varepsilon Q_i$ is finite.

(b) (Elastic media connected by springs) Let E be as above and let

$$\mu(B) = \frac{1}{n+1}\Big(|B| + \mathcal{H}^{n-1}(E \cap B)\Big)$$

$$\mu_\varepsilon(B) = \frac{1}{n+1}\Big(|B| + \varepsilon\mathcal{H}^{n-1}((\varepsilon E) \cap B)\Big).$$

In this case the functions in $W^{1,p}_{\mu_\varepsilon}(\Omega; \mathbf{R}^m)$ are functions whose restriction to each $\varepsilon Q_i \cap \Omega$ belongs to $W^{1,p}(\varepsilon Q_i \cap \Omega; \mathbf{R}^m)$, and such that the difference of the traces on both sides of $\partial(\varepsilon Q_i) \cap \partial(\varepsilon Q_j) \cap \Omega$ is p-summable for every $i, j \in \mathbf{Z}^n$. The functionals F_ε take the form

$$\int_\Omega f\left(\frac{x}{\varepsilon}, \frac{dDu}{dx}\right) dx + \varepsilon \int_{\Omega \cap \varepsilon E} g\left(\frac{x}{\varepsilon}, \frac{1}{\varepsilon}\frac{dDu}{d\mathcal{H}^{n-1}}\right) d\mathcal{H}^{n-1},$$

up to a normalization factor.

While no assumption had to be made on measures μ to prove the Homogenization Theorem in Section 24.1, in the case of Sobolev spaces with respect to a measure, some requirements have to be made in order to obtain in the limit functionals which admit an integral representation on $W^{1,p}(\Omega; \mathbf{R}^m)$.

Definition 24.10 *A σ-finite 1-periodic positive Borel measure μ on $\mathbf{R}^n$ is p-homogenizable if the following properties hold:*

(i) (existence of cut-off functions) there exist $K > 0$ and $\delta > 0$ such that for all $\varepsilon > 0$, for all pairs U, V of open subsets of $\mathbf{R}^n$ with $U \subset\subset V$, and $\mathrm{dist}\,(U, \partial V) \geq \delta\varepsilon$, and for all $u \in W^{1,p}_{\mu_\varepsilon}(V)$ there exists $\phi \in W^{1,\infty}_{\mu_\varepsilon}(V)$ with $0 \leq \phi \leq 1$, $\phi = 1$ on U, $\phi = 0$ in a neighbourhood of ∂V, such that

$$\int_V \left|\frac{dD\phi}{d\mu_\varepsilon}\tilde{u}\right|^p d\mu_\varepsilon \leq \frac{K}{(\mathrm{dist}\,(U, \partial V))^p}\int_{V\setminus U}|u|^p\,dx; \qquad (24.26)$$

(ii) (existence of periodic test functions) for all $i = 1, \ldots, n$, there exists $z_i \in W^{1,p}_{\mu,\mathrm{loc}}(\mathbf{R}^n)$ such that $x \mapsto z_i(x) - x_i$ is 1-periodic.

Remark 24.11 The measures of Example 24.9 are p-homogenizable for all $p \geq 1$ (see Exercises 24.2 and 24.3).

If $\mathcal{K}$ is a periodic set defined as in Chapter 20, then the measure μ defined by $\mu(B) = |B \setminus \mathcal{K}|$ is p-homogenizable for all $p \geq 1$. The proof of property (i) is contained in Proposition 20.5, and the proof of property (ii) is contained in Proposition 20.8.

The Homogenization Theorem for functionals in (24.25) takes the following form.

Theorem 24.12 *Let μ be a p-homogenizable measure, and for every bounded open subset Ω of $\mathbf{R}^n$ let $F_\varepsilon(\cdot, \Omega)$ be defined on $\mathrm{L}^p(\Omega; \mathbf{R}^m)$ by (24.25). Then the Γ-limit*

$$F_{\mathrm{hom}}(u, \Omega) = \Gamma(\mathrm{L}^p)\text{-}\lim_{\varepsilon \to 0} F_\varepsilon(u, \Omega) \qquad (24.27)$$

exists for all bounded open subsets Ω with Lipschitz boundary and for all $u \in \mathrm{W}^{1,p}(\Omega; \mathbf{R}^m)$, and it can be represented as

$$F_{\mathrm{hom}}(u, \Omega) = \int_\Omega f_{\mathrm{hom}}(Du) \, dx, \qquad (24.28)$$

where the homogenized integrand satisfies the asymptotic formula

$$f_{\mathrm{hom}}(A) = \lim_k \inf \Big\{ \frac{1}{k^n} \int_{[0,k)^n} f\left(x, \frac{dDu}{d\mu}\right) d\mu :$$
$$u \in \mathrm{W}^{1,p}_{\mu,\mathrm{loc}}(\mathbf{R}^n; \mathbf{R}^m), \ u - Ax \ k\text{-}periodic \Big\}. \qquad (24.29)$$

Furthermore, if f is convex then the cell-problem formula holds

$$f_{\mathrm{hom}}(A) = \inf \Big\{ \int_{[0,1)^n} f\left(x, \frac{dDu}{d\mu}\right) d\mu :$$
$$u \in \mathrm{W}^{1,p}_{\mu,\mathrm{loc}}(\mathbf{R}^n; \mathbf{R}^m), \ u - Ax \ 1\text{-}periodic \Big\} \qquad (24.30)$$

for all $A \in \mathbf{M}^{m \times n}$.

Remark 24.13 In formulae (24.29) and (24.30) we cannot replace the sets $[0, k)^n$ and $[0, 1)^n$ by the sets $(0, k)^n$ and $(0, 1)^n$, respectively, if μ charges $[0, 1)^n \setminus (0, 1)^n$. Note, moreover, that if μ is not a p-homogenizable measure, then in general f_{hom} may be equal to $+\infty$ for all non-zero matrices A (see Exercise 24.4).

The proof of the Homogenization Theorem will be obtained at the end of the section, as a consequence of the following propositions, which adapt to this case the methods for the homogenization in Chapter 14.

Proposition 24.14 *For every $A \in \mathbf{M}^{m \times n}$ there exists $z_A \in \mathrm{W}^{1,p}_{\mu,\mathrm{loc}}(\mathbf{R}^n; \mathbf{R}^m)$ such that $z_A - Ax$ is 1-periodic and satisfies*

$$\int_{[0,1)^n} \left| \frac{dDz_A}{d\mu} \right|^p d\mu \leq c|A|^p. \qquad (24.31)$$

Proof Define $z_A = \sum_{i=1}^{m} \sum_{j=1}^{n} A_{ij} e_i z_j$, where z_j are as in Definition 24.10(ii). Inequality (24.31) is trivial. $\qquad\square$

From now on Ω will be a fixed, bounded open subset of $\mathbf{R}^n$ with Lipschitz boundary.

Proposition 24.15 *For every $\sigma > 0$ there exist ε_σ and $M > 0$ such that for all U, U', V open subsets of Ω with $U' \subset U$ and $\operatorname{dist}(U', V \setminus U) > 0$, for all $\varepsilon < \varepsilon_\sigma \operatorname{dist}(U', V \setminus U)$ and for all $u \in \mathrm{W}^{1,p}_{\mu_\varepsilon}(\Omega; \mathbf{R}^m)$, $v \in \mathrm{W}^{1,p}_{\mu_\varepsilon}(\Omega; \mathbf{R}^m)$ there exists a cut-off function between U' and U, $\phi \in \mathrm{W}^{1,\infty}_{\mu_\varepsilon}(U \cup V)$, such that*

$$F_\varepsilon(\phi u + (1 - \phi)v, U' \cup V) \le (1 + \sigma)(F_\varepsilon(u, U) + F_\varepsilon(v, V))$$
$$+ \frac{M}{\left(\operatorname{dist}(U', V \setminus U)\right)^p} \int_{(U \cap V) \setminus U'} |u - v|^p dx + \sigma \mu_\varepsilon((U \cap V) \setminus U'). \quad (24.32)$$

Proof Let δ be given by Definition 24.10(b), let $N \in \mathbf{N}$ be such that $N\delta\varepsilon \le \operatorname{dist}(U', V \setminus U)$, and let $U_k = \{x \in U : N\operatorname{dist}(x, U') < k \operatorname{dist}(U', V \setminus U)\}$, $U_0 = U'$. For each $k = 1, \ldots, N$ let ϕ_k be a cut-off function between U_{k-1} and U_k, satisfying (24.26), which exists since $\operatorname{dist}(U_{k-1}, \partial U_k) \ge \delta\varepsilon$. We have, using Remark 24.8(b), (14.2) and (24.26)

$$F_\varepsilon(\phi_k u + (1 - \phi_k)v, U' \cup V)$$
$$= \int_{U' \cup V} f\left(\frac{x}{\varepsilon}, \tilde{\phi}_k \frac{dDu}{d\mu_\varepsilon} - (1 - \tilde{\phi}_k)\frac{dDv}{d\mu_\varepsilon} + (\tilde{u} - \tilde{v}) \otimes \frac{dD\phi_k}{d\mu_\varepsilon}\right) d\mu_\varepsilon$$
$$\le \int_U f\left(\frac{x}{\varepsilon}, \frac{dDu}{d\mu_\varepsilon}\right) d\mu_\varepsilon + \int_V f\left(\frac{x}{\varepsilon}, \frac{dDv}{d\mu_\varepsilon}\right) d\mu_\varepsilon$$
$$+ 4^p \beta \int_{(U_k \setminus U_{k-1}) \cap V} \left(1 + \left|\frac{dDu}{d\mu_\varepsilon}\right|^p + \left|\frac{dDv}{d\mu_\varepsilon}\right|^p\right) d\mu_\varepsilon$$
$$+ 4^p \beta \int_{(U_k \setminus U_{k-1}) \cap V} \left|(\tilde{u} - \tilde{v}) \otimes \frac{dD\phi_k}{d\mu_\varepsilon}\right|^p d\mu_\varepsilon$$
$$\le F_\varepsilon(u, U) + F_\varepsilon(v, V)$$
$$+ 4^p \beta \int_{(U_k \setminus U_{k-1}) \cap V} \left(1 + \left|\frac{dDu}{d\mu_\varepsilon}\right|^p + \left|\frac{dDv}{d\mu_\varepsilon}\right|^p\right) d\mu_\varepsilon$$
$$+ 4^p \beta \frac{cKN^p}{\left(\operatorname{dist}(U', V \setminus U)\right)^p} \int_{(U_k \setminus U_{k-1}) \cap V} |u - v|^p dx$$

where K is the constant appearing in (24.26).

Choose k such that

$$\int_{(U_k \setminus U_{k-1}) \cap V} \left(1 + \left|\frac{dDu}{d\mu_\varepsilon}\right|^p + \left|\frac{dDv}{d\mu_\varepsilon}\right|^p\right) d\mu_\varepsilon$$
$$+ \frac{cKN^p}{\left(\operatorname{dist}(U', V \setminus U)\right)^p} \int_{(U_k \setminus U_{k-1}) \cap V} |u - v|^p dx$$

$$\leq \frac{1}{N}\left(\int_{(U\cap V)\setminus U'}\left(1+\left|\frac{dDu}{d\mu_\varepsilon}\right|^p+\left|\frac{dDv}{d\mu_\varepsilon}\right|^p\right)d\mu_\varepsilon\right.$$
$$\left.+\frac{cKN^p}{\left(\mathrm{dist}\,(U',V\setminus U)\right)^p}\int_{(U\cap V)\setminus U'}|u-v|^p\,dx\right).$$

Then, taking into account (14.2) also,

$$F_\varepsilon(\phi_k u+(1-\phi_k)v,U'\cup V)$$
$$\leq F_\varepsilon(u,U)+F_\varepsilon(v,V)$$
$$+\frac{4^p\beta}{N\alpha}\left(\int_{(U\cap V)\setminus U'}f\left(\frac{x}{\varepsilon},\frac{dDu}{d\mu_\varepsilon}\right)d\mu_\varepsilon+\int_{(U\cap V)\setminus U'}f\left(\frac{x}{\varepsilon},\frac{dDv}{d\mu_\varepsilon}\right)d\mu_\varepsilon\right)$$
$$+4^p\beta\frac{cKN^{p-1}}{\left(\mathrm{dist}\,(U',V\setminus U)\right)^p}\int_{(U\cap V)\setminus U'}|u-v|^p dx+\frac{4^p\beta}{N}\mu_\varepsilon((U\cap V)\setminus U')$$
$$\leq\left(1+\frac{4^p\beta}{N\alpha}\right)\left(F_\varepsilon(u,U)+F_\varepsilon(v,V)\right)$$
$$+4^p\beta\frac{cKN^{p-1}}{\left(\mathrm{dist}\,(U',V\setminus U)\right)^p}\int_{(U\cap V)\setminus U'}|u-v|^p dx+\frac{4^p\beta}{N}\mu_\varepsilon((U\cap V)\setminus U').$$

We can choose ε_σ satisfying

$$\frac{4^p\beta}{\sigma\min\{1,\alpha\}}+1=\frac{1}{\delta\varepsilon_\sigma},$$

so that we can choose N, depending only on σ and on the constants of the problem, in such a way that (24.32) holds, with $M=c4^pK\beta N^{p-1}$ $\qquad\square$

Using Proposition 7.9 and a diagonal procedure, we fix an infinitesimal sequence (ε_j) along which the Γ-limit

$$\Gamma(\mathrm{L}^p)\text{-}\lim_j F_{\varepsilon_j}(u,U)=F(u,U)$$

exists for all $u\in L^p(\Omega;\mathbf{R}^m)$ and for all sets U in the countable family $\mathcal{R}$ of all finite unions of open rectangles of Ω with rational vertices. Moreover, we define

$$F'(u,U)=\Gamma(\mathrm{L}^p)\text{-}\liminf_j F_{\varepsilon_j}(u,U)$$

$$F''(u,U)=\Gamma(\mathrm{L}^p)\text{-}\limsup_j F_{\varepsilon_j}(u,U)$$

for all $u\in L^p(\Omega;\mathbf{R}^m)$, and for all open subsets U of Ω.

Proposition 24.16 *We have*

$$F''(u,U)\leq c\int_U(1+|Du|^p)dx$$

for all $u\in\mathrm{W}^{1,p}(\Omega;\mathbf{R}^m)$ and for all open subsets U of Ω with $|\partial U|=0$.

Proof *Step* 1: *we have* $F''(Ax, U) \leq c|U|(1 + |A|^p)$ *for all* $A \in \mathbf{M}^{m \times n}$ *and for all* $U \in \mathcal{A}(\Omega)$ *with* $|\partial U| = 0$.

Let z_A be given by Proposition 24.14. The functions $z_A^\varepsilon(x) = \varepsilon z_A(\frac{x}{\varepsilon})$ converge in $L_{\mathrm{loc}}^p(\mathbf{R}^n; \mathbf{R}^m)$ to Ax, so that

$$F''(Ax, U) \leq \limsup_{\varepsilon \to 0+} \int_U f\left(\frac{x}{\varepsilon}, \frac{dDz_A^\varepsilon}{d\mu_\varepsilon}\right) d\mu_\varepsilon$$

$$\leq c \limsup_{\varepsilon \to 0+} \int_U \left(1 + \left|\frac{dDz_A^\varepsilon}{d\mu_\varepsilon}\right|^p\right) d\mu_\varepsilon \leq c|U|(1 + |A|^p).$$

Step 2: *we have* $F''(u, U) \leq c \int_U (1 + |Du|^p) dx$ *for all piecewise affine functions* $u \in \mathrm{W}^{1,p}(\Omega; \mathbf{R}^m)$ *and for all* $U \in \mathcal{A}(\Omega)$ *with* $|\partial U| = 0$.

We write $u = \sum_{i=1}^N \chi_{U_i} u_i$, where $U_1, \ldots, U_N$ are disjoint open subsets of U such that $|U \setminus \bigcup_i U_i| = 0$ and $|\overline{U}_i| = |U_i|$, and $u_i(x) = A_i x + c_i$ for some $A_i \in \mathbf{M}^{m \times n}$ and $c_i \in \mathbf{R}^m$. For each i we set $u_i^\varepsilon(x) = z_{A_i}^\varepsilon(x) + c_i$, as from Step 1.

We will prove Step 2 by finite induction. First, we give an estimate on $U_1 \cup U_2$. For all $\varepsilon > 0$ sufficiently small, we can apply Proposition 24.15 with $U' = U_2$,

$$U = U_2^\eta = \{x \in U : \ \mathrm{dist}\,(x, U_2) < \eta\},$$

$V = U_1$, where $\eta = \eta_\varepsilon > 0$ will be determined later, $\sigma = 1$, $u = u_2^\varepsilon$ and $v = u_1^\varepsilon$. We then obtain a cut-off function $\phi = \phi_\varepsilon$ between U_2 and U_2^η such that

$$F_\varepsilon(\phi_\varepsilon u_2^\varepsilon + (1 - \phi_\varepsilon)u_1^\varepsilon, U_1 \cup U_2) \leq 2(F_\varepsilon(u_1^\varepsilon, U_1) + F_\varepsilon(u_2^\varepsilon, U_2^\eta))$$

$$+ \frac{M}{\eta^p} \int_{U_1 \cap U_2^\eta} |u_2^\varepsilon - u_1^\varepsilon|^p \, dx + \mu_\varepsilon(U_1 \cap U_2^\eta). \qquad (24.33)$$

The constant M is the one given by Proposition 24.15 with $\sigma = 1$.

Taking into account that

$$\lim_{\varepsilon \to 0} \int_{U_1 \cap U_2^\eta} |u_2^\varepsilon - u_1^\varepsilon|^p \, dx = \int_{U_1 \cap U_2^\eta} |u_2 - u_1|^p \, dx \leq c\|Du\|_\infty^p \eta^{p+1}$$

since u_i are affine and $u_2 = u_1$ on $\partial U_1 \cap \partial U_2$, we can choose $\eta = \eta_\varepsilon$, tending to 0 as $\varepsilon \to 0$, in such a way that

$$\lim_{\varepsilon \to 0} \frac{1}{\eta_\varepsilon^p} \int_{U_1 \cap U_2^{\eta_\varepsilon}} |u_2^\varepsilon - u_1^\varepsilon|^p \, dx = 0.$$

If we define $w_1^\varepsilon = \phi_\varepsilon u_2^\varepsilon + (1 - \phi_\varepsilon)u_1^\varepsilon$, we have $w_1^\varepsilon \to u$ in $L^p(U_1 \cup U_2; \mathbf{R}^m)$ and, by (24.33),

$$\limsup_{\varepsilon \to 0} F_\varepsilon(w_1^\varepsilon, U_1 \cup U_2) \leq c \int_{U_1 \cup U_2} (1 + |Du|^p) \, dx$$

as in the proof of Step 1.

We can now proceed by induction, repeating at each step the previous argument replacing U_1 by $U_1 \cup \cdots \cup U_j$, U_2 by U_{j+1}, u_1^ε by the w_j^ε constructed in the preceding step, and u_2^ε by u_{j+1}^ε.

To conclude the proof of the proposition, it suffices to recall that $F''(\cdot, U)$ is weakly lower semicontinuous, and that piecewise affine functions are dense in $W^{1,p}(\Omega; \mathbf{R}^m)$. $\qquad\square$

Proposition 24.17 *For all open subsets U of Ω there exists the Γ-limit*

$$\Gamma\text{-}\lim_j F_{\varepsilon_j}(u, U) = F(u, U),$$

and there exists a function $\varphi : \mathbf{M}^{m \times n} \to \mathbf{R}$ such that

$$F(u, U) = \int_U \varphi(Du)\,dx$$

for all $u \in W^{1,p}(\Omega; \mathbf{R}^m)$ and $U \in \mathcal{A}(\Omega)$ with $|\partial U| = 0$.

Proof First of all, observe that for all $U \in \mathcal{A}(\Omega)$ with $|\partial U| = 0$ and $u \in W^{1,p}(\Omega; \mathbf{R}^m)$ we have

$$F''(u, U) = \sup\{F''(u, V) : \ V \subset\subset U, \ V \text{open}\},$$

$$F'(u, U) = \sup\{F'(u, V) : \ V \subset\subset U, \ V \text{open}\}.$$

In fact, it suffices to apply Proposition 11.6 with $q = p$, as the fundamental estimate is proven in Proposition 24.15. Note that this implies that $F'(u, U) = F''(u, U)$, as both supremums above can be performed on open sets in $\mathcal{R}$.

Next, by Theorem 10.3, and by Proposition 11.5 and Theorem 10.2, we obtain that the Γ-limit $\Gamma\text{-}\lim_j F_{\varepsilon_j}(u, U) = F(u, U)$ exists for all $U \in \mathcal{A}(\Omega)$ with $|\partial U| = 0$, and for all $u \in W^{1,p}(\Omega; \mathbf{R}^m)$ the function $U \mapsto \sup\{F''(u, V) : \ V \subset\subset U, \ V \text{open}\}$ is the restriction to $\mathcal{A}(\Omega)$ of a Borel measure on Ω.

Eventually, the existence of $\varphi : \mathbf{M}^{m \times n} \to \mathbf{R}$ such that

$$F(u, U) = \int_U \varphi(Du)\,dx$$

for all $u \in W^{1,p}(\Omega; \mathbf{R}^m)$ and $U \subseteq \Omega$ with $|\partial U| = 0$ follows from the Integral Representation Theorem 9.2, observing that translation invariance in x can be obtained as in Proposition 14.3. $\qquad\square$

Proposition 24.18 *For all $A \in \mathbf{M}^{m \times n}$ there exists the limit in (24.29) and we have $\varphi(A) = f_{\mathrm{hom}}(A)$.*

Proof Up to a translation we can suppose that $\mu([0,1)^n \setminus (0,1)^n) = 0$, in order to simplify the proof of formula (24.29). For all $A \in \mathbf{M}^{m \times n}$ and $k \in \mathbf{N}$ we define

$$g_k(A) = \inf\left\{\frac{1}{k^n} \int_{(0,k)^n} f\left(x, \frac{dDu}{d\mu}\right) d\mu : u \in W^{1,p}_{\mu,\mathrm{loc}}(\mathbf{R}^n; \mathbf{R}^m), \ u - Ax \ k\text{-periodic}\right\}.$$

Fix $A \in \mathbf{M}^{m \times n}$, let $u \in W^{1,p}_{\mu,\text{loc}}(\mathbf{R}^n; \mathbf{R}^m)$ with $u - Ax$ k-periodic, and define the sequence $u_j(x) = \varepsilon_j u(x/\varepsilon_j)$. Note that $u_j \to Ax$ in $L^p_{\text{loc}}(\mathbf{R}^n; \mathbf{R}^m)$. We then have

$$\varphi(A) = F(Ax, (0,1)^n) \le \liminf_j F_{\varepsilon_j}(u_j, (0,1)^n) = \frac{1}{k^n} \int_{(0,k)^n} f\left(x, \frac{dDu}{d\mu}\right) d\mu.$$

The last equality follows from Example 2.7 with $Y = (0,k)^n$. Hence, $\varphi(A) \le g_k(A)$, so that

$$\varphi(A) \le \liminf_k g_k(A). \tag{24.34}$$

Conversely, let $w_j \to Ax$ in $L^p((0,1)^n; \mathbf{R}^m)$ such that

$$\varphi(A) = F(Ax, (0,1)^n) = \lim_j F_{\varepsilon_j}(w_j, (0,1)^n).$$

Let $\sigma > 0$. Let $T_j = 1/\varepsilon_j$ and let $u_j(x) = T_j w_j(x/T_j)$. We use the notation $K_j = [T_j] + 1$.

If j is large enough and $N > 4$, we can use Proposition 24.15 with $\varepsilon = 1$, $U = (0, T_j)^n$, $V = (0, K_j)^n \setminus (2T_j/N, T_j - 2(T_j/N))^n$, $U' = (T_j/N, T_j - (T_j/N))^n$, $u = u_j$, and $v = z_A$. We then get

$$\begin{aligned}
&F_1(\phi u + (1 - \phi)v, (0, K_j)^n) \\
&= F_1(\phi u + (1 - \phi)v, U' \cup V) \\
&\le (1 + \sigma)(F_1(u, U) + F_1(v, V)) \\
&\quad + MN^p T_j^{-p} \int_{(U \cap V) \setminus U'} |u - v|^p dx + \sigma\mu((U \cap V) \setminus U'). \tag{24.35}
\end{aligned}$$

The function $\phi u + (1 - \phi)v - Ax$ can be extended to a K_j-periodic function. We then obtain

$$\begin{aligned}
&K_j^n g_{K_j}(A) \\
&\le (1 + \sigma)(F_1(u_j, (0, T_j)^n) + F_1(z_A, V)) \\
&\quad + MN^p T_j^{-p} \int_{(0,T_j)^n \setminus (T_j/N, T_j - (T_j/N))^n} |u_j - z_A|^p dx + \sigma\mu((U \cap V) \setminus U') \\
&\le (1 + \sigma)\left(T_j^n F_{\varepsilon_j}(w_j, (0,1)^n) + c\frac{K_j^n}{N}(1 + |A|^p)\right) \\
&\quad + MN^p T_j^n \int_{(0,1)^n} |w_j - z_j|^p dx + \sigma c K_j^n,
\end{aligned}$$

where $z_j(x) = T_j^{-1} z_A(T_j x)$. Note that $z_j \to Ax$ in $L^p((0,1)^n; \mathbf{R}^m)$; hence

$$\lim_j \int_{(0,1)^n} |w_j - z_j|^p dx = 0.$$

Dividing by K_j^n and letting first $j \to +\infty$ and then $\sigma \to 0$ and $N \to +\infty$ we get

$$\limsup_j g_{K_j}(A) \le \varphi(A). \tag{24.36}$$

By (24.34) and (24.36) we then obtain

$$\varphi(A) = \liminf_k g_k(A) = \limsup_j g_{K_j}(A).$$

The first equality shows that φ is independent of the sequence (ε_j). Eventually, repeating the reasoning with a sequence (ε_j) such that

$$\lim_j g_{K_j}(A) = \limsup_k g_k(A),$$

the proof is completed. $\qquad\square$

Proof of Theorem 24.12 The previous propositions show that the limit in (24.27) exists and admits the integral representation (24.28) with φ satisfying formula (24.29). In order to conclude the proof it remains to remark that formula (24.30) in the convex case follows as in Theorem 14.7. $\qquad\square$

24.3 Homogenization of periodic thin structures

We begin by defining the full-dimensional domains in $\mathbf{R}^{n+N}$ which will tend to a n-dimensional set. Let $\gamma : \mathbf{R}^n \to \mathbf{R}$ be a bounded, strictly positive and 1-periodic lower semicontinuous function. We set $c_\gamma = \min \gamma$ and suppose, without loss of generality, that $\sup \gamma = 1$. We consider a bounded open set $\Omega \subset \mathbf{R}^n$ and define the domains

$$\Omega_\varepsilon = \left\{ (x,y) \in \mathbf{R}^{n+N} : x \in \Omega, \ |y| < \varepsilon\, \gamma\!\left(\frac{x}{\varepsilon}\right) \right\}. \tag{24.37}$$

For example, if $n = 1$ and $N = 2$ Ω_ε model a rod with vanishing modulated thickness, while if $n = 2$ and $N = 1$ they can be viewed as thin films with an oscillating profile.

We consider a Borel function $f : \mathbf{R}^n \times \mathbf{M}^{m \times (n+N)} \to [0, +\infty)$, 1-periodic in the first variable, and satisfying the standard growth condition of order $p > 1$

$$\alpha |A|^p \le f(x, A) \le \beta(1 + |A|^p) \tag{24.38}$$

for all $x \in \mathbf{R}^n$ and $A \in \mathbf{M}^{m \times (n+N)}$. The 'elastic free energy' of the thin structure at size $\varepsilon > 0$, with energy density f, is given by

$$I_\varepsilon(u) = \int_{\Omega_\varepsilon} f\!\left(\frac{x}{\varepsilon}, Du\right) dx\, dy \tag{24.39}$$

for all $u \in \mathrm{W}^{1,p}(\Omega_\varepsilon; \mathbf{R}^m)$.

After normalizing I_ε, multiplying by a factor ε^{-N}, and a change of variables, we obtain a scaled version of this energy defined on functions $u \in \mathrm{W}^{1,p}(\tilde{\Omega}_\varepsilon; \mathbf{R}^m)$, where

$$\tilde{\Omega}_\varepsilon = \Big\{(x,y): \ x \in \Omega, \ |y| < \gamma\Big(\frac{x}{\varepsilon}\Big)\Big\},$$

as follows:

$$F_\varepsilon(u) = \int_{\tilde{\Omega}_\varepsilon} f\Big(\frac{x}{\varepsilon}, D_x u, \frac{1}{\varepsilon} D_y u\Big)\, dx\, dy. \qquad (24.40)$$

Let B_t denote the ball in $\mathbf{R}^N$ of centre 0 and radius t. In order to have a common domain for all ε, we extend the definition of F_ε to the whole space $\mathrm{L}^p(\Omega \times B_1; \mathbf{R}^m)$ by taking $F_\varepsilon(u)$ as above if the restriction $u_{|\tilde{\Omega}_\varepsilon}$ belongs to $\mathrm{W}^{1,p}(\tilde{\Omega}_\varepsilon; \mathbf{R}^m)$, and $F_\varepsilon(u) = +\infty$ otherwise.

In order to state the Homogenization Theorem, we define the space

$$\mathcal{V} = \{u \in \mathrm{W}^{1,p}(\Omega \times B_1; \mathbf{R}^m): \ D_y u = 0\},$$

which can be identified with $\mathrm{W}^{1,p}(\Omega; \mathbf{R}^m)$.

Theorem 24.19 (Homogenization of thin structures) *The Γ-limit*

$$F_{\mathrm{hom}}(u) = \Gamma\text{-}\lim_{\varepsilon \to 0+} F_\varepsilon(u)$$

in the $\mathrm{L}^p(\Omega \times B_1; \mathbf{R}^m)$-topology exists for all $u \in \mathrm{L}^p(\Omega \times B_1; \mathbf{R}^m)$. Moreover, if we set

$$
\begin{aligned}
f_{\mathrm{hom}}(A) = \lim_{T \to +\infty} \inf\Big\{ &\frac{1}{T^n |B_1|} \int_{((0,T)^n \times \mathbf{R}^N) \cap \{|y| < \gamma(x)\}} f(x, A + D_x u, D_y u)\, dx\, dy \\
&u \in \mathrm{W}^{1,p}(((0,T)^n \times \mathbf{R}^N) \cap \{|y| < \gamma(x)\}; \mathbf{R}^m), \\
&u = 0 \ on \ ((\partial(0,T)^n) \times \mathbf{R}^N) \cap \{|y| < \gamma(x)\}\Big\}, \qquad (24.41)
\end{aligned}
$$

the function f_{hom} satisfies

$$c_\gamma^N |A|^p \le f_{\mathrm{hom}}(A) \le \beta(1 + |A|^p) \qquad (24.42)$$

for all $x \in \Omega$ and $A \in \mathbf{M}^{m \times n}$, and we have

$$F_{\mathrm{hom}}(u) = \int_{\Omega \times B_1} f_{\mathrm{hom}}(D_x u)\, dx\, dy$$

for all $u \in \mathcal{V}$, and $F_{\mathrm{hom}}(u) = +\infty$ if $u \notin \mathcal{V}$.

Before proving Theorem 24.19, we state a compactness result for more general structures.

Theorem 24.20 (Compactness of thin structures) *For every $\varepsilon > 0$ let E_ε be an open subset of $\Omega \times B_1$ and let $f_\varepsilon : \Omega \times B_1 \times \mathbf{M}^{m \times (n+N)} \to [0, +\infty)$ be a Borel function satisfying the standard growth condition of order $p > 1$ (24.38). Set*

$$F_\varepsilon(u) = \int_{E_\varepsilon} f_\varepsilon\left(x, y, D_x u, \frac{1}{\varepsilon} D_y u\right) dx\, dy \qquad (24.43)$$

if $u_{|E_\varepsilon} \in W^{1,p}(E_\varepsilon; \mathbf{R}^m)$, and $F_\varepsilon(u) = +\infty$ otherwise. Then for every sequence (ε_j) of positive numbers converging to 0 there exist a subsequence (not relabelled) and a function $f_0 : \Omega \times \mathbf{M}^{m\times n} \to [0, +\infty)$, satisfying

$$0 \le f_0(x, A) \le \beta(1 + |A|^p) \qquad (24.44)$$

for all $x \in \Omega$ and $A \in \mathbf{M}^{m\times n}$, such that the Γ-limit in the $L^p(\Omega \times B_1; \mathbf{R}^m)$-topology

$$F_0(u) = \Gamma\text{-}\lim_j F_{\varepsilon_j}(u) \qquad (24.45)$$

exists for all $u \in \mathcal{V}$, and

$$F_0(u) = \int_{\Omega \times B_1} f_0(x, D_x u)\, dx\, dy \qquad (24.46)$$

for all $u \in \mathcal{V}$.

Proof We can follow the direct methods of Γ-convergence described in Chapters 9–11. We begin by 'localizing the functionals in the x-space' by setting for all $U \in \mathcal{A}(\Omega)$

$$F_\varepsilon(u, U) = \int_{(U \times \mathbf{R}^N) \cap E_\varepsilon} f_\varepsilon\left(x, y, D_x u, \frac{1}{\varepsilon} D_y u\right) dx\, dy$$

if $u_{|E_\varepsilon} \in W^{1,p}(E_\varepsilon; \mathbf{R}^m)$, $F(u, U) = +\infty$ otherwise. The functionals F_ε satisfy the following variant of the L^p-fundamental estimate: for all U, U' and V open subsets of Ω with $U' \subset\subset U$ and for all $\sigma > 0$ there exists M_σ such that for all $u, v \in W^{1,p}(\Omega; \mathbf{R}^m)$ there exists a cut-off function $\varphi = \varphi(x)$ between U' and U such that

$$F_\varepsilon(\varphi u + (1 - \varphi)v, U' \cup V) \le (1 + \sigma)(F_\varepsilon(u, U) + F_\varepsilon(v, V)) \qquad (24.47)$$

$$+ M_\sigma \int_{((U\cap V)\setminus U') \times B_1} |u - v|^p \, dx\, dy + \sigma.$$

The proof of this estimate follows exactly the one of Proposition 12.2, after noting that $D_y(\varphi(u - v)) = \varphi D_y(u - v)$.

Using (24.47) and the fact that for $u \in \mathcal{V}$ we trivially have

$$\Gamma\text{-}\limsup_{\varepsilon \to 0+} F_\varepsilon(u, U) \le \beta \int_{U \times B_1} (1 + |Du|^p)\, dx, \qquad (24.48)$$

it can be proved, exactly as in Chapters 10 and 11, that for every sequence (ε_j) there exists a subsequence (not relabelled) along which the Γ-limit $F_0(u, U)$ of

$F_{\varepsilon_j}(u, U)$ exists for all $u \in \mathcal{V}$ and $U \in \mathcal{A}(\Omega)$ (proceeding as in Proposition 11.6 and Theorem 10.3, using the L^p-estimate as above when needed). Moreover, the functional F_0, after identifying $\mathcal{V}$ with $W^{1,p}(\Omega; \mathbf{R}^m)$, satisfies properties (i)–(v) of Theorem 9.1, which are an immediate consequence of the definition of F_0, except for the measure property (ii), which follows from the L^p-fundamental estimate, Proposition 11.6 and Theorem 10.2. Hence, there exists a Carathéodory function $\tilde{f}_0$ defined on $\Omega \times \mathbf{M}^{m \times n}$, quasiconvex in the second variable, such that

$$F_0(u, U) = \int_U \tilde{f}_0(x, D_x u)\, dx$$

for all $u \in W^{1,p}(\Omega; \mathbf{R}^m)$. In particular, taking $U = \Omega$, and identifying the space $W^{1,p}(\Omega; \mathbf{R}^m)$ with $\mathcal{V}$ we obtain (24.46), with $f_0 = \tilde{f}_0 |B_1|^{-1}$. $\hfill\square$

Proof of Theorem 24.19 We begin by observing that if (ε_j) is a sequence of positive numbers converging to 0, and if $u_j \to u$ in $L^p(\Omega \times B_1; \mathbf{R}^m)$ with $\sup_j F_{\varepsilon_j}(u_j) \leq +\infty$, then $u \in \mathcal{V}$. To see this, first remark that from the growth conditions (24.38) we have $u \in W^{1,p}(\Omega \times B_{c_\gamma}; \mathbf{R}^m)$ and

$$\lim_j \int_{\tilde{\Omega}_{\varepsilon_j}} |D_y u_j|^p \, dx\, dy = 0. \tag{24.49}$$

Fix $0 < t < 1$, and define

$$\omega_j = \left\{ x \in \Omega : \ \gamma\!\left(\frac{x}{\varepsilon_j}\right) \geq t \right\}.$$

We have $\chi_{\omega_j} \rightharpoonup c_t$ weakly* in $L^\infty(\Omega)$, where $c_t > 0$ is a constant. We set $S_j = \omega_j \times B_t$. Note that $|y| < \gamma(x/\varepsilon_j)$ on S_j, $\chi_{S_j}(x, y) = \chi_{\omega_j}(x)$ on $\Omega \times B_t$, and $\chi_{S_j} \rightharpoonup c_t$ weakly* in $L^\infty(\Omega \times B_t)$. Let $\varphi \in C_0^\infty(\Omega \times B_t)$. We have

$$\begin{aligned}
c_t \int_{\Omega \times B_t} u D_y \varphi \, dx\, dy &= \lim_j \int_{\Omega \times B_t} u_j \chi_{S_j} D_y \varphi \, dx\, dy \\
&= -\lim_j \int_{\Omega \times B_t} \varphi \chi_{S_j} D_y u_j \, dx\, dy \\
&= -\lim_j \int_{S_j} \varphi D_y u_j \, dx\, dy = 0,
\end{aligned}$$

where we used (24.49). Hence $D_y u = 0$ in the sense of distributions on $\Omega \times B_t$, and by the arbitrariness of $t \in (0, 1)$ we conclude that $D_y u = 0$ in $\Omega \times B_1$, so that $u \in \mathcal{V}$. Note that this implies the last statement of the theorem.

By Theorem 24.20, applied with $E_\varepsilon = \tilde{\Omega}_\varepsilon$ and $f_\varepsilon(x, y, A) = f(x/\varepsilon, y, A)$, from every sequence (ε_j) we can extract a subsequence (not relabelled) such that the Γ-limit of $F_{\varepsilon_j}(\cdot, U)$ exists for all bounded open sets U in $\mathbf{R}^n$ (where now $F_{\varepsilon_j}(\cdot, U)$ is defined in the same way as F_ε, replacing Ω by U). Note that the inequality

$$F_0(u, U) \geq \alpha \int_{U \times B_{c_\gamma}} |D_x u|^p \, dx \, dy$$

implies the first inequality in (24.42) for f_0.

By the periodicity assumptions we obtain, repeating the proof of Proposition 14.3, that the function f_0 in Theorem 24.20 can be chosen to be independent of the variable x. We have to show now that for all $A \in \mathbf{M}^{m \times n}$ we have $f_0(A) = f_{\mathrm{hom}}(A)$ given by formula (24.41). Let

$$g(T) = \inf\left\{ \frac{1}{T^n |B_1|} \int_{((0,T)^n \times \mathbf{R}^N) \cap \{|y| < \gamma(x)\}} f(x, A + D_x u, D_y u) \, dx \, dy \right.$$
$$\left. u = 0 \text{ on } ((\partial(0,T)^n) \times \mathbf{R}^N) \cap \{|y| < \gamma(x)\} \right\}.$$

On one hand we know (following word for word the proof of Proposition 11.7 using the L^p-estimate as stated above) that there exists a sequence u_j converging to 0 in $L^p((0,1)^n \times B_1; \mathbf{R}^m)$ such that $u_j = 0$ on $((\partial(0,1)^n) \times \mathbf{R}^N) \cap \{|y| < \gamma(\frac{x}{\varepsilon_j})\}$, and

$$|B_1| f_0(A) = F_0(Ax, (0,1)^n) = \lim_j F_{\varepsilon_j}(Ax + u_j, (0,1)^n).$$

This shows, considering

$$v_j(x, y) = \frac{1}{\varepsilon_j} u_j(\varepsilon_j x, y)$$

as test function in the minimum problem defining $g(\frac{1}{\varepsilon_j})$, that

$$f_0(A) \geq \limsup_j g\left(\frac{1}{\varepsilon_j}\right). \tag{24.50}$$

On the other hand, for all $T > 0$ we can choose u_T, an admissible test function in the definition of $g(T)$, such that

$$\int_{((0,T)^n \times \mathbf{R}^N) \cap \{|y| < \gamma(x)\}} f(x, A + D_x u_T, D_y u_T) \, dx \, dy \leq T^n |B_1| g(T) + 1.$$

We can extend this function by 0 on $(0, [T+1])^n \times B_1$, and then by $[T+1]$-periodicity in the first variable on the whole $\mathbf{R}^n \times B_1$, still denoting this extension by u_T. We can consider

$$v_j(x, y) = \varepsilon_j \, u_T\left(\frac{x}{\varepsilon_j}, y\right),$$

obtaining

$$\liminf_j F_{\varepsilon_j}(Ax + v_j, (0,1)^n) \leq g(T)|B_1| + o\left(\frac{1}{T}\right).$$

Moreover $v_j \to 0$ in $L^p(E \times B_1; \mathbf{R}^m)$ for all E bounded open subsets of $\mathbf{R}^n$. Hence, by the Γ-convergence of F_{ε_j},

$$f_0(A)|B_1| = F_0(Ax, (0,1)^n) \le \liminf_j F_{\varepsilon_j}(Ax + v_j, (0,1)^n),$$

and, by the arbitrariness of T,

$$f_0(A) \le \liminf_{T \to +\infty} g(T). \tag{24.51}$$

Note that the right-hand side of the previous expression is less than or equal to $\liminf_j g(\frac{1}{\varepsilon_j})$; hence, (24.50) and (24.51) show that

$$f_0(A) = \liminf_{T \to +\infty} g(T) = \lim_j g\left(\frac{1}{\varepsilon_j}\right).$$

The first equality shows that indeed the function f_0 is independent of (ε_j), and by the arbitrariness of (ε_j) the second one implies also that the limit of $g(T)$ as $T \to +\infty$ exists, concluding thus the proof. $\qquad\square$

24.4 Exercises

Exercise 24.1 Prove the L^∞-fundamental estimate as $\varepsilon \to 0$ for the functionals in (24.6).

Hint: proceed as in Proposition 12.2, integrating with respect to μ_ε in place of dx, noting that $\int |u - v|^p \, d\mu_\varepsilon$ can be estimated by $c\|u - v\|_\infty^p$ if ε is small enough.

Exercise 24.2 Prove that the measure μ in Example 24.9(a) is p-homogenizable for all $p \ge 1$.

Hint: to prove (i) let $\delta = 5\sqrt{n}$. Denote $|x - y|_\infty = \max_{1 \le i \le n} |x_i - y_i|$. Fix $\varepsilon > 0$, and set $U_\varepsilon = \bigcup \{\varepsilon Q_i : \varepsilon Q_i \cap U \ne \emptyset\}$. Note that $U_\varepsilon \subset\subset V$. Choose

$$\phi(x) = 1 - \left(\frac{1}{C}\left[\frac{1}{\varepsilon}\inf\{|x - y|_\infty : y \in U_\varepsilon\}\right] \wedge 1\right),$$

where $C = \left[\frac{1}{\varepsilon}\inf\{|x - y|_\infty : x \in U_\varepsilon,\ y \in \partial V\}\right] - 2$. Note that $|dD\phi/d\mu_\varepsilon| \le n/(C\varepsilon) \le c/\mathrm{dist}\,(U, \partial V)$ for some constant c independent of U and V. Moreover, if $u \in W^{1,p}_{\mu_\varepsilon}(V)$ then u is equal to a constant u_i on each cube εQ_i such that $D\phi \ne 0$ on $\partial \varepsilon Q_i$. Hence, for two such cubes

$$\varepsilon \int_{\partial \varepsilon Q_i \cap \partial \varepsilon Q_j} |\tilde{u}|^p \, d\mathcal{H}^{n-1} \le \varepsilon \int_{\partial \varepsilon Q_i \cap \partial \varepsilon Q_j} (|u_i|^p + |u_j|^p) \, d\mathcal{H}^{n-1} = \int_{\varepsilon Q_i \cup \varepsilon Q_j} |u|^p \, dx$$

so that

$$\int_V \left|\frac{dD\phi}{d\mu_\varepsilon} \tilde{u}\right|^p d\mu_\varepsilon \le \varepsilon \frac{c^p}{(\mathrm{dist}\,(U, \partial V))^p} \int_{(V \setminus U) \cap \varepsilon E \cap \mathrm{spt}\, D\phi} |\tilde{u}|^p \, d\mathcal{H}^{n-1}$$
$$\le 2n \frac{c^p}{(\mathrm{dist}\,(U, \partial V))^p} \int_{V \setminus U} |u|^p \, dx.$$

The proof of (i) is then complete. To verify (ii) simply take $z_i(x) = [x_i]$.

Exercise 24.3 Prove that the measure μ in Example 24.9(b) is p-homogenizable for all $p \geq 1$.

Hint: the proof of (i) and (ii) is trivial since μ is absolutely continuous with respect to the Lebesgue measure.

Exercise 24.4 Let μ be defined as in (24.13). Prove that if $f_{\mathrm{hom}}(A)$ is defined as in (24.29) then $f_{\mathrm{hom}}(A) = +\infty$ if $A \neq 0$.

Hint: note that test functions u must be constant on a periodic connected component of $\mathbf{R}^n$.

Exercise 24.5 Let μ and E be defined as in Example 24.9(a), $m = 1$ and $f(x, A) = |A|^2$. Prove that $f_{\mathrm{hom}}(A) = n|A|^2$.

Hint: the only functions admissible in the minimum problem defining $f_{\mathrm{hom}}(z)$ are of the form $\sum_i z_i[x_i] + c$, with c an arbitrary constant.

Exercise 24.6 Prove that if f is convex then the formula for f_{hom} in Theorem 24.19 simplifies to

$$f_{\mathrm{hom}}(A) = \inf\left\{ \frac{1}{|B_1|} \int_{((0,1)^n \times \mathbf{R}^N) \cap \{|y| < \gamma(x)\}} f(x, A + D_x u, D_y u)\, dx\, dy \right.$$
$$\left. u \text{ 1-periodic in the } x\text{-direction}\right\}. \tag{24.52}$$

Hint: argue as in Section 14.3.1.

Exercise 24.7 Take in Theorem 24.19 $\gamma = 1$ and $f(x, A) = f(A)$ to be locally Lipschitz on $\mathbf{M}^{m \times (n+N)}$. Prove that in this case the function f_{hom} coincides with $Q f_0$, where $f_0(A) = \min\{f(A, B) : B \in \mathbf{M}^{m \times N}\}$ for $A \in \mathbf{M}^{m \times n}$, and Q indicates the operation of quasiconvexification.

Hint: it suffices to prove that $f_{\mathrm{hom}} \leq Q f_0$. Fix $\eta > 0$ and find 1-periodic functions $u \in C^1(\mathbf{R}^n; \mathbf{R}^m)$ and $B \in C^1(\mathbf{R}^n; \mathbf{M}^{m \times N})$ such that

$$\int_{(0,1)^n} f(x, A + Du(x), B(x))\, dx \leq Q f_0(A) + \eta.$$

To give an estimate for $f_{\mathrm{hom}}(A)$, use $u_\varepsilon(x, y) = Ax + \sqrt{\varepsilon}\, u(\frac{x}{\sqrt{\varepsilon}}) + \varepsilon B(\frac{x}{\sqrt{\varepsilon}}) y$ in the lim inf inequality for $F_{\mathrm{hom}}(Ax, (0,1)^n)$.

Exercise 24.8 Let f_{hom} be as in Theorem 24.19, and let $f(x, A) = f(A)$ be locally Lipschitz on $\mathbf{M}^{m \times (n+N)}$. Prove that we have the estimate

$$f_{\mathrm{hom}}(A) \leq \left(\int_{(0,1)^n} \gamma\, dx \right) Q f_0(A) \tag{24.53}$$

for all $A \in \mathbf{M}^{m \times n}$ (f_0 as in the previous exercise). Prove that if $\gamma \neq 1$ then this inequality may not be sharp even in the convex case.

Hint: to prove the inequality use the previous exercise. To check that it is not sharp, let $n = N = m = 1$, $f(A, B) = g(A) + |B|^p$ with g a convex function satisfying the required growth conditions, and

$$g(1) = 0, \qquad g(-1) = 2, \qquad g(0) = 1.$$

Choose γ simply as

$$\gamma(x) = \begin{cases} 1 & \text{if } 0 < x < 1/2, \\ 1/2 & \text{if } 1/2 \leq x \leq 1 \end{cases}$$

(extended by periodicity). In this case $f_0 = g = Qf_0$. Prove that (24.53) gives $f_{\text{hom}}(0) \leq 3/4$, while $f_{\text{hom}}(0) \leq 1/2$. To check this, choose

$$u(x) = \begin{cases} x & \text{if } 0 < x \leq 1/2, \\ \frac{1}{2} - x & \text{if } 1/2 < x \leq 1 \end{cases}$$

(extended by periodicity) as test function in (24.52).

Part V

Appendices

APPENDIX A

ALMOST-PERIODIC FUNCTIONS

Uniformly almost-periodic functions

Definition A.1 *Let $(X, \| \ \|)$ be a complex Banach space. We say that a measurable function $v : \mathbf{R}^N \to X$ is* uniformly almost periodic *(u.a.p. for short), and we write $v \in UAP(\mathbf{R}^N; X)$, if it is the uniform limit of a sequence of trigonometric polynomials on X, i.e.*

$$\lim_k \| P_k(\cdot) - v(\cdot) \|_\infty = 0$$

for some functions of the form

$$P_k(y) = \sum_{j=1}^{r_k} \mathbf{x}_j^k \, e^{i(\lambda_j^k, y)},$$

with $\mathbf{x}_j^k \in X$, $\lambda_j^k \in \mathbf{R}^N$, and $r_k \in \mathbf{N}$. The definition easily extends to real Banach spaces. If $X = \mathbf{R}$, this definition is the usual definition of uniformly almost-periodic functions in the sense of Bohr.

Definition A.2 *If $u : \mathbf{R}^N \to \mathbf{R}$ is an $\mathrm{L}^1_{\mathrm{loc}}$ function, we define the* mean value *of u (over $\mathbf{R}^N$) as*

$$\fint u \, dx = \limsup_{T \to +\infty} \frac{1}{(2T)^N} \int_{[-T,T]^N} u(x) \, dx.$$

Proposition A.3 *If $v \in UAP(\mathbf{R}^N; X)$ then the limit*

$$\lim_{T \to +\infty} \frac{1}{(2T)^N} \int_{a+[-T,T]^N} u(x) \, dx = \fint u \, dx < +\infty$$

exists uniformly in $a \in \mathbf{R}^N$.

Proof Clearly, the statement is true for trigonometric polynomials and passes to the limit for uniform convergence. $\qquad\qquad\square$

Definition A.4 *Let $u : \mathbf{R}^N \to X$ be a function, and let $\eta > 0$. We say that $\tau \in \mathbf{R}^N$ is an η-almost period if*

$$|u(x + \tau) - u(x)| < \eta \qquad \text{for every } x \in \mathbf{R}^N.$$

Definition A.5 *A set $T \subset \mathbf{R}^N$ is* relatively dense *in $\mathbf{R}^N$ if there exists an* inclusion length *$L > 0$ such that $T + [0, L)^N = \mathbf{R}^N$, i.e. for every $z \in \mathbf{R}^N$ there exists $\tau \in T \cap (z + [0, L)^N)$.*

The following theorem states that uniformly almost-periodic functions have relatively dense sets of η-almost periods for all $\eta > 0$.

Theorem A.6 *Let $g \in UAP(\mathbf{R}^N; X)$; then*

(i) *for every $\eta > 0$ the set*

$$T_\eta = \{\tau \in \mathbf{R}^N \ : \ |g(x + \tau) - g(x)| < \eta \ \text{for every} \ x \in \mathbf{R}^N\}$$

is relatively dense in $\mathbf{R}^N$;

(ii) *for every $y \in \mathbf{R}^N$ and $\eta > 0$ the set*

$$T_\eta^y = \{\tau \in \mathbf{R} \ : \ |g(x + \tau y) - g(x)| < \eta \ \text{for every} \ x \in \mathbf{R}^N\}$$

is relatively dense in $\mathbf{R}$.

Proof It is easy to see that (i) and (ii) hold for trigonometric polynomials (see e.g. Besicovitch (1932), Theorems 11 and 12, p. 5), and that these properties pass to the limit for uniform convergence. $\square$

Remark A.7 The converse of Theorem A.6 also holds: if g is a continuous function satisfying (i) then $g \in UAP(\mathbf{R}^N; X)$. A proof of this fact can be easily deduced from Besicovitch (1932) pp. 59–66, where the case $N = 2$ and $X = \mathbf{R}$ is dealt with. From this remark we can deduce that if a is uniformly almost periodic and $|a| \geq c > 0$ then $\frac{1}{a}$ is uniformly almost periodic.

Remark A.8 Note that a function which is uniformly almost periodic in each variable need not be uniformly almost periodic itself, as the function $\cos xy$ shows.

Quasiperiodic functions

Definition A.9 *A function $v : \mathbf{R}^N \to X$ is* quasiperiodic *if it is a diagonal function of a continuous periodic function of a larger number of variables $V :$ $\mathbf{R}^{NM} \to X$, i.e. $v(x) = V(x, \ldots, x)$.*

Remark A.10 An equivalent definition of quasiperiodicity can be given: f is quasiperiodic if there exist $m_1, \ldots, m_n \in \mathbf{N}$ and a continuous function $F : \mathbf{R}^M \to \mathbf{R}$, where $M = m_1 + \cdots + m_N$, $M \geq N$, such that

$$f(x_1, \ldots, x_N) = F(\underbrace{x_1, \ldots, x_1}_{m_1}, \underbrace{x_2, \ldots, x_2}_{m_2}, \ldots, \underbrace{x_N, \ldots, x_N}_{m_N}),$$

and F is periodic in each variable of periods $2\pi/\lambda_1^1, 2\pi/\lambda_1^2, \ldots, 2\pi/\lambda_1^{m_1}, 2\pi/\lambda_2^1, \ldots,$ $2\pi/\lambda_2^{m_2}, \ldots, 2\pi/\lambda_n^{m_N}$, respectively, with $\lambda_r^l \in (0, +\infty)$. It is not restrictive to assume that the frequencies $\lambda_r^1, \ldots, \lambda_r^{m_r}$ are linearly independent on $\mathbf{Z}$ for every $r = 1, \ldots, N$. This will be done constantly in the sequel. Under this assumption, Kronecker's lemma (see below) guarantees that F is uniquely determined by f.

Given $m_1, \ldots, m_N$ as above and given $\lambda \in \mathbf{R}^{m_1} \times \cdots \times \mathbf{R}^{m_N}$, $\lambda = (\lambda_1, \ldots, \lambda_N)$ with $\lambda_r = (\lambda_r^1, \ldots, \lambda_r^{m_r})$ for $r = 1, \ldots, N$, we denote by $QP(\lambda)$ the set of all quasiperiodic functions $f : \mathbf{R}^n \to X$ with frequencies λ. Furthermore, by $Trig(\lambda)$ we indicate the set of all trigonometric polynomials with frequencies λ, i.e. finite sums of terms of the form

$$P(x) = \mathrm{Re}\left(\mathbf{x} \exp\left(i \sum_{l,r} k_r^l \lambda_r^l x_r\right)\right),$$

where $k \in \mathbf{Z}^M$, $\mathbf{x} \in X$.

Remark A.11 Trigonometric polynomials are obviously quasiperiodic functions. Moreover, using the density of trigonometric polynomials in the set of continuous periodic functions, it is easy to prove that every function f of $QP(\lambda)$ is the uniform limit of a sequence of trigonometric polynomials of $Trig(\lambda)$. In particular, quasiperiodic functions are uniformly almost periodic.

It is useful to consider the function $j : \mathbf{R}^N \to \mathbf{R}^M$ defined by

$$j(x) = (\underbrace{x_1, \ldots, x_1}_{m_1}, \underbrace{x_2, \ldots, x_2}_{m_2}, \ldots, \underbrace{x_n, \ldots, x_n}_{m_N}).$$

We also set

$$T = \prod_{l,r}\left(0, \frac{2\pi}{\lambda_r^l}\right) \subset \mathbf{R}^M, \tag{A.1}$$

and we denote by $Trig(T)$ the set of all trigonometric polynomials in $\mathbf{R}^M$ with period T.

The following lemma is a well-known result on Diophantine approximations.

Lemma A.12 (Kronecker's lemma) *The set of vectors which are equivalent modulo T to vectors of the form $j(x)$, with $x \in \mathbf{R}^N$, is dense in $\mathbf{R}^M$.*

The following theorem states that the diagonal is uniformly dense in the periodicity torus T. For simplicity, we state and prove it in the case $X = \mathbf{R}$.

Theorem A.13 (Birkhoff's Ergodic Theorem for quasiperiodic functions) *Let $f : \mathbf{R}^n \to \mathbf{R}$ be a quasiperiodic function in $QP(\lambda)$. Then*

$$\fint f(x)\,dx = \frac{1}{|T|}\int_T F(y)\,dy,$$

where T is defined by (A.1).

Proof Let $W = j(\mathbf{R}^N)$ and $Z = W^\perp$. We identify $\mathbf{R}^M = W \oplus Z$. Moreover, we introduce

$$P_s^M(0) = j(Q_s(0)) \times B_s^Z(0),$$

where $B_s^Z(z)$ is the ball in Z of radius s and centre z, and $Q_s(z)$ is the cube of side-length s and centre $z \in \mathbf{R}^n$. Define

$$S = \{z \in B_s^Z(0) : \ z \text{ is equivalent modulo } T \text{ to some vector } j(\tau) \text{ with } \tau \in \mathbf{R}^N\}.$$

By Kronecker's lemma the set S is dense in $B_s^Z(0)$. Then, given $z \in S$ there exists $\tau = \tau(z) \in \mathbf{R}^N$ such that

$$F(j(x) + z) = f(x + \tau)$$

for every $x \in \mathbf{R}^N$. Since the limit

$$\lim_{s \to +\infty} \frac{1}{s^n} \int_{Q_s(\tau)} f(x)\,dx = \fint f(x)\,dx$$

exists uniformly with respect to τ, given $\eta > 0$ we have

$$\left| \frac{1}{s^n} \int_{Q_s(0)} f(x + \tau)\,dx - \fint f(x)\,dx \right| < \eta$$

for every $\tau \in \mathbf{R}^N$ and for s sufficiently large. Hence, if $z \in S$

$$\left| \frac{1}{s^n} \int_{Q_s(0)} F(j(x) + z)\,dx - \fint f(x)\,dx \right| < \eta \tag{A.2}$$

for s sufficiently large. By the uniform continuity of F and the density of S, the estimate (A.2) holds for every $z \in B_s^Z(0)$. By Fubini's Theorem we have

$$\frac{1}{|P_s^M(0)|} \int_{P_s^M(0)} F(y)\,dy = \frac{1}{s^n} \int_{Q_s(0)} \left(\frac{1}{|B_s^Z(0)|} \int_{B_s^Z(0)} F(j(x) + z)\,dz \right) dx$$

$$= \frac{1}{|B_s^Z(0)|} \int_{B_s^Z(0)} \left(\frac{1}{s^n} \int_{Q_s(0)} F(j(x) + z)\,dx \right) dz,$$

which by (A.1) yields that

$$\left| \frac{1}{|P_s^M(0)|} \int_{P_s^M(0)} F(y)\,dy - \fint f(x)\,dx \right| < \eta.$$

This implies, together with the periodicity of F, that

$$\frac{1}{|T|} \int_T F(y)\,dy = \lim_{s \to +\infty} \frac{1}{|P_s^M(0)|} \int_{P_s^M(0)} F(y)\,dy = \fint f(x)\,dx$$

and concludes the proof. $\qquad\square$

APPENDIX B

CONSTRUCTION OF EXTENSION OPERATORS

In this appendix we include the proof of the existence of suitable extension operators, that are fundamental tools for proving the compactness of the solutions to the homogenization problems considered in Chapter 19. The results of this appendix are due to Acerbi *et al.* (1992).

In this appendix $1 \le p < +\infty$, and $Q = (-\frac{1}{2}, \frac{1}{2})^n$ is the open unit cube of $\mathbf{R}^n$ centred at the origin. If $A \subset \mathbf{R}^n$ is any open set and $\lambda > 0$, we use the notation $A(\lambda)$ for the *retracted set* $\{x \in A : \text{dist}\,(x, \partial A) > \lambda\}$.

Definition B.1 *An open set $E \subseteq \mathbf{R}^n$ has* Lipschitz boundary at $x \in \partial E$ *if ∂E is locally the graph of a Lipschitz function, in the sense that there exist a coordinate system $(y_1, \ldots, y_n)$, a Lipschitz function Φ of $n - 1$ variables, and an open rectangle U_x in the y-coordinates, centred at x, such that $E \cap U_x = \{y : y_n = \Phi(y_1, \ldots, y_{n-1})\}$ and that ∂E splits U_x into two connected sets, $E \cap U_x$ and $U_x \setminus \overline{E}$. If this property holds for every $x \in \partial E$ with the same Lipschitz constant, we say that E has* Lipschitz boundary. *We also say that E has the* cone property *if there exists a finite open cone C such that each point $x \in E$ is the vertex of a finite cone C_x contained in E and congruent to C. It is clear that an open set with Lipschitz boundary has the cone property.*

The main result of this appendix is the following theorem.

Theorem B.2 *Let E be a periodic, connected, open subset of $\mathbf{R}^n$, with Lipschitz boundary, and let $E_\varepsilon = \varepsilon E$. Given a bounded open set $\Omega \subset \mathbf{R}^n$, and a real number $\varepsilon > 0$, there exist a linear and continuous extension operator $T_\varepsilon : W^{1,p}(\Omega \cap E_\varepsilon) \to W_{\text{loc}}^{1,p}(\Omega)$ and three constants $k_0, k_1, k_2 > 0$, such that*

$$T_\varepsilon u = u \quad \text{a.e. in } \Omega \cap E_\varepsilon, \tag{B.1}$$

$$\int_{\Omega(\varepsilon k_0)} |T_\varepsilon u|^p dx \le k_1 \int_{\Omega \cap E_\varepsilon} |u|^p dx, \tag{B.2}$$

$$\int_{\Omega(\varepsilon k_0)} |D(T_\varepsilon u)|^p dx \le k_2 \int_{\Omega \cap E_\varepsilon} |Du|^p dx, \tag{B.3}$$

for every $u \in W^{1,p}(\Omega \cap E_\varepsilon)$. The constants k_0, k_1, k_2 depend on E, n, p, but are independent of ε and Ω.

Note that in general it is not possible to construct a family of extension operators $T_\varepsilon : W^{1,p}(\Omega \cap E_\varepsilon) \to W^{1,p}(\Omega)$ satisfying (B.1)–(B.3) with $\Omega(\varepsilon k_0)$

replaced by Ω, since we do not have any control of the behaviour of E_ε near $\partial\Omega$. In particular, even with $\Omega = Q$, it may happen that $\Omega \cap E_\varepsilon$ is disconnected for every $0 < \varepsilon < 1$, so that (B.3) cannot hold with Ω in place of $\Omega(\varepsilon k_0)$.

The proof of this result will be given after some preliminary lemmas. In order to have an idea of the proof, consider the particular case where the set $E \cap 2Q$ is connected and has Lipschitz boundary. These properties yield the existence of an extension operator $\tau : W^{1,p}(E \cap 2Q) \to W^{1,p}(2Q)$ (see Lemma 3 in Cioranescu and Saint Jean Paulin (1979)) that has separate estimates for the gradients. To construct a global extension operator first we can consider the family $\tau^\alpha : W^{1,p}(E \cap (\alpha + 2Q)) \to W^{1,p}(\alpha + 2Q)$ of extension operators obtained by translating τ by an integer vector $\alpha \in \mathbf{Z}^n$. The next step consists in gluing together the extension operators $(\tau_\alpha)_\alpha$ by means of a periodic partition of unity, and in proving that an estimate between the gradients still holds.

In the general case, $E \cap 2Q$ is neither connected nor Lipschitz. The firsts two lemmas B.3 and B.4 below will take care of the fact that the set $E \cap 2Q$ may be disconnected.

Lemma B.3 *Let A, ω, ω' be open subsets of $\mathbf{R}^n$. Assume that ω, ω' are bounded, with $\omega \subset\subset \omega'$, and that A has Lipschitz boundary at each point of $\partial A \cap \overline{\omega}$. Then the number of connected components of $A \cap \omega'$ that intersect $A \cap \omega$ is finite.*

Proof Since A has Lipschitz boundary at each point of $\partial A \cap \overline{\omega}$, for every $x \in \partial A \cap \overline{\omega}$ there exists an open neighbourhood U_x of x, contained in ω', such that $U_x \cap A$ is connected. For every $x \in A \cap \overline{\omega}$, let U_x be any open ball centred at x, contained in $A \cap \omega'$. Since $\overline{A} \cap \overline{\omega}$ is compact, there exist $x_1, \dots, x_N \in \overline{A}$ such that $\overline{A} \cap \overline{\omega} \subseteq \bigcup_{i=1}^N U_{x_i}$. In particular, $A \cap \omega \subseteq \bigcup_{i=1}^N (U_{x_i} \cap A)$. Now, if C is a connected component of $A \cap \omega'$ that intersects $A \cap \omega$, then C intersects $U_{x_{i_0}} \cap A$ for some $i_0 = 1, \dots, N$. But, since $U_{x_{i_0}} \cap A$ is connected, then $U_{x_{i_0}} \cap A \subseteq C$, and this implies that the number of connected components of $A \cap \omega'$ which intersect $A \cap \omega$ is at most N. $\qquad\square$

Lemma B.4 *Let E be a connected open subset of $\mathbf{R}^n$ with Lipschitz boundary. Then, there exists $k \in \mathbf{N}$, $k \geq 3$, such that $2Q \cap E$ is contained in a single connected component of $kQ \cap E$.*

Proof By applying Lemma B.3 with $A = E$, $\omega = 2Q$, $\omega' = 3Q$, we can assume that $2Q \cap E = \bigcup_{i=1}^N C_i$, where $C_i = C_i' \cap (2Q \cap E)$, and C_i' are connected components of $3Q \cap E$. Fix $x_i \in C_i$, $i = 1, \dots, N$; since E is connected, for every $i, j \in \{1, \dots, N\}$ there exists a continuous map $\gamma_{i,j} : [0,1] \to E$, such that $\gamma_{i,j}(0) = x_i$, $\gamma_{i,j}(1) = x_j$. Moreover, for every i, j, there exists $k_{i,j} \in \mathbf{N}$ such that $\gamma_{i,j}([0,1]) \subseteq k_{i,j}Q$. By taking $k = \max_{i,j} k_{ij}$, we obtain that $2Q \cap E$ is contained in a single connected component of $kQ \cap E$. $\qquad\square$

In the next lemmas B.5 and B.6 the difficulties due to the lack of regularity of the boundary of $E \cap 2Q$ are overcome.

Lemma B.5 *Let A, ω be open subsets of $\mathbf{R}^n$. Assume that ω is bounded and that A is connected and has Lipschitz boundary at each point of $\partial A \cap \overline{\omega}$. Then* *there exists a bounded and connected open set $B \subseteq A$, with the cone property and having Lipschitz boundary at each point of $\partial B \cap \overline{\omega}$, such that $\omega \cap B = \omega \cap A$.*

Proof Since the set $U = \{x \in \partial A : A \text{ has Lipschitz boundary at } x\}$ is open in the relative topology of ∂A, and $\partial A \cap \overline{\omega}$ is a compact subset of U, there exist two bounded open subsets ω_1, ω_2 of $\mathbf{R}^n$, such that $\omega \subset\subset \omega_1 \subset\subset \omega_2$ and A has Lipschitz boundary at the points of $\partial A \cap \overline{\omega_2}$. By Lemma B.3, let $C_1, \ldots, C_N$ be the connected components of $A \cap \omega_2$ that intersect $A \cap \omega_1$, and let $x_i \in C_i$, for every $i = 1, \ldots, N$. Since A is connected, for every $i, j \in \{1, \ldots, N\}$ there exists a continuous map $\gamma_{i,j} : [0,1] \to A$, such that $\gamma_{i,j}(0) = x_i$, $\gamma_{i,j}(1) = x_j$. We denote by S the bounded connected set

$$S = \left(\bigcup_{i=1}^{N} C_i \right) \cup \left(\bigcup_{i,j=1}^{N} \gamma_{ij}([0,1]) \right),$$

and by $K = \overline{S}$ its closure, which is a compact connected subset of $\overline{A}$. Since $\partial A \cap K \subseteq \partial A \cap \overline{\omega_2}$, for every $x \in \partial A$ there exists an open neighbourhood U_x of x, such that $U_x \cap A$ satisfies the cone property. Moreover, for every $x \in A \cap K$, let U_x be an open ball centred at x and contained in A. Since K is compact, there exist $x_1, \ldots, x_M$, such that $K \subseteq \bigcup_{i=1}^{M} U_{x_i}$.

Define $B = \bigcup_{i=1}^{M}(U_{x_i} \cap A)$; by our construction, B is a bounded open subset of A. Moreover, the cone property holds for B, since it holds for each set $U_{x_i} \cap A$. It is easy to check that B is connected, since for every i the set $U_{x_i} \cap A$ is connected and has non-empty intersection with the connected set S. We now prove that $\omega_1 \cap B = \omega_1 \cap A$. The inclusion $\subseteq$ is trivial, since $B \subseteq A$. On the other hand, if $x \in \omega_1 \cap A$, then $x \in C_{i_0}$ for some $i_0 = 1, \ldots, N$. Hence $x \in K \cap A \subseteq \bigcup_{i=1}^{M}(U_{x_i} \cap A) = B$, i.e. $x \in \omega_1 \cap B$. Since $\omega \subset\subset \omega_1$, the conclusions of the lemma follow immediately. $\qquad\square$

Lemma B.6 *Let B, ω be open subsets of $\mathbf{R}^n$. Assume that ω is bounded and that B has Lipschitz boundary at each point of $\partial B \cap \overline{\omega}$. Then, there exists a linear and continuous operator $S : \mathrm{W}^{1,p}(B) \to \mathrm{W}^{1,p}(\omega)$ such that for every $u \in \mathrm{W}^{1,p}(B)$*

$$Su = u \quad \text{a.e. in } B \cap \omega, \tag{B.4}$$

$$\|Su\|_{\mathrm{L}^p(\omega)} \leq c\|u\|_{\mathrm{L}^p(B)}, \tag{B.5}$$

$$\|Su\|_{\mathrm{W}^{1,p}(\omega)} \leq c\|u\|_{\mathrm{W}^{1,p}(B)}, \tag{B.6}$$

where $c = c(n, p, B, \omega)$.

Proof The problem can be localized by means of a partition of unity, and the proof can be obtained by applying the standard reflection technique (see for instance Adams (1975), Theorems 4.26, 4.28 and Section 4.29 for details) to the neighbourhoods of the points of $\partial B \cap \overline{\omega}$. $\qquad\square$

The following lemma states the existence of an extension operator that has separate estimates in terms of the gradients.

Lemma B.7 *Let A, ω be open subsets of $\mathbf{R}^n$. Assume that ω is bounded and that A is connected and has Lipschitz boundary at each point of $\partial A \cap \overline{\omega}$. Then, there exists a linear and continuous operator $\tau : \mathrm{W}^{1,p}(A) \to \mathrm{W}^{1,p}(\omega)$ such that, for every $u \in \mathrm{W}^{1,p}(A)$,*

$$\tau u = u \quad \text{a.e. in } A \cap \omega, \tag{B.7}$$

$$\int_\omega |\tau u|^p dx \le c_1 \int_A |u|^p dx, \tag{B.8}$$

$$\int_\omega |D(\tau u)|^p dx \le c_2 \int_A |Du|^p dx, \tag{B.9}$$

where c_1, c_2 depend only on n, p, A, ω.

Proof By Lemma B.5, there exists a bounded and connected open set $B \subseteq A$, with the cone property, such that $\omega \cap B = \omega \cap A$. Moreover, B has Lipschitz boundary at each point of $\partial B \cap \overline{\omega}$. Then, by applying Lemma B.6 to the sets B, ω, there exists a linear and continuous operator $S : \mathrm{W}^{1,p}(B) \to \mathrm{W}^{1,p}(\omega)$ satisfying (B.4)–(B.6). For every $u \in \mathrm{W}^{1,p}(A)$ we set

$$\tau u = S(u_{|B} - (u)_B) + (u)_B,$$

where $u_{|B}$ denotes the restriction of u to the set B, and $(u)_B = |B|^{-1} \int_B u \, dx$. From the properties of S we have that $\tau u \in \mathrm{W}^{1,p}(\omega)$ and $\tau u = u$ a.e. in $B \cap \omega = A \cap \omega$, i.e. (B.7). Moreover, by (B.5) and Hölder's inequality, we have

$$\int_\omega |\tau u|^p dx = \int_\omega |S(u - (u)_B) + (u)_B|^p dx$$

$$\le c \int_\omega |S(u - (u)_B)|^p dx + c \int_\omega |(u)_B|^p dx$$

$$\le c \int_B |u|^p dx \le c \int_A |u|^p dx,$$

where, for simplicity, the letter c denotes a positive constant that depends only on n, p, B, ω, A, and can change from line to line. Condition (B.8) is then completely proven.

To show (B.9), note that, since B is a bounded open set satisfying the cone property, the imbedding of $\mathrm{W}^{1,p}(B)$ into $\mathrm{L}^p(B)$ is compact by Rellich's Theorem. By taking (B.6) and Poincaré's inequality into account, we have finally

$$\int_\omega |D(\tau u)|^p dx = \int_\omega |D(S(u - (u)_B))|^p dx$$

$$\le c \int_B |u - (u)_B|^p dx + c \int_B |Du|^p dx$$

$$\le c \int_B |Du|^p dx \le c \int_A |Du|^p dx.$$

The proof of Lemma B.7 is then complete. $\qquad\square$

In the rest of the appendix we use the following notation. For every set $A \subseteq \mathbf{R}^n$, for every $\alpha \in \mathbf{Z}^n$, and for every real number $h > 0$ we denote by

$$A_h^\alpha = \alpha + hA \tag{B.10}$$

the translated image of the set $hA = \{hx : x \in A\}$ by the integer vector α. Moreover, we indicate by $\pi_h^\alpha : \mathbf{R}^n \to \mathbf{R}^n$ the invertible affine map defined by

$$\pi_h^\alpha(x) = \alpha + hx, \tag{B.11}$$

for every $x \in \mathbf{R}^n$. When $h = 1$, we simply write $A_1^\alpha = A^\alpha$ and $\pi_1^\alpha = \pi^\alpha$, while for $\alpha = 0 \in \mathbf{Z}^n$ we put $A_h^0 = A_h$ and $\pi_h^0 = \pi_h$.

In the next lemma we glue together the family of all integer translations of the extension operator defined above, by means of a periodic partition of unity.

Lemma B.8 *Let E be a periodic connected open subset of $\mathbf{R}^n$ with Lipschitz boundary. Let Ω, Ω' be open subsets of $\mathbf{R}^n$ such that $\Omega' \subset\subset \Omega$ and $\mathrm{dist}\,(\Omega', \partial\Omega) > 2\sqrt{n}k$, where $k \ge 3$ is the integer given by Lemma B.4. Then there exist two positive constants k_1, k_2, and a linear continuous operator $L : \mathrm{W}^{1,p}(\Omega \cap E) \to \mathrm{W}^{1,p}(\Omega')$, such that*

$$Lu = u \quad \text{a.e. in } \Omega' \cap E, \tag{B.12}$$

$$\int_{\Omega'} |Lu|^p dx \le k_1 \int_{\Omega \cap E} |u|^p dx, \tag{B.13}$$

$$\int_{\Omega'} |D(Lu)|^p dx \le k_2 \int_{\Omega \cap E} |Du|^p dx, \tag{B.14}$$

for every $u \in \mathrm{W}^{1,p}(\Omega \cap E)$. The constants k_1, k_2 depend only on E, n and p, but are independent of Ω and Ω'.

Proof Let C be the connected component of $kQ \cap E$ containing $2Q \cap E$, given by Lemma B.4. Since E has Lipschitz boundary, the set C has Lipschitz boundary at each point of $\partial C \cap 2\overline{Q}$ and we can apply Lemma B.7 to the sets $A = C$ and $\omega = 2Q$. Therefore, there exists a linear and continuous operator $\tau : \mathrm{W}^{1,p}(C) \to \mathrm{W}^{1,p}(2Q)$ such that, for every $u \in \mathrm{W}^{1,p}(C)$,

$$\tau u = u \quad \text{a.e. in } 2Q \cap C = 2Q \cap E, \tag{B.15}$$

$$\int_{2Q} |\tau u|^p dx \le c_1 \int_C |u|^p dx \le c_1 \int_{kQ \cap E} |u|^p dx, \tag{B.16}$$

$$\int_{2Q} |D(\tau u)|^p dx \le c_2 \int_C |Du|^p dx \le c_2 \int_{kQ \cap E} |Du|^p dx, \tag{B.17}$$

where c_1, c_2 depend only on n, p, E.

Now, consider the open cover of $\mathbf{R}^n$ given by the cubes $(Q_2^\alpha)_{\alpha \in \mathbf{Z}^n}$ (we use the notation (B.10)), and for every set $A \subseteq \mathbf{R}^n$ define $I(A) = \{\alpha \in \mathbf{Z}^n : Q_2^\alpha \cap A \neq \emptyset\}$. Under our assumptions, for every $\alpha \in I(\Omega')$ we have $Q_{2k}^\alpha \subseteq \Omega$.

For every $\alpha \in I(\Omega')$ we define by $\tau^\alpha : \mathrm{W}^{1,p}(C^\alpha) \to \mathrm{W}^{1,p}(Q_2^\alpha)$ the extension operator obtained by translating the operator τ, i.e. for every $u \in \mathrm{W}^{1,p}(C^\alpha)$

$$\tau^\alpha u = (\tau(u \circ \pi^\alpha)) \circ \pi^{-\alpha} \tag{B.18}$$

(we use the notation (B.11)). For simplicity, if $u \in \mathrm{W}^{1,p}(\Omega \cap E)$ we denote by u^α the function

$$u^\alpha = \tau^\alpha(u_{|C^\alpha}) \in \mathrm{W}^{1,p}(Q_2^\alpha). \tag{B.19}$$

In order to define a global extension operator $L : \mathrm{W}^{1,p}(\Omega \cap E) \to \mathrm{W}^{1,p}(\Omega')$, we consider a partition of unity $(\varphi^\alpha)_\alpha$ associated to the open cover $(Q_2^\alpha)_\alpha$, such that $\varphi^\beta = \varphi^\alpha \circ \pi^{\alpha-\beta}$, for every $\alpha, \beta \in \mathbf{Z}^n$. A partition of unity of this kind can be constructed, for instance, by choosing a non-negative function $\varphi \in C_0^\infty(\mathbf{R}^n)$, with spt $\varphi \subset\subset (2Q)$ and $\varphi > 0$ on $\frac{3}{2}Q$, and by setting

$$\varphi^\alpha(x) = \frac{\varphi(x - \alpha)}{\sum_{\beta \in \mathbf{Z}^n} \varphi(x - \beta)} \tag{B.20}$$

for every $\alpha \in \mathbf{Z}^n$ and for every $x \in \mathbf{R}^n$. In this way $\sum_{\alpha \in I(\Omega')} \varphi^\alpha(x) = 1$ for every $x \in \Omega'$, and there exists a positive constant M such that

$$(|\varphi^\alpha(x)| + |D\varphi^\alpha(x)|) \leq M, \tag{B.21}$$

for every $\alpha \in \mathbf{Z}^n$ and for every $x \in \mathbf{R}^n$.

Now, for every $u \in \mathrm{W}^{1,p}(\Omega \cap E)$ we set

$$Lu = \sum_{\alpha \in I(\Omega')} u^\alpha \varphi^\alpha, \tag{B.22}$$

with u^α, φ^α given by (B.19), (B.20), respectively. It is clear that L is a linear operator from $\mathrm{W}^{1,p}(\Omega \cap E)$ to $\mathrm{W}^{1,p}(\Omega')$ and that condition (B.12) is satisfied, since by (B.19), (B.18) and (B.15) we have

$$\sum_{\alpha \in I(\Omega')} u^\alpha(x)\varphi^\alpha(x) = \sum_{\alpha \in I(\Omega')} u(x)\varphi^\alpha(x) = u(x)$$

for a.e. $x \in \Omega' \cap E$. To prove (B.13), note that, by (B.21), for every $u \in \mathrm{W}^{1,p}(\Omega \cap E)$ and for every $\beta \in I(\Omega')$

$$\int_{Q_2^\beta} |Lu|^p dx \leq N^{p-1} M^p \sum_{\alpha \in I(Q_2^\beta)} \int_{Q_2^\alpha} |u^\alpha|^p dx, \tag{B.23}$$

where, from now on, N denotes the cardinality of the set $I(Q_2^\beta)$. Since by (B.19) and (B.16) we have for every $\alpha \in I(Q_2^\beta)$

$$\int_{Q_2^\alpha} |u^\alpha|^p dx \le c_1 \int_{Q_k^\alpha \cap E} |u|^p dx,$$

and since $\alpha \in I(Q_2^\beta)$ implies $Q_k^\alpha \subseteq Q_{2k}^\beta$, then from (B.23) we obtain

$$\int_{Q_2^\beta} |Lu|^p dx \le N^p M^p c_1 \int_{Q_{2k}^\beta \cap E} |u|^p dx.$$

By taking the sum over $\beta \in I(\Omega')$ in the previous inequality, we have

$$\int_{\Omega'} |Lu|^p dx \le \sum_{\beta \in I(\Omega')} \int_{Q_2^\beta} |Lu|^p dx$$
$$\le N^p M^p c_1 \sum_{\beta \in I(\Omega')} \int_{Q_{2k}^\beta \cap E} |u|^p dx \le N^p M^p c_1 c(k) \int_{\Omega \cap E} |u|^p dx,$$

$$\text{(B.24)}$$

where $c(k)$ is a constant depending only on k and n, such that each point $x \in \mathbf{R}^n$ is contained in at most $c(k)$ cubes of the form $(Q_{2k}^\beta)_{\beta \in \mathbf{Z}^n}$. Hence (B.13) is proven with $k_1 = N^p M^p c_1 c(k)$.

In order to prove (B.14), for every $u \in W^{1,p}(\Omega \cap E)$ and for every $\beta \in I(\Omega')$ we write

$$\int_{Q_2^\beta} |D(Lu)|^p dx \le 2^{p-1} \int_{Q_2^\beta} \Big| \sum_{\alpha \in I(Q_2^\beta)} Du^\alpha \varphi^\alpha \Big|^p dx$$
$$+ 2^{p-1} \int_{Q_2^\beta} \Big| \sum_{\alpha \in I(Q_2^\beta)} u^\alpha D\varphi^\alpha \Big|^p dx. \qquad \text{(B.25)}$$

By (B.17), the first term can be estimated as follows:

$$\int_{Q_2^\beta} \Big| \sum_{\alpha \in I(Q_2^\beta)} Du^\alpha \varphi^\alpha \Big|^p dx \le N^{p-1} \sum_{\alpha \in I(Q_2^\beta)} \int_{Q_2^\beta \cap Q_2^\alpha} |Du^\alpha \varphi^\alpha|^p dx$$
$$\le N^{p-1} M^p c_2 \sum_{\alpha \in I(Q_2^\beta)} \int_{Q_k^\alpha \cap E} |Du|^p dx$$
$$\le N^{p-1} M^p c_2 \int_{Q_{2k}^\beta \cap E} |Du|^p dx.$$

Since $\sum_{\alpha \in I(Q_2^\beta)} D\varphi^\alpha(x) = 0$ in Q_2^β, the second integrand on the right-hand side of (B.25) can be written as

$$\sum_{\alpha \in I(Q_2^\beta)} u^\alpha D\varphi^\alpha = \sum_{\alpha \in I(Q_2^\beta)} \left((u^\alpha - u^\beta) D\varphi^\alpha + u^\beta D\varphi^\alpha \right) = \sum_{\alpha \in I(Q_2^\beta)} (u^\alpha - u^\beta) D\varphi^\alpha$$

for a.e. $x \in Q_2^\beta$. Hence, by (B.23) we have

$$\int_{Q_2^\beta} |\sum_{\alpha \in I(Q_2^\beta)} u^\alpha D\varphi^\alpha|^p dx = \int_{Q_2^\beta} |\sum_{\alpha \in I(Q_2^\beta)} (u^\alpha - u^\beta) D\varphi^\alpha|^p dx$$

$$\leq N^{p-1} M^p \sum_{\alpha \in I(Q_2^\beta)} \int_{Q_2^\alpha \cap Q_2^\beta} |u^\alpha - u^\beta|^p dx.$$

Since $u^\alpha - u^\beta = 0$ a.e. in $Q_2^\alpha \cap Q_2^\beta \cap E$, the Poincaré inequality on $Q_2^\alpha \cap Q_2^\beta$ yields

$$\int_{Q_2^\alpha \cap Q_2^\beta} |u^\alpha - u^\beta|^p dx \leq c \int_{Q_2^\alpha \cap Q_2^\beta} |Du^\alpha - Du^\beta|^p dx$$

for a suitable constant c depending only on n, p and E, which together with (B.17) implies that

$$\int_{Q_2^\beta} |\sum_{\alpha \in I(Q_2^\beta)} u^\alpha D\varphi^\alpha|^p dx \leq c(n, p, E, M, N)c_2 \int_{Q_{2k}^\beta \cap E} |Du|^p dx, \qquad \text{(B.26)}$$

for every $u \in W^{1,p}(\Omega \cap E)$ and for every $\beta \in I(\Omega')$. From (B.25)–(B.26) we get finally

$$\int_{Q_2^\beta} |D(Lu)|^p dx \leq c(n, p, E) \int_{Q_{2k}^\beta \cap E} |Du|^p dx.$$

Now, to conclude the proof of (B.14) it is enough to sum over $\beta \in I(\Omega')$ in the previous inequality, and the conclusion follows as in (B.24). $\qquad\square$

Proof of Theorem B.2 Let $\varepsilon > 0$ and $k_0 = 4\sqrt{n}k$, where k is the integer given by Lemma B.4. By means of Lemma B.8 we can construct a linear operator $L_\varepsilon : W^{1,p}(\Omega \cap E_\varepsilon) \to W^{1,p}(\Omega(\varepsilon k_0/2))$ such that

$$L_\varepsilon u = u \quad \text{a.e. in } \Omega(\varepsilon k_0/2) \cap E_\varepsilon, \qquad \text{(B.27)}$$

$$\int_{\Omega(\varepsilon k_0/2)} |L_\varepsilon u|^p dx \leq k_1 \int_{\Omega \cap E_\varepsilon} |u|^p dx, \qquad \text{(B.28)}$$

$$\int_{\Omega(\varepsilon k_0/2)} |D(L_\varepsilon u)|^p dx \leq k_2 \int_{\Omega \cap E_\varepsilon} |Du|^p dx, \qquad \text{(B.29)}$$

for every $u \in W^{1,p}(\Omega \cap E_\varepsilon)$. In fact, since $\text{dist}\,(\frac{1}{\varepsilon}\Omega(\varepsilon k_0/2), \partial(\frac{1}{\varepsilon}\Omega)) > k_0/2 = 2\sqrt{n}k$, and $u \circ \pi_\varepsilon \in W^{1,p}(\frac{1}{\varepsilon}\Omega \cap E)$ if $u \in W^{1,p}(\Omega \cap E_\varepsilon)$, Lemma B.8 ensures the existence of a linear operator $L : W^{1,p}(\frac{1}{\varepsilon}\Omega \cap E) \to W^{1,p}(\frac{1}{\varepsilon}\Omega(\varepsilon k_0/2))$, such that $Lv = v$ a.e. in $\frac{1}{\varepsilon}\Omega(\varepsilon k_0/2) \cap E$, for every $v \in W^{1,p}(\frac{1}{\varepsilon}\Omega \cap E)$. Moreover, the following estimates

$$\int_{\frac{1}{\varepsilon}\Omega(\varepsilon k_0/2)} |Lv|^p dx \le k_1 \int_{\frac{1}{\varepsilon}\Omega \cap E} |v|^p dx,$$

$$\int_{\frac{1}{\varepsilon}\Omega(\varepsilon k_0/2)} |D(Lv)|^p dx \le k_2 \int_{\frac{1}{\varepsilon}\Omega \cap E} |Dv|^p dx,$$

hold for every $v \in W^{1,p}(\frac{1}{\varepsilon}\Omega \cap E)$, where k_1, k_2 are the constants given in Lemma B.8 and, in particular, are independent of ε. Define $L_\varepsilon u = (L(u \circ \pi_\varepsilon)) \circ \pi_{(1/\varepsilon)}$. It is clear that $L_\varepsilon u \in W^{1,p}(\Omega(\varepsilon k_0/2))$ and that it satisfies (B.27)–(B.29).

To complete the proof of Theorem B.2, we have to construct an extension operator $T_\varepsilon : W^{1,p}(\Omega \cap E_\varepsilon) \to W^{1,p}_{\mathrm{loc}}(\Omega)$. To this end we choose a locally finite open cover $(A_i)_{i \in \mathbf{N}}$ of Ω such that $A_0 = \Omega(\varepsilon k_0/2)$, $A_i \subset\subset \Omega$, $A_i \cap \overline{\Omega(\varepsilon k_0)} = \emptyset$ for every $i \ne 0$. Let $(\varphi_i)_{i \in \mathbf{N}}$ be a partition of unity associated to the sequence (A_i), i.e. a sequence of functions $\varphi_i \in C_0^\infty(\mathbf{R}^n)$, with spt $\varphi_i \subset\subset A_i$, and $\sum_{i=0}^\infty \varphi_i(x) = 1$ for every $x \in \Omega$. In particular we have $\varphi_0(x) \equiv 1$ in $\Omega(\varepsilon k_0)$. Now, for every $\varepsilon > 0$, $i \in \mathbf{N} \setminus \{0\}$, by Lemma B.6 applied to $B = \Omega \cap E_\varepsilon$ and $\omega = A_i$, there exists a linear and continuous operator $L_{\varepsilon i} : W^{1,p}(\Omega \cap E_\varepsilon) \to W^{1,p}(A_i)$ such that $L_{\varepsilon i} u = u$ a.e. in $A_i \cap E_\varepsilon$, and for which estimates of the type (B.5)–(B.6) hold. For $i = 0$ we choose $L_{\varepsilon 0} = L_\varepsilon$, where L_ε is the operator satisfying (B.27)–(B.29). Finally, for every $u \in W^{1,p}(\Omega \cap E_\varepsilon)$ we set

$$T_\varepsilon u = \sum_{i=0}^\infty (L_{\varepsilon i} u) \varphi_i$$

where the function $(L_{\varepsilon i} u)\varphi_i \in W_0^{1,p}(A_i)$ is extended to the whole set Ω by the constant 0. It is easy to check that $T_\varepsilon u \in W^{1,p}_{\mathrm{loc}}(\Omega)$, that T_ε is linear and continuous from $W^{1,p}(\Omega \cap E_\varepsilon)$ into $W^{1,p}_{\mathrm{loc}}(\Omega)$, and that (B.1) is satisfied. Moreover, since $\varphi_0 \equiv 1$ in $\Omega(\varepsilon k_0)$ and $\Omega(\varepsilon k_0) \cap$ spt $\varphi_i = \emptyset$ for every $i \ge 1$, we have $T_\varepsilon u = L_\varepsilon u$ a.e. in $\Omega(\varepsilon k_0)$; hence (B.2) and (B.3) follow immediately from (B.28) and (B.29). $\qquad\square$

APPENDIX   C

SOME REGULARITY RESULTS

In this appendix we present some regularity results that are used in the book.

L^p estimates for the Laplace operator

We recall a well-known estimate for the Laplace operator, which is used in Chapters 6 and 18 (for a proof see Morrey (1966), Stein (1970)).

Theorem C.1 *Let $p > 1$ and Ω be a bounded open subset of $\mathbf{R}^n$. Let $f \in \mathrm{W}^{1,p}(\Omega; \mathbf{R}^m)$, and let $u \in \mathrm{W}_0^{1,2}(\Omega; \mathbf{R}^m)$ be a weak solution to the problem*

$$\begin{cases} -\Delta u = f \\ u \in \mathrm{W}_0^{1,2}(\Omega; \mathbf{R}^m). \end{cases}$$

Then $u \in \mathrm{W}_0^{1,p}(\Omega; \mathbf{R}^m)$ and

$$\|Du\|_{\mathrm{L}^p(\Omega; \mathbf{M}^{m\times n})} \leq c\|f\|_{\mathrm{W}^{-1,p}(\Omega; \mathbf{R}^m)},$$

where c is a positive constant.

Meyers regularity result for non-linear elliptic systems

The following partial regularity result for non-linear elliptic systems due to Meyers and Elcrat (1975) is used in Chapters 17 and 23.

Theorem C.2 *Let Ω be a bounded open subset of $\mathbf{R}^n$ and let $1 < p < +\infty$. For every $j = 1, \ldots, n$ consider Carathéodory functions*

$$A_j : \Omega \times \mathbf{M}^{m\times n} \to \mathbf{R}^m, \qquad A_0 : \Omega \times \mathbf{R}^m \to \mathbf{R}^m$$

Let $u \in \mathrm{W}_0^{1,p}(\Omega; \mathbf{R}^m)$, and assume

(i) $\displaystyle\sum_{j=1}^{n}\langle A_j(x, Du(x)), D_j u(x)\rangle \geq |Du|^p - a(x)$ *a.e., where $a \in L^\gamma(\Omega)$, $\gamma > 1$;*

(ii) $|A_j(x, Du(x))| \leq a_1|Du(x)|^{p-1} + b_1(x)$ *a.e., where $a_1 \geq 0$, $b_1 \in L^{\beta_1}(\Omega)$, $\beta_1 > \max\{1, p'\}$;*

$|A_0(x, u(x))| \leq b_0(x)$ *a.e., with $b_0 \in L^{\beta_0}(\Omega)$, $\beta_0 > \max\{1, np'/(n + p')\}$;*

(iii) $\displaystyle-\sum_{j=1}^{n} D_j(A_j(x, Du(x))) + A_0(x, u) = 0$ *in the sense of distributions.*

Then there exists $\eta > 0$ such that

$$\|u\|_{W^{1,p+\eta}(\Omega;\mathbf{R}^m)} \leq c\Big(\|a\|_{L^\gamma(\Omega)}^{1/p}$$

$$+ \sum_{k=0}^{1} \|b_k\|_{L^{\beta_k}(\Omega)}^{1/p} \|u\|_{W^{1,p}(\Omega;\mathbf{R}^m)}^{1/p} + \|u\|_{W^{1,p}(\Omega;\mathbf{R}^m)}\Big),$$

where η depends only on $\{n,m,p,\gamma,a_1,b_0,b_1\}$, while the constant c depends in addition on Ω.

Remark C.3 We can deduce from the previous result a proof for Theorem 17.1, for which we adapt the notation.

We first assume that $f(x,\cdot)$ is of class C^2. Then the minimum points u_λ of $G_\lambda(u)$ on $W_0^{1,p}(B;\mathbf{R}^m)$ satisfy the Euler equations

$$-\mathrm{div}\left(\frac{\partial f}{\partial \xi}(x, Du_\lambda + D\overline{u})\right) + \lambda p u_\lambda |u_\lambda|^{p-2} = 0.$$

Set $A(x, Du) = (A_1(x, Du), \ldots, A_n(x, Du)) = \frac{\partial f}{\partial \xi}(x, Du + D\overline{u})$ and $A_0(x, u) = \lambda p u |u|^{p-2}$. By the assumptions on f, the functions A_j and A_0 satisfy the hypotheses of Theorem C.2 and, taking $\Omega = B$, we get that there exists $\eta > 0$ such that

$$\|u_\lambda\|_{W^{1,p+\eta}(B;\mathbf{R}^m)} \leq C.$$

If f is not of class C^2, by a convolution argument we can take a sequence (f^k) of quasiconvex functions, twice differentiable in the second variable, converging uniformly to f. Setting

$$G_\lambda^k(u) = \int_B f^k(x, Du + D\overline{u})\, dx + \lambda \int_B |u|^p\, dx$$

we have that $G_\lambda(u) = \Gamma(\mathrm{L}^p)\text{-}\lim_k G_\lambda^k(u)$ for every $W_0^{1,p}(B;\mathbf{R}^m)$, so that if u_λ is a minimum point of G_λ, there exists a sequence (u_λ^k) (u_λ^k minimum point of G_λ^k) converging weakly to u_λ in $W^{1,p}(B;\mathbf{R}^m)$. We have

$$\|u_\lambda^k\|_{W^{1,p+\eta}(B;\mathbf{R}^m)} \leq C,$$

so that, up to a subsequence, (u_λ^k) converges weakly to u_λ in $W^{1,p+\eta}(B;\mathbf{R}^m)$. It follows by the lower semicontinuity of the norm that

$$\|u_\lambda\|_{W^{1,p+\eta}(B;\mathbf{R}^m)} \leq \liminf_k \|u_\lambda^k\|_{W^{1,p+\eta}(B;\mathbf{R}^m)} \leq C,$$

which concludes the proof.

A theorem on rigid motions

The following theorem can be found in Reshetnyak (1967). It is used in Chapter 14.

Theorem C.4 *Let Ω be a connected open subset in $\mathbf{R}^n$ and $u \in \mathrm{W}^{1,\infty}(\mathbf{R}^n; \mathbf{R}^n)$. If*

$$Du^t Du = I \qquad and \ \det Du > 0$$

a.e. in Ω then u is a rigid motion.

Higher integrability of gradients

The following lemma (see Fonseca *et al.* (1998)), that allows to pass from bounded sequences to sequences with equi-integrable p-th power of the gradient, can be sometimes helpful. It is used in Chapter 22.

Lemma C.5 *For every bounded sequence (u_j) in $\mathrm{W}^{1,p}(\Omega; \mathbf{R}^m)$ there exists a subsequence (not relabelled) and a sequence (v_j) in $\mathrm{W}^{1,p}(\Omega; \mathbf{R}^m)$ such that*

$$\lim_j |\{u_j \neq v_j\} \cup \{Du_j \neq Dv_j\}| = 0$$

and $(|Dv_j|^p)$ is equi-integrable.

Remark C.6 Let $f_j \in \mathcal{F}(\alpha, \beta, p)$ (as in Chapter 12) and let (u_j) be a bounded sequence in $\mathrm{W}^{1,p}(\Omega; \mathbf{R}^m)$ with $u_j \to u$ in $\mathrm{L}^p(\Omega; \mathbf{R}^m)$. Then there exist a subsequence, still denoted by (u_j), and a sequence (v_j) in $\mathrm{W}^{1,p}(\Omega; \mathbf{R}^m)$ such that $v_j \to u$ in $\mathrm{L}^p(\Omega; \mathbf{R}^m)$, $(|Dv_j|^p)$ is equi-integrable and

$$\limsup_j \int_\Omega f_j(x, Dv_j)\,dx \leq \limsup_j \int_\Omega f_j(x, Du_j)\,dx.$$

To check this, choose (v_j) as in Lemma C.5, so that $v_j \to u$ and

$$\limsup_j \int_\Omega f_j(x, Du_j)\,dx \geq \limsup_j \int_{\{u_j = v_j\} \cap \{Du_j = Dv_j\}} f_j(x, Du_j)\,dx$$
$$= \limsup_j \int_\Omega f_j(x, Dv_j)\,dx$$

the last equality following from the equi-integrability of $(|Dv_j|^p)$, and the growth conditions on f_j.

REFERENCES

1. Acerbi, E. and Buttazzo, G. (1983) On the limits of periodic Riemannian metrics. *J. Anal. Math.* **43**, 183–201.
2. Acerbi, E., Chiadò Piat, V., Dal Maso, G. and Percivale, D. (1992) An extension theorem from connected sets, and homogenization in general periodic domains. *Nonlinear Anal.* **18**, 481–496.
3. Acerbi, E. and Fusco, N. (1984) Semicontinuity problems in the calculus of variations. *Arch. Ration. Mech. Anal.* **86**, 125–145.
4. Adams, R.A. (1975) *Sobolev Spaces.* Academic Press, New York.
5. Allaire, G. (1992) Homogenization and two-scale convergence. *SIAM J. Math. Anal.* **23**, 1482–1518.
6. Ambrosio, L. and Braides, A. (1990) Functionals defined on partitions of sets of finite perimeter, II: semicontinuity, relaxation and homogenization. *J. Math. Pures Appl.* **69**, 307–333.
7. Ambrosio, L., Buttazzo, G. and Fonseca, I. (1996) Lower semicontinuity problems in Sobolev spaces with respect to a measure. *J. Math. Pures Appl.* **75**, 211–224.
8. Ansini, N., Braides, A. and Chiadò Piat, V. (1997) Homogenization of periodic multi-dimensional structures, *Boll. Un. Mat. Ital.*, to appear.
9. Attouch, H. (1984) *Variational Convergence for Functions and Operators.* Pitman, Boston.
10. Avellaneda, M. (1987) Iterated homogenization, differential effective medium theory and applications. *Commun. Pure Appl. Math.* **40**, 527–556.
11. Bakhvalov, N.S. and Panasenko, G.P. (1989) *Homogenisation: Averaging Processes in Periodic Media.* Kluwer, Dordrecht.
12. Ball, J.M. (1977) Convexity conditions and existence theorems in nonlinear elasticity. *Arch. Ration. Mech. Anal.* **63**, 337–403.
13. Ball, J.M. and Murat, F. (1984) $W^{1,p}$-quasiconvexity and variational problems for multiple integrals. *J. Funct. Anal.* **58**, 225–253.
14. Bendsøe, M.P. and Kikuchi, N. (1988) Generating optimal topologies in structural design using a homogenization method. *Comput. Meth. Appl. Mech. Eng.* **71**, 197–224.
15. Bensoussan, A., Lions, J.L. and Papanicolaou, G. (1978) *Asymptotic Analysis of Periodic Structures.* North-Holland, Amsterdam.
16. Besicovitch, A. (1932) *Almost Periodic Functions.* Cambridge University Press, Cambridge.
17. Bouchitté, G. (1985) Homogénéisation sur $BV(\Omega)$ de fonctionnelles intégrales à croissance linéaire. *C. R. Acad. Sci. Paris Sér. I Math.* **301**, 785–788.
18. Bouchitté, G., Buttazzo, G. and Seppecher, P. (1997) Energies with respect to a measure and applications to low dimensional structures. *Calc. Var.* **5**,

37–54.

19. Braides, A. (1983) Omogeneizzazione di integrali non coercivi. *Ric. Mat.* **32**, 347–368.

20. Braides, A. (1985) Homogenization of some almost periodic functional. *Rend. Accad. Naz. Sci. XL* **103**, 313–322.

21. Braides, A. (1986) A homogenization theorem for weakly almost periodic functionals. *Rend. Accad. Naz. Sci. XL* **104**, 261–281.

22. Braides, A. (1991) Correctors for the homogenization of almost periodic monotone operators. *Asymptotic Anal.* **5**, 47–74.

23. Braides, A. (1992) Almost periodic methods in the theory of homogenization. *Appl. Anal.* **47**, 259–277.

24. Braides, A. (1994) Loss of polyconvexity by homogenization. *Arch. Ration. Mech. Anal.* **127**, 183–190.

25. Braides, A. and Chiadò Piat, V. (1994) Remarks on the homogenization of connected media. *Nonlinear Anal.* **22**, 391–407.

26. Braides, A., Chiadò Piat, V. and Defranceschi, A. (1992) Homogenization of almost periodic monotone operators. *Ann. Inst. H. Poincaré, Anal. Non Linéaire* **9**, 399–432.

27. Braides, A., Defranceschi, A. and Vitali, E. (1996) Homogenization of free discontinuity problems. *Arch. Ration. Mech. Anal.* **135**, 297–356.

28. Braides, A. and Garroni, A. (1995) Homogenization of nonlinear media with soft and stiff inclusions. *Math. Mod. Methods Appl. Sci.* **5**, 543–564.

29. Braides, A. and Lukkassen, D. (1997) Reiterated homogenization of integral functionals, *Math. Mod. Methods Appl. Sci.*, to appear.

30. Bruno, O. P. (1991) The effective conductivity of strongly heterogeneous composites. *Proc. R. Soc. London A*, **433**, 353–381.

31. Bruno, O.P. and Leo, P.H. (1992) On the stiffness of materials containing a disordered array of microscopic holes or hard inclusions. *Arch. Ration. Mech. Anal.* **121**, 303–338.

32. Buttazzo, G. (1989) *Semicontinuity, Relaxation and Integral Representation in the Calculus of Variations.* Pitman, London.

33. Buttazzo, G. and Dal Maso, G. (1978) Γ-limit of a sequence of non-convex and non-equi-Lipschitz integral functionals. *Ric. Mat.* **27**, 235-251.

34. Buttazzo, G., Dal Maso, G. and Mosco, U. (1987) A derivation theorem for capacities with respect to a Radon measure. *J. Funct. Anal.* **71**, 263–278.

35. Cherkaev, A.V. and Kohn, R.V. (eds) (1997) *Topics in the Mathematical Modeling of Composite Materials.* Birkhäuser, Boston.

36. Ciarlet, P.G. (1988) *Mathematical Elasticity.* North-Holland, Amsterdam.

37. Cioranescu, D. and Murat, F. (1982) Un terme etrange venu d'ailleur. *Nonlinear Partial Differential Equations and Their Applications.* Res. Notes in Math. **60**, Pitman, London, 98–138.

38. Cioranescu, D. and Saint Jean Paulin, J. (1979) Homogenization in open sets with holes. *J. Math. Anal. Appl.* **71**, 590–607.

39. Corbo Esposito, A. and De Arcangelis, R. (1992) The Lavrentieff phenomenon and different processes of homogenization. *Commun. Partial Dif-*

fer. Equations **17**, 1503–1538.

40. Crandall, M.G., Ishii, H. and Lions, P.L. (1992) User's guide to viscosity solutions of second order partial differential equations. *Bull. AMS* **27**, 1–67.

41. Dacorogna, B. (1989) *Direct Methods in the Calculus of Variations.* Springer Verlag, Berlin.

42. Dal Maso, G. (1993) *An Introduction to Γ-convergence.* Birkhäuser, Boston.

43. Dal Maso, G. (1995) Problemi di semicontinuità e rilassamento nel calcolo delle variazioni. *Quaderni dell'Unione Matematica Italiana* **39**, Pitagora, Bologna, 145–196.

44. Dal Maso, G. and Dell'Antonio, G.F. (eds) (1991) *Composite Media and Homogenization Theory.* Birkhäuser, Boston.

45. Dal Maso, G. and Dell'Antonio, G.F. (eds) (1995) *Composite Media and Homogenization Theory. Proceedings of the second Workshop.* World Scientific, Singapore.

46. Dal Maso, G. and Defranceschi, A. (1990) Correctors for the homogenization of monotone operators. *Differ. Integral Equations* **3**, 1151–1166.

47. Dal Maso, G. and Mosco, U. (1987) Wiener's criterion and Γ-convergence. *Appl. Math. Optim.* **15**, 15–63.

48. De Giorgi, E. (1975) Sulla convergenza di alcune successioni di integrali del tipo dell'area. *Rend. Mat.* **8**, 277–294.

49. De Giorgi, E. and Franzoni, T. (1975) Su un tipo di convergenza variazionale. *Atti Accad. Naz. Lincei Rend. Cl. Sci. Mat.* **58**, 842–850.

50. De Giorgi, E. and Letta, G. (1977) Une notion générale de convergence faible pour des fonctions croissantes d'ensemble. *Ann. Scuola Norm. Sup. Pisa Cl. Sci.* **4**, 61–99.

51. De Giorgi, E. and Spagnolo, S. (1973) Sulla convergenza degli integrali dell'energia per operatori ellittici del secondo ordine. *Boll. Un. Mat. Ital.* **8**, 391–411.

52. Dellacherie, C. and Meyer, P.-A. (1975) *Probabilités et potentiel.* Hermann, Paris.

53. Donaldson, T.K. and Trudinger, N.S. (1971) Orlicz-Sobolev spaces and imbedding theorems. *J. Funct. Anal.* **8**, 52–75.

54. E, W. (1991) A class of homogenization problems in the calculus of variations. *Commun. Pure Appl. Math.* **44**, 733–759.

55. Ekeland, I. and Temam, R. (1976) *Convex Analysis and Variational Problems.* North-Holland, Amsterdam.

56. Evans, L.C. (1989) The perturbed test function method for viscosity solutions of nonlinear PDE. *Proc. R. Soc. Edinburgh A* **111**, 359–375.

57. Evans, L.C. (1990) *Weak Convergence Methods in Nonlinear PDEs.* AMS, Providence.

58. Evans, L.C. and Gariepy, R.F. (1992) *Measure Theory and Fine Properties of Functions.* CRC Press, Boca Raton.

59. Fonseca, I., Müller, S. and Pedregal, P. (1998) Analysis of concentration and oscillation effects generated by gradients. *SIAM J. Math. Anal.* **29**, 736–756.

60. Francfort, G. and Müller, S. (1994) Combined effect of homogenization and singular perturbations in elasticity. *J. reine angew. Math.* **454**, 1–35.

61. Francfort, G. and Murat, F. (1986) Homogenization and optimal bounds in linear elasticity. *Arch. Ration. Mech. Anal.* **94**, 307–334.

62. Fusco, N. (1983) On the convergence of integral functionals depending on vector-valued functions. *Ric. Mat.* **32**, 321–339.

63. Geymonat, G., Müller, S. and Tryantafillidis, N. (1993) Bifurcation and macroscopic loss of rank-one convexity. *Arch. Ration. Mech. Anal.* **122**, 231–290.

64. Gossez, J.-P. (1974) Nonlinear elliptic boundary value problems for equations with rapidly (or slowly) increasing coefficients. *Trans. AMS* **190**, 163–204.

65. Hashin, Z. and Strikman, S. (1963) A variational approach to the theory of the elastic behavior of multiphase materials. *J. Mech. Phys. Solids*, **11**, 127–140.

66. Kohn, R.V. and Strang, G. (1986) Optimal design and relaxation of variational problems I–III. *Commun. Pure Appl. Math.* **39**, 113–137, 139–182, 353–377.

67. Kozlov, S. (1989) Geometric aspects of averaging. *Russ. Math. Surv.* **44**, 91–144.

68. Krasnoselskii, M.A. and Rutickii, Ya.B. (1961) *Convex Functions and Orlicz Spaces*. P. Noordhoff, Groningen.

69. Lions, P.L. (1982) *Generalized solutions of Hamilton-Jacobi equations*. Pitman, Boston.

70. Lions, P.L., Papanicolaou, G. and Varadhan, S.R.S. (1987) Homogenization of Hamilton-Jacobi equations. Unpublished note.

71. Lurie, K.A. and Cherkaev, A.V. (1984) G-closure of a set of anisotropically conductivity media in the two-dimensional case. *J. Optim. Theory Appl.* **42**, 283–304.

72. Lurie, K.A., Cherkaev, A.V. and Fedorov, A. (1982) Regularization of optimal design problems for bars and plates I, II. *J. Optim. Theory Appl.* **39**, 499–521, 523–543.

73. Marcellini, P. (1978) Periodic solutions and homogenization of nonlinear variational problems. *Ann. Mat. Pura Appl.* **117**, 481–498.

74. Meyers, N. and Elcrat, A. (1975) Some results on regularity for solutions of nonlinear elliptic systems and quasiregular functions. *Duke Math. J.* **42**, 121–136.

75. Migorski, S., Mortola, S. and Traple, J. (1992) Homogenization of first order differential operators. *Rend. Accad. Naz. Sci. XL* **110**, 259–276.

76. Milton, G.W. (1990) On characterizing the set of possible effective tensors of composites: the variational method and the translation method. *Commun. Pure Appl. Math.* **43**, 63–125.

77. Milton, G.W. (1992) Composite materials with Poisson's ratios close to -1. *J. Mech. Phys. Solids* **40**, 1105–1137.

78. Morrey, C.B. (1952) Quasiconvexity and the semicontinuity of multiple integrals. *Pacific J. Math.* **2**, 25–53.

79. Morrey, C.B. (1966) *Multiple Integrals of the Calculus of Variations.* Springer Verlag, Berlin.

80. Müller, S. (1987) Homogenization of nonconvex integral functionals and cellular elastic materials. *Arch. Ration. Mech. Anal.* **99**, 189–212.

81. Müller, S. (1998) Reiterated homogenization. Manuscript.

82. Murat, F. (1978) Compacité par compensation. *Ann. Scuola Norm. Sup. Pisa Cl. Sci.* **5**, 489–507.

83. Murat, F. and Tartar, L. (1985) Calcul des variations et homogénéisation. *Les Méthodes de l'Homogénéisation: Theorie et Applications en Physique*, Eyrolles, 319–369.

84. Murat, F. and Tartar, L. (1985a) Optimality conditions and homogenization. *Proceedings of 'Nonlinear variational problems', Isola d'Elba 1983*, Res. Notes in Math. **127**, Pitman, London, 1–8.

85. Reshetnyak, Yu.G. (1967) On the stability of conformal mappings in multidimensional spaces. *Siberian J. Math.* **8**, 69–85.

86. Sanchez-Palencia, E. (1980) *Non-Homogeneous Media and Vibration Theory.* Lecture Notes in Physics 127, Springer Verlag, Berlin.

87. Shu,Y.C. (1998) Heterogeneous thin films of martensitic materials.

88. Spagnolo, S. (1968) Sulla convergenza delle soluzioni di equazioni paraboliche ed ellittiche. *Ann. Scuola Norm. Sup. Pisa Cl. Sci.* **22**, 571–597.

89. Stein, E.M. (1970) *Singular Integrals and Differentiability Properties of Functions.* Princeton University Press, Princeton.

90. Struwe, M. (1990) *Variational Methods.* Springer Verlag, Berlin.

91. Šverák, V. (1991) Quasiconvex functions with subquadratic growth. *Proc. R. Soc. London* **433**, 725–733.

92. Šverák, V. (1992) Rank-one convexity does not imply quasiconvexity. *Proc. R. Soc. Edinburgh A* **120**, 185–189.

93. Tartar, L. (1975) Problèmes de contrôle des coefficients dans des équations aux dérivées partielles. *Control Theory, Numerical Methods and Computer Systems Modelling.* Lect. Notes in Econom. and Math. Systems **107**, Springer Verlag, Berlin, 420–426.

94. Tartar, L. (1979) Compensated compactness and applications to partial differential equations. Nonlinear analysis and mechanics. *Heriot-Watt Symposium vol. IV.* Res. Notes in Math. **39**, Pitman, London, 136–211.

95. Tartar, L. (1985) Estimations fines des coefficients homogénéises. *Ennio De Giorgi Colloquium* (P. Krée ed.). Res. Notes in Math. **125**, Pitman, London, 168–187.

96. Tartar, L. (1990) H-measures, a new approach for studying homogenization, oscillations and concentration effects in partial differential equations. *Proc. R. Soc. Edinburgh A* **115**, 193–230.

97. Willis, J.R. (1983) The overall elastic response of composite materials. *J. Appl. Mech.* **33**, 1202–1209.

98. Yosida, K. (1965) *Functional Analysis.* Springer Verlag, Berlin.

99. Zhikov, V.V. (1995) Lavrentiev phenomenon and homogenization for some variational problems. *Composite Media and Homogenization Theory*, World Scientific, Singapore, 273–288.
100. Zhikov, V.V., Kozlov, S.K. and Oleinik, O.A. (1994) *Homogenization of Differentiable Operators and Integral Functionals*. Springer Verlag, Berlin.
101. Ziemer, W. (1989) *Weakly Differentiable Functions*. Springer Verlag, Berlin.

The reference list takes into account only papers and books directly referred to in the book, and has no claim on completeness. For a wider bibliography, we refer to the book by Dal Maso (1993); see also the books by Bensoussan *et al.* (1978), Sanchez-Palencia (1980), Bakhvalov and Panasenko (1989), Zhikov *et al.* (1994) and Cherkaev and Kohn (1997).

Notes to references

Part I

Chapter 1: more information about l.s.c. functions can be found in the books by Ekeland and Temam (1976), Attouch (1984) and Dal Maso (1993). Chapter 2: the canonical reference for Sobolev spaces is the book by Adams (1975) (see also the book by Ziemer (1989)); information about weak convergence can be found, for example, in the books by Yosida (1965) and by Evans (1990). Chapter 3: the proof of Proposition 3.3 is adapted from Ekeland and Temam (1976), where a deeper study of integrands is carried out. Other good sources are the introduction to the book by Struwe (1990), and the second part of the book by Ciarlet (1988). Chapter 4: much of this chapter is adapted from the paper by Ball and Murat (1984). Chapter 5: the convex case is dealt with in many books and papers with many different proofs (e.g. Ekeland and Temam (1976), Dal Maso (1995)). Other proofs for the polyconvex and quasiconvex case can be found in the book by Dacorogna (1989), and in the paper by Acerbi and Fusco (1984). Chapter 6: another proof of the properties of quasiconvexifications can be found in Dacorogna (1989). The counterexamples are adapted from the papers by Šverák (1990) and (1991).

Part II

Chapter 7: all the results on Γ-limits can be found in the original paper by De Giorgi and Franzoni (1975), and in the book by Dal Maso (1993), for the case of functions defined on topological spaces. Chapter 8: some results on the convergence of problems of the type (8.5) can be found in the book by Bensoussan *et al.* (1978). Chapter 9: a collection of integral representation results can be found in the book by Buttazzo (1989). Chapter 10: the 'measure property criterion' Theorem 10.2 is due to De Giorgi and Letta (1977). Chapters 11 and 12: the version of the fundamental estimate in Definition 11.2 is due to Braides and Garroni (1995), and is a generalization of the previous version by Dal Maso (1993). An example of an application where the fundamental estimate is stated as in Remark 11.3 is given in Ambrosio and Braides (1990). The propositions in these chapters generalize similar results that can be found in

Dal Maso (1993). The proof of the Compactness Theorem in the vector-valued case is given by Fusco (1983). The same techniques give a compactness theorem in Orlicz-Sobolev spaces satisfying the Δ_2-property (for the necessary information on Orlicz spaces and related imbedding theorems we refer to the book by Krasnoselskii and Rutickii (1961), and to the papers by Donaldson and Trudinger (1971) and Gossez (1974)).

Part III

Chapter 14: the proof of the Homogenization Theorem 14.5 follows the one given in Braides (1985). An independent proof can be found in the paper by Müller (1987). The convex scalar case is dealt with by Marcellini (1978) and the convex scalar non-coercive case by Braides (1983); see also Willis (1983). The same method gives an integral representation for the homogenization of functionals with a singular perturbation, of the form $\int_\Omega f(\frac{x}{\varepsilon}, Du)\, dx + \varepsilon^\gamma \int_\Omega |\Delta u|^2\, dx$. The form of the homogenization formula for such functionals, which depends on γ, has been described by Francfort and Müller (1994). The asymptotic behaviour of the linearized systems related to homogenization has been studied by Geymonat *et al.* (1993). A generalization of Theorem 14.5 to functionals with surface energies can be found in Braides *et al.* (1996), where applications to fracture mechanics are given. If we do not have a growth condition 'from below', as in Theorem 14.8, then the homogenized functional may be finite on the space of functions of bounded variation (see Bouchitté (1985)). The counterexample of the cell-problem formula is due to Müller (1987). For a comprehensive list of references on the homogenization of elliptic operators we refer to Dal Maso (1993). Chapter 15: almost-periodic homogenization techniques are illustrated by Braides (1986) and (1992). The results of this chapter generalize the paper by E (1991), where 'periodic techniques' are used, and earlier homogenization results for functionals $\int f(u/\varepsilon, Du)\, dx$ by Buttazzo and Dal Maso (1978), from which the last section is adapted. Section 16.1 is an elaboration of some results from Acerbi and Buttazzo (1983). Section 16.2 contains results by E (1991) and Braides (1992), which draw some ideas from Lions *et al.* (1987). An introduction to viscosity solutions is given in the review paper by Crandall *et al.* (1992). Chapter 17: the proof of the results adapts those by Braides (1986) and (1992). Chapter 18 improves the results of Braides (1994).

Part IV

Chapter 19: the homogenization of connected non-linear media has been considered by Acerbi *et. al.* (1992) in the scalar case, and by Braides and Chiadò Piat (1994) in the vector-valued case, whose proof is reproduced here. Chapter 20: part of the results are from Braides and Garroni (1995) (see also Bruno (1991) and Bruno and Leo (1992)). For the treatment of stiff inclusions in the linear case we refer to Zhikov *et al.* (1994). The results of Chapter 21 are a generalization of the results of Corbo Esposito *et al.* (1992), who treated the convex case, and to whom we owe Example 21.15. Non-standard integrals have been dealt with also by Kozlov (1989) and Zhikov (1995), who presents more examples not included here. Chapter 22: the proof of the Iterated Homogenization Theorem follows the lines of the one in S. Müller (1998). An iterated homogenization theorem for the linear case can be found in Bensoussan *et al.* (1978); see also Avellaneda (1987). An interesting example of a material with extremal properties obtained by iteration is given by Milton (1992). An application of the Non-linear Iterated Homogenization Theorem can be found in Braides and Lukkassen (1997), where a simplified proof in the convex case is presented. Chapter 23: the result in this chapter is taken from Dal Maso and Defranceschi (1990). Correctors in the almost-periodic case are dealt with by Braides (1991); see also Braides *et al.* (1992). Chapter 24: integral functionals with respect to measures μ depending on smooth functions have been studied by Bouchitté *et al.* (1997). Some related homogenization results have been obtained by Zhikov (1995) when μ is the restriction of the Lebesgue measure to a periodic set. Energies defined on Sobolev spaces with respect to a measure have been studied by Ambrosio *et al.* (1996). Their homogenization is dealt with by Ansini *et al.* (1997). Section 24.3 contains results obtained in collaboration with I. Fonseca and G. Francfort. The homogenization of thin films with flat profile (i.e. $\gamma = 1$) is dealt with in detail by Shu (1998).

Further reading

Related techniques for the homogenization. G-convergence: Spagnolo (1968), De Giorgi and Spagnolo (1973). Compensated compactness: Murat (1978), Tartar (1979). The perturbed test-function method: Evans (1989). H-measures: Tartar (1990). Two-scale convergence: Allaire (1992).

Some applications of homogenization. Optimal design problems: Tartar (1975), Lurie *et al.* (1982), Murat and Tartar (1985), Kohn and Strang (1986), Bendsøe and Kikuchi (1988). Optimal bounds for composites: Hashin and Strikman (1963), Lurie and Cherkaev (1984), Murat and Tartar (1985a), Tartar (1985), Francfort and Murat (1986), Milton (1990). Relaxed Dirichlet problems (by Γ-convergence techniques): Buttazzo *et al.* (1987), Dal Maso and Mosco (1987) (see also Cioranescu and Murat (1982)).

Bibliography: Dal Maso and Dell'Antonio (eds) (1991) and (1995), Cherkaev and Kohn (eds) (1997).

INDEX